Organic Superconductivity

Organic Superconductivity

Edited by

Vladimir Z. Kresin

Lawrence Berkeley Laboratory
University of California, Berkeley
Berkeley, California

and

William A. Little

Stanford University
Stanford, California

PLENUM PRESS • NEW YORK AND LONDON

Library of Congress Cataloging in Publication Data

International Conference on Organic Superconductivity (1990: South Lake Tahoe, Calif.)

Organic superconductivity / edited by Vladimir Z. Kresin and William A. Little.

p. cm.

"Proceedings of the International Conference on Organic Superconductivity, held May 20–24, 1990, in South Lake Tahoe, California"—T.p. verso.

Includes bibliographical references and indexes.

ISBN 0-306-43730-9

1. Organic superconductors—Congresses. I. Kresin, Vladimir Z. II. Little, William A., 1930- .III. Title.

QC611.98.074I57 1990 90-22264

537.6'233—dc20 CIP

Proceedings of the International Conference on Organic Superconductivity, held May 20–24, 1990, at Stanford Conference Center, South Lake Tahoe, California

ISBN 0-306-43730-9

A Division of Plenum Publishing Corporation
233 Spring Street, New York, N.Y. 10013

Printed in the United States of America

INTERNATIONAL WORKSHOP ON ORGANIC SUPERCONDUCTIVITY

CHAIRMAN

W.A. Little, Stanford University

INTERNATIONAL COMMITTEE

D. Jerome (France)
H.J. Keller (Germany)
I.F. Schegoler (USSR)
G. Saito (Japan)
T. Timusk (Canada)
J. Williams (USA)

PROGRAM COMMITTEE

J.P. Collman, Stanford University
E. Edelsack, Cryogenics
R.L. Greene, University of Maryland
V.Z. Kresin, Lawrence Berkeley Laboratory

PROCEEDINGS

V.Z. Kresin
W.A. Little

SPONSORS

Department of Energy
Office of Naval Research

PREFACE

This book contains papers presented at the International Conference on Organic Superconductivity which was held May 20-24, 1990, at the Stanford Sierra Conference Center, South Lake Tahoe, California.

In the twenty years since the First Conference on Organic Superconductivity was held (Hawaii, 1969), there has been remarkable progress in the field. At present, development is accelerating with contributions from many groups in many countries worldwide.

The discovery of high T_c superconductivity by G. Bednorz and K. Muller in 1986 and subsequent developments in the ceramic superconductors have had an enormous impact on the field of superconductivity as a whole. This discovery occurred in an area entirely different from that of conventional superconductivity, underscoring the importance of the search for and study of novel materials of all kinds. We believe that the organics, with their wide range of structural, chemical, and physical properties, belong in this category of novel materials.

This book reflects the efforts of researchers from various disciplines: physicists, chemists, and materials scientists. It addresses the normal and superconducting properties of organic materials, as well as the search for new compounds and new syntheses. We are pleased to note that one of these papers reports on the discovery of a new organic superconductor with a record high T_c in this class. One chapter is devoted to a comparison of organic superconductors and the cuprates, another, to the prospects of discovering other novel conducting or superconducting compounds.

We are grateful to the Department of Energy and the Office of Naval Research for their support. We thank Dr. Edgar Edelsack, who provided support and encouragement for the First Conference in 1969 and initiated work for the most recent meeting as well. We also thank C. Caley, J. Collman, D. Cowan, and J. Williams for guidance and advice on organizing the conference and the final manuscript of the meeting. We are very grateful to Bertha O'Keefe for careful editing of the manuscript and many improvements. Finally, we are grateful to the staff of the Stanford Sierra Conference Center for their help and hospitality.

Vladimir Z. Kresin and William A. Little

CONTENTS

THEORY

SEARCH FOR NEW MATERIALS, STRUCTURE AND T_c, SYNTHESIS

ORGANIC SUPERCONDUCTIVITY: THE FIRST QUARTER CENTURY

W. A. Little

Physics Department, Stanford University
Stanford, California, USA

This, the International Conference on Organic Superconductors commemorates the twenty-first anniversary of the first such Conference[1] which was held in the Fall of 1969, in Honolulu. The intent of that meeting was to bring together theoretical and experimental physicists, and theoretical, organic and inorganic chemists, and introduce them to the challenge of designing and synthesizing polymeric, organic compounds which would superconduct at high temperatures[2]. That conference served to catalyze work in this area and within the next decade the first 2D-superconductors intercalated with organic compounds were prepared[3]; the first polymeric superconductor was synthesized[4] in 1975; and, in 1980, the first organic superconductor was discovered[5]. Much progress has been made since then[6], with more than thirty organic superconductors now known, and some with transition temperatures as high as 11K. On the other hand, during this period, the search for high temperature superconductivity appeared more elusive, with little progress made until the discoveries[7] of the high Tc cuprates in 1986. These discoveries have established the validity of the early arguments that such high transition temperatures were possible, although the exact[8] mechanism responsible for the superconductivity in these materials has not yet been fully identified. They have provided encouragement for the study of non-conventional conducting materials including such materials as the oxides and the organics. The latter continue to offer a rational route to the synthesis of compounds which may yet yield materials with transition temperatures comparable to, or higher, than those of the cuprates.

In the light of these developments, Ed Edelsack of the Office of Naval Research, Washington suggested to me about a year ago that it would be appropriate on this anniversary occasion to bring together once again, a similar group of physicists and chemists, as we had done previously, to review the progress that has been made in recent years in the organics, and to consider new approaches to the synthesis of other classes of conducting, or possibly superconducting compounds. My enthusiasm for doing this was tempered by the thought of the work and organization that would be required to make it happen, but when Vladimir Kresin of the Lawrence Berkeley Laboratory volunteered to share the burden, we agreed to do it together. Our proposal received the enthusiastic support of the United States Department of Energy and the Office of Naval Research. These agencies also endorsed our proposal to include a substantial number of students among the attendees at the meeting, recognising in so doing, the long term nature of the field, and the need to involve both the present and future generations of scientists in it.

In preparing for this meeting, I reviewed the program and topics which were covered in the 1969, Hawaii meeting. The coverpage of that program was decorated with illustrations of phenomena from superconductivity, and some elementary concepts of organic chemistry - an early effort to establish some common ground between chemist and physicist. Great progress has been made in the past two decades in furthering this endeavor, with many groups of chemists and physicists now collaborating, and each conversant with the terminology and thinking of the other's discipline.

Organic Superconductivity, Edited by V.Z. Kresin and W.A. Little, Plenum Press, New York, 1990

It was interesting to note also, the breadth of coverage at that meeting, and at the same time to appreciate the distance we have traveled since then. Among the compounds which were discussed were charge transfer salts, the TCNQ salts, ferrocene polymers, phthalocyanines, polyynes, polyacetylene, SN_x (not known at that time as a superconductor), porphyrins, the Krogmann salts (pointed out to us by Professor J. Collman), metal oxalates and glyoximes and the possible bio-synthesis of diamagnetic and superconducting compounds. It was striking to note that the theoretical chemists were much more familiar, than were the physicists at that time, with the existence of Charge Density Waves, Spin Density Waves, and the Peierls distortion, and the distinctions among them. The key role of packing forces in establishing the crystal structure and through this, the conductive behavior of the material, was also recognised. At that time, it had already been recognised by the physicists that fluctuations were not nearly as destructive in linear, or one-dimensional systems than earlier considerations had led one to believe; and, the role of the dimensionality of the system was beginning to be appreciated.

I have also found it useful to review for this meeting, the history of the organics and to recognise the major milestones along the road. Probably the earliest discussion of organic *semiconductors* was that of Akamatu and Inokuchi[9] in 1950, followed in 1960 by the publication in Physical Review Letters by Kepler[10], of the behavior of an organic metal containing TCNQ. The first suggestion of the possibility of synthesizing an organic, polymeric *superconductor* and the consideration of dimensional effects in superconductivity, was my 1964 paper [2] in Physical Review, followed by Ginzburg's application[11] of these ideas to two-dimensional systems in a discussion of surface superconductivity later that year. Fred Wudl's introduction of the donor[12], TTF in 1970, and the discovery of the properties of doped polyacetylene by Shirakawa[13], were other key events in the period preceeding the discovery of the first organic superconductor.

Notable experimental discoveries in the past several years since the Hawaii meeting, were the discovery by F. Gamble et al., of the possibility of intercalating the two-dimensional compound, TaS_2 with organic compunds[3]; the discovery of superconductivity in polymeric, SN_x by R. L. Greene[4] in 1975; the discovery of the first organic superconductor, TMTSF-PF_6 by D. Jerome et al.[5], in 1980; the synthesis of β-$(BEDT\text{-}TTF)_2I_3$ by several groups[14] and the preparation of $(BEDT\text{-}TTF)_2Cu(NCS)_2$ by Urayama et al.[15] with a transition temperature above 10K, in 1987. One might also include the discovery of the high T_c cuprates by Bednorz and Müller[7] in 1986 because of its dramatic impact upon the field of non-conventional superconductors, and in particular the recognition of the role of dimensionality in these superconductors.

At the Hawaii meeting, Felix Bloch opened the meeting with a review of the history of superconductivity, and introduced it with a story of a visit of a friend of his to Washington, DC. While driving around the city in a taxi cab admiring the grand buildings, this friend noticed an inscription above the portals of a building containing the archives of the great. It read,"The Past is but a Prologue". His friend turned to the cab-driver and asked him what it meant, and the driver replied, "Sir, it means that you ain't seen nothing yet". That was Bloch's opinion of superconductivity in 1969! I believe it is equally true today. We ain't seen nothing yet!

For the meeting at Lake Tahoe, the program was organized to include both review talks in each of several key area, whose purpose was to serve as an introduction to those new to the field, and presentations of original, recent contributions by those most active in these areas. These two were meant to provide a basis for the main focus of the meeting, which was to stimulate the participants to consider new and different materials to synthesize. This it was hoped, would lead eventually to the discovery of new classes of superconductors, or materials with other novel properties. The program was organized in such a way as to keep the physicists and chemists together, interleaving recent theoretical and experimental work, with work on the synthesis or strategic design of chemical syntheses; rather than separating the sessions into those that would be predominantly of interest to physicists, and others of predominant interest to chemists. Ample time was set aside for discussion and interaction among the participants, by limiting the formal sessions to the mornings and late afternoons, leaving the afternoons free.

The rapid developments in the field make it likely that new organic superconductors, and new classes of such superconductors are likely to be discovered in the near future, pushing the value of the highest transition temperature beyond that of the metallic alloys, and, perhaps, up to the cuprates and beyond. At the same time, in view of the rich structure of the phase diagrams that many of these compounds exhibit, one can expect to see some new discoveries in the magnetic, thermal and conductivebehavior of related materials which may not even superconduct. The field appears to have a rich and lustrous future.

REFERENCES

1. W. A. Little, Proceedings of the International Conference on Organic Superconductors, Journal of Polymer Science, Part C, POLYMER SYMPOSIA, **29** (1970).
2. W. A. Little, Phys. Rev. **A134**, 1416 (1964)
3. F. Gamble, F. J. DiSalvo, R. A. Klemm and T. H. Geballe, Science **168**, 568 (1970)
4. R. L. Greene, G. B. Street and L. J. Suter, Phys. Rev. Lett. **34**, 577 (1975)
5. D. Jerome, A. Mazaud, M. Ribault, and K. Bechgaard, J. Phys. Lett. **41**, L-95 (1980)
6. T. Ishiguro, Physica **C 153**, 1055 (1988)
7. J. G. Bednorz and K. A. Müller, Z. Phys. **B64**, 189 (1986)
8. W. A. Little, Science **242**, 1390 (1988)
9. H. Akamatu and H. Inokuchi, J. Chem. Phys., **18**, 810 (1950)
10. R. G. Kepler, P. E. Bierstedt, R. E. Merrifield, Phys. Rev. Lett. **5**, 503 (1960)
11. V. L. Ginzburg, Zh. Eksp. Teor. Fiz. **47**, 2318 (1964); [Soviet Phys. - JETP **20**, 1549 (1965)].
12. F. Wudl, G. M. Smith, and E. J. Hüfnagel, J. Chem. Soc. Chem. Commun. 1425 (1970).
13. H. ShirakawaE, J, Lewis, A. G. MacDiarmid, C. K. Chian, and A. J. Heeger: J. Chem. Soc. Chem. Commun. 578 (1977).
14. K. Murata, M. Tokumoto, H. Anzai, H. Bando, G. Saito, K. Kajimura, T. Ishiguro: J. Phys. Soc. Jpn **54**, 1236 (1985); V. N. Laukhin, E .E. Kostyuchenko, Yu. V. Sushko, I. F. Schegolev, E. B. Yagubskii: JETP Lett. **41**, 81, (1985); F. Creuzet, G. Creuzet, D. Jerome, D. Schweitzer and H. J. Keller: J. Phys. Lett. **46**, L1079 (1985).
15. H. Urayama, H. Yumochi, G. Saito, K. Nozawa,T. Sugano, M. Kinoshita, S. Sato, K. Oshima, A. Kawamoto, and J. Tanaka: Chem. Lett. **55**, (1988)

ORGANIC SUPERCONDUCTORS VS HIGH HIGH T_c OXIDES

ORGANIC AND OXIDE SUPERCONDUCTORS: AN EXPERIMENTAL COMPARISON

Richard L. Greene

Center for Superconductivity Research†
Department of Physics
University of Maryland
College Park, MD

ABSTRACT

The recent discovery of superconductivity in an organic metal with transition temperature (T_c) near 12K has led to renewed interest in the physics and chemistry of organic metals. Organic metals and superconductors have many properties that are strikingly similar to the high-T_c copper oxides. This paper will give a brief experimental comparison of these two classes of materials.

INTRODUCTION

In the mid 1960's Little and afterwards Ginzburg suggested that some one-dimensional or two-dimensional metals might be superconductors at temperatures much higher than conventional three dimensional metals. These ideas stimulated an enhanced research effort into the synthesis and properties of low dimensional (anisotropic) materials. The dramatic discovery of the high-T_c copper oxides by Müller and Bednorz in 1986 is the most spectacular success of this research effort, although many other anisotropic materials with novel properties have been found during the past 25 years. One of the most interesting of these classes of materials is the organic metals (including polymeric metals). They have been extensively studied since 1973 when TTF-TCNQ was synthesized. The first organic superconductors were the Bechgaard salts, $(TMTSF)_2X$ (X= PF_6, $C\ell O_4$, etc.), discovered in 1980 by Jerome and coworkers[1], with a $T_c \lesssim 1K$. Another class of organic metals $(BEDT\text{-}TTF)_2X$, or for abbreviation, $(ET)_2X$, were synthesized in 1983 and one member $(ET)_2ReO_4$ was found to be superconducting below ≈ 2.5K by a group at IBM Almaden[2]. Since then many more superconductors in the $(ET)_2X$ class have been found as well as four new classes of organic superconductors. To date, the highest T_c organics have been found in the $(ET)_2X$ class by Saito and coworkers[3] (X = $Cu(SCN)_2$, $T_c \approx 10.5K$) and Williams et al[4] (X = $Cu[N(CN)_2]Br$, $T_c \approx 12K$). This latter result — just reported at this conference - raises T_c to the level of $Ba(Pb,Bi)O_3$, the compound discovered by Sleight et al[5] which stimulated much of the research on superconducting oxides prior to the cuprates.

---

†Also at IBM Research, Yorktown Hts., N.Y. 10598

Organic Superconductivity, edited by V.Z. Kresin and W.A. Little, Plenum Press, New York, 1990

Although nobody can foresee if the T_c of organics will follow that of the copper oxides there are enough similarities between these two classes of superconductors that the organizers of this conference felt that a lecture comparing the properties of these materials should be given. In hindsight, this was impossible to do in any detail because of the enormous literature of research that has been devoted to these materials in recent years (well over 10,000 papers). In this brief summary I will compare some of the properties that are likely to be relevant to understanding the mechanism of superconductivity in these materials. Needless to say, the nature of superconducting state and normal state in these two systems is not yet understood. However, I will not discuss any of the many theories that have been proposed to explain these properties (or T_c). Instead I will discuss the experimental facts which I believe are the most reliable. The rapid improvement in the quality of the materials means that some of these results may change in the future. It is also impossible to adequately reference all the important prior work - I apologize to all those authors whose work is omitted. I will refer the reader primarily to review articles or conference proceedings[6,7]. Because of time constraints I have not included any figures here — the reader will find these in the references. In spite of these limitations I hope that the reader will find this short aritcle useful for getting an overview of the similarities and differences of these two novel classes of superconductors, and perhaps some insight into future research directions.

Some basic features of the copper oxides and the $(ET)_2X$ organics illustrate their similarities. Both have two dimensional (2D) crystal structures with a 2D electronic band structure (closed Fermi surface) derived from Cu-O interactions and ET-ET interactions in the 2D plane. The coupling between the 2D layers is typically ten times smaller than within the plane. The carrier density is low ($n \sim 10^{21} cm^{-3}$) and the carriers are holes with the exception of the $(Nd,Ce)_2CuO_4$ system ("214"). The order of magnitude reduction in carrier density when compared to traditional superconductors orignates from a large unit cell volume in the organics and small doping levels in the oxides. The low carrier density implies less screening and in both systems Coulomb interactions (e-e correlations) are important, however, their exact role is not yet well understood. The undoped parent oxides, e.g. La_2CuO_4, $YBa_2Cu_3O_6$ are antiferromagnetic and the $(TMTSF)_2X$ compounds have SDW transitions in close proximity to the superconducting state. The $(ET)_2X$ compounds have not shown direct evidence of magnetism but Coulomb interactions (represented by the Hubbard U) are of the same order or larger than the bandwidth ($4t \sim 0.5eV$). The small n and large U lead to other gross similarities such as a low Fermi velocity $V_F \sim 10^7 cm/sec$, a small Fermi energy, $E_F \sim 0.1eV$ and a large in-plane effective mass, $m^*/m_e \gtrsim 5$. Although all the oxide and organic superconductors have small n they are not all 2D in structure. The $(Ba,K)BiO_3$, $Ba(Pb,Bi)O_3$ system is 3D with a maximum T_c of order 30K while the $(TMTSF)_2X$ system is quasi-1D with a maximum T_c of order 1K. We will not discuss these latter systems here (see the reviews) although they have many properties that are similar to the 2D materials, and possibly the same pairing mechanism in the superconducting state.

NORMAL STATE PROPERTIES

Both the copper oxides and the organic metals can be classified as anomalous metals. Electron-electron correlations are significant and the present band structure calculations are probably not adequate. The extent to which conventional Fermi liquid theory is applicable is also a subject of intense debate, especially for the oxides. A good summary of the normal state properties of the oxides has been given by Batlogg in reference 7c —

a theoretical overview by Lee can be found in the same book. An experimental overview of the $(ET)_2X$ materials is given in references 6a and 6b.

Perhaps the most unusual feature of the copper oxides is the linear temperature dependence of the resistivity ($\rho \propto T$) and the inverse linear dependence of the Hall coefficient ($R_H^{-1} \propto T$) for hole transport in the 2D plane. This is rigorously true only for $YBa_2Cu_3O_7$ ("123") but is found to some extent in the Bi (2212 and 2201) and $T\ell$ superconductors as well. In Bi-2201 the linear $\rho(T)$ extends from 10K to 700K — a remarkable behavior not easily explained by electron-phonon or election-electron scattering. In the best oxide samples the resistivity extrapolates to zero at T=0 suggesting no residual ρ_o and a long mean free path. Since the measured coherence length is small these superconductors are in the clean limit. The electron-like $(Nd,Ce)CuO_4$ system seems to be different ($\rho \propto T^2$ below ~100K) but sample homogeneity is still a problem (even in crystals) so the intrinsic behavior may not yet be determined.

In the $(ET)_2X$ superconductors one finds $\rho = A + BT^2$ below 100K, a behavior attributed[6a] to electron-electron scattering (Fermi liquid). However, at room temperature (300K) $\rho \approx 4000 u\Omega$-cm is larger than the estimated maximum metallic resistivity so the metallic T dependence of ρ is surprising. Certainly near 300K the carriers must be localized. The Hall effect is also T dependent[9] (but not linear) and cannot be explained by any simple model. At low temperature the organics appear to be very clean since Shubnikov-de Haas oscillations have been observed in high fields below 1K.

The resistivity perpendicular to the 2D plane is 10 to 100 times larger for organics and oxides. In both cases this would suggest the carriers are localized to the planes and move between planes only by tunneling or hopping. The temperature dependence of $\rho_\perp$ is "metallic-like" for the organics and "semiconductor-like" for the oxides. This behavior is not understood for either class of materials.

The optical properties give information about the frequency dependent conductivity $\sigma(\omega)$. For both organics and copper oxides $\sigma(\omega)$ is anomalous and cannot be described by the usual Drude model. As yet there is not complete agreement on the interpretation of $\sigma(\omega)$. One interesting approach is to analyze the data using a generalized Drude model with frequency dependent scattering time (τ) and effective mass (m^*). This method was used to consistently interpret $\sigma(\omega)$ in the heavy fermion superconductors (where $1/\tau \propto \omega^2$ and $\rho \propto T^2$). For $\omega \leq 2000 cm^{-1}$ this analysis gives[10,11] $\hbar/\tau \simeq \pi kT + \hbar\omega$ in both "123" and $(ET)_2[Cu(SCN)_2]$ and $m^*/m_e \gtrsim 5$ at $\omega=0$. This suggests that the carriers scatter inelastically off a broad spectrum of excitations (of unknown origin). For a Fermi liquid one anticipates $1/\tau \sim A[kT^2 + \hbar\omega^2]$ with A much less than 1, so in this sense the organics and oxides are anomalous. Inelastic light scattering[12] in "123" directly indicates a broad continuum of excitations up to $\sim 8000 cm^{-1}$ (100meV) in agreement with the analysis of $\sigma(\omega)$. To my knowledge light scattering experiments have not yet been done in the $(ET)_2X$ compounds.

Both systems have a magnetic susceptibility which is roughly temperature independent above T_c. This has been interpreted in terms of a Pauli susceptibility with an effective mass ($m^*/m_e \sim 5$) enhanced by correlations. Correlation effects also are found in NMR experiments where the nuclear relaxation rate ($1/T_1$) is enhanced over the rate expected from the band theory density of states. The Korringa law ($T_1T \approx$ constant) is obeyed in the organics and in "123" for the oxygen sites and chain Cu site (but not for Cu sites in the plane).

Much experimental and theoretical work needs to be done before it can be said that we understand the normal state properties of organic and oxide superconductors. The brief summary above illustrates some of the important properties which are anomalous. Although the organics and copper oxides have many similar normal state properties it remains to be seen if they originate from a similar mechanism.

SUPERCONDUCTING STATE

While we have seen that the normal state is rather anomalous the superconducting state in the oxides and organics seems almost normal (i.e. conventional), except for the high T_c. A good discussion of the superconducting properties of the oxides and the possible nature of the pairing state is given by Annett et al in Volume II of reference 7a. A similar review does not exist for the organics but most of the relevant information can be found in references 6a,b,c and in this article.

The nature of the superconducting state is best revealed by the symmetry, wave vector ($\tilde{k}$) and temperature dependence of the order parameter or gap function $\Delta(\tilde{k},T)$. Many experiments, e.g. tunneling, NMR, Raman and infrared, are influenced by $\Delta(k,T)$ below T_c. In anisotropic materials with conventional pairing one might expect the energy gap to be anisotropic but still nonzero everywhere in $\tilde{k}$ space. We call this "s-wave" pairing. If the gap goes to zero somewhere in $\tilde{k}$ space the pairing is unconventional (non s-wave) and we expect power law temperature dependencies as $T \rightarrow 0$ for many properties, e.g. specific heat, T_1^{-1}, and penetration depth $\lambda(T)$. Another important question is the nature of the excitations responsible for the pairing and whether they are strongly or weakly coupled.

One of the most unusual properties of the oxides and organics is that superconductivity only occurs in a very narrow (perhaps single) composition range. All the organic superconductors have two cations for each anion (X). Because of complete charge transfer this yields a 3/4 filled conduction band and an effective charge of 0.5 holes per cation. In the copper oxides the maximum T_c occurs at a composition between 0.15 and 0.25 holes (or elections) per plane Cu. The difficulty in preparing homogeneous, stoichiometric, single phase compounds has caused much confusion in this regard. In my opinion, it is not yet determined if T_c can be varied by changing composition within a given system, e.g. Sr in $La_{2-x}Sr_xCuO_4$, or if bulk superconductivity occurs only at one fixed composition, e.g. $La_{1.85}Sr_{0.15}CuO_4$. This is an issue to be resolved in the future. It would also be interesting to vary the band filling in the organics to see if T_c can be varied.

The most direct probe of the energy gap is electron tunneling. This has proven to be a very difficult experiment in both oxides and organics. Although no conclusive statements can be made at this time the consensus of the experiments[13] on "123" suggest an anisotropic gap with $2\Delta/kT_c \sim 6$ in the ab plane and $2\Delta/kT_c \sim 3.5$ in the c-direction. The tunneling data for the organics is not as extensive as for the oxides but it appears that the in plane gap is large $2\Delta/kT_c \gtrsim 6$, suggestive of strong coupling.[14,6a]

Far-infrared reflectivity (absorption or transmission) is another classic method to determine the gap. The results on the oxides have been quite controversial. The most recent reflectivity data on untwinned "123" crystals[10] suggests a large gap ($2\Delta/kT_c \sim 8$) in the ab plane with a smaller gap ($2\Delta/kT_c \sim 3.5$) in the c direction[15]. This gap anisotropy agrees with the tunneling data. Absorption data[16] on twinned "123" and transmission data[17] on untwinned Bi-2212 give evidence for states below the gap and no clear signature of a gap for energy up to $2\Delta/kT \sim 10$. However, the absorption data seems to agree with the prior reflectivity results on twinned "123" so it is likely that the absorption below the gap comes from excitations associated with the Cu-O chains. Absorption measurements on untwinned "123" should resolve this issue. The Bi-2212 transmission result is anomalous and may result from a strong pair breaking interaction or from the possibility of being in the extreme clean limit (i.e. $1/\tau \ll 2\Delta$). In either case the gap would be difficult to observe in transmission.

A large gap implies strong coupling and associated structure above the gap in both far-IR and tunneling in the conventional Eliashberg theory (from $\alpha^2F(\omega)$). This structure has not been observed.[18] Raman scattering experiments[12,7a] on twinned "123" crystals also show evidence for low energy excitations in the gap region and no clear evidence for a gap. The present consensus from all these experiments indicates that there are excitations (states) in the gap. These may arise from spurious effects, or may originate from the Cu-O chains, or may be an indication of a non s-wave order parameter. The latter possibility seems to be ruled out by penetration depth experiments (to be discussed below).

To my knowledge only one far-IR experiment has been attempted in the $(ET)_2X$ system. Kornelsen et al (this conference) did a careful absorption experiment on the X = $Cu(SCN)_2$ salt which gave no evidence for a gap in the energy region up to $2\Delta/kT_c \sim 10$. In this case it is likely that the sample is in the clean limit, making the gap unobservable, but more experiments are needed. Measurements[6d] in the $(TMTSF)_2X$ materials have been controversial but suggest a strong-coupling value for the gap.

NMR is an important probe of the superconducting state via the Knight shift (K) and the relaxation rate (T_1^{-1}). In a conventional superconductor (s-wave pairing) one finds a Hebel-Slichter coherence peak in $1/T_1$ at T_c, an activated behavior $1/T_1 \sim \exp(-\Delta/kT)$ for $T < T_c$ and a spin contribution to K which goes to zero at T=0 (with a T dependence given by the BCS weak coupling Yosida function). In both oxides[7a,c] and organics[19,20] the coherence peak is absent and $1/T_1$ follows a power law ($\sim T^3$) below T_c. This behavior is non-conventional and is similar to that found in the heavy fermion superconductors (believed to have non s-wave pairing). In $(ET)_2Cu(SCN)_2$ a large peak in T_1^{-1} has been seen well below T_c.[20] This cannot be the usual coherence peak and its origin is not understood. Knight shift experiments have so far only been done[21] on "123". The shift for the chain site Cu follows the Yosida function (s-wave pairing) while the shift for the plane site Cu does not. The Knight shift anisotropy appears to rule out triplet (p-wave) pairing (see Annett et al in 7a). The shift for the Cu plane site has been interpreted as due to either strong coupling s-wave pairing or d-wave pairing.

The important characteristic lengths in a superconductor are the Pippard coherence length $\xi \propto V_F/T_c$ and the London penetration depth $\lambda_L \propto \left(\frac{m^*}{n_s}\right)^{1/2}$. The small n_s, low V_F and high T_c implies that $\lambda_L \gg \xi$ and thus both the oxides and organics are extreme type-II superconductors. In the oxides ξ is very small, estimated ~15Å in the plane and ~3Å perpendicular. Thus, the coherence volume contains only a few Cooper pairs and fluctuations will be significant. In the $(ET)_2X$ organics the in-plane ξ is of order 200Å so fluctuation effects are expected to be less important than in the oxides. The temperature dependence of $\lambda_L(T)$ can give information on the pairing mechanism. In conventional superconductors $\lambda(T)$ varies as $\exp(-\Delta/T)$ for $T \ll T_c$ whereas for unconventional pairing one gets a power law $\lambda \propto T^n$ (because of the non-zero gap on parts of the Fermi surface). Recent experiments on "123" crystals[22] and $(ET)_2Cu(SCN)_2$ crystals[23] give a temperature dependence for $\lambda_L(T)$ consistent with conventional (s-wave) pairing. These results are in obvious contradiction with the NMR experiments — further experimental work is needed to resolve the differences. The temperature variation of $\lambda(T)$ seems to follow the two fluid model (similar to strong coupling BCS) rather than the weak coupling BCS dependence.

A few other properties of the superconducting state are worth mentioning. The specific heat jump at T_c has been measured and the ratio $\Delta C(T_c)/\gamma T_c$ evaluated. For "123" Phillips et al[24] find $\Delta C/\gamma T_c = 4.8$ (compared to 1.43 for weak coupling) while in $(ET)_2Cu(SCN)_2$ Andraka et al[25] find $\Delta C/\gamma T_c \gtrsim 2.8$. These results suggest strong coupling in both systems. The excitations involved in the coupling are not known and they may not be the same for oxides and organics. The biggest difference between these materials is in the hydrostatic pressure dependence of T_c. In the organics T_c is strongly decreased with pressure while in the oxides T_c is increased or unchanged by pressure. For the organics this suggests that electron-phonon coupling is important for determining T_c whereas in the oxides coupling to other excitations must dominate. The rather small isotope effect found in "123" would seem to confirm this although a larger isotope effect has been found in the "214" system. In the organics the substitution of deuterium for hydrogen on the ET molecule gives almost no change in T_c (if anything an inverse isotope effect). This appears inconsistent with the pressure results but it may be that other phonon modes are involved in the pairing interaction. Further work is needed here.

CONCLUSIONS

From this brief overview it is clear that the 2D organic and copper oxide superconductors are very similar. The origin of the superconductivity has not yet been determined in either case. Important issues, such as the symmetry of the pairing, the excitations involved in the coupling and the strength of the coupling need to be resolved. The normal state is also anomalous, perhaps more so in the oxides, and a detailed understanding is lacking. There is no doubt that an understanding of the normal state will help solve the problem of the superconducting state. Since this is a conference proceedings I will take some license and give a few of my present opinions. The superconductivity most likely originates from a conventional (s-wave) pairing mechanism with an anisotropic gap. I believe the excitations (states) observed below the gap in Raman, tunneling, IR and NMR will turn out to be artifacts of the samples or the experimental technique, i.e. not intrinsic to the superconductivity. This has proven to be the case for the linear term in the specific heat of some oxides which was once thought to be intrinsic but has now been shown[24] to be non-intrinsic. The best present data suggest that a large gap exists in the 2D plane ($2\Delta/kT_c \sim 8$). This probably results from strong coupling but the absence of $\alpha^2F(\omega)$ structure in the IR and tunneling is puzzling. It is likely that phonons are involved in the coupling in the organics but the nature of the excitations involved in the copper oxides is still anybody's guess. In the normal state, electron correlations certainly play an important role but this is not understood in detail.

Our understanding of these novel superconductors will improve in direct proportion to the improvement in sample quality. I believe many of the experiments are still influenced by "dirt" effects which are quite difficult to eliminate. Both our basic understanding and technological applications will depend on the skill of the materials people. In comparison to the oxides very little work has been done on the organics. The similarities between the two systems suggests that a larger research effort devoted to organic metals and superconductors is likely to lead to important discoveries.

REFERENCES

1. D. Jerome, A. Mazaud, M. Ribault, K. Bechgaard, J. Phys. Lett. (Paris) 41, L95 (1980).
2. S. S. P. Parkin, et. al, Phys. Rev. Lett. 50, 270 (1983).
3. H. Urayama, et. al, Chem. Lett. 55 (1988).
4. J. Williams, et. al, this volume
5. A. W. Sleight, J. L. Gillson and P. E. Bierstedt, Solid State Comm. 17, 27 (1975).
6. Some reviews of organic superconductors are:
   a) L. N. Bulaevskii, Adv. Phys. 37, 443 (1988).
   b) G. Saito, et. al, Synth. Met. 27, A331 (1988).
   c) "The Physics and Chemistry of Organic Superconductors", G. Saito and S. Kagoshima, ed., Springer-Verlag (1989).
   d) D. Jerome, H. L. Schulz, Adv. Phys. 31, 299 (1982).
   e) R. L. Greene, P. M. Chaikin, Physica 126B, 431 (1984).
7. Some reviews of oxide superconductors are:
   a) "Physical Properties of High Temperature Superconductors I and II", D. M. Ginsberg, ed., World Scientific (1989,1990).
   b) Proceedings of M^2S-HTSC, Stanford (1989); Physica C, 162-164 (1989).
   c) "High Temperature Superconductivity — Los Alamos Symposium Proceedings", edited by K. S. Bedell, et. al, Addison-Wesley (1990).
   d) IBM J. Res. Devel. 33, 197-404 (1989).
8. S. Martin, et. al, Phys. Rev. B 41, 846 (1990).
9. B. Korin-Hamzic, L. Forro and J. R. Cooper, Phys. Rev B41 (June 1990).
10. Z. Schlesinger, et. al, Phys. Rev. Lett. (1990).
11. K. Kornelsen, et. al, Solid State Comm. 72, 475 (1989).
12. S. L. Cooper, et. al, Phys. Rev. B38, 11934 (1988).
13. J. S. Tsai, et. al, in "Mechanisms of High Temperature Superconductivity," H. Kamimura and A. Oshiyama, ed., Springer (1989); J. F. Annett, et. al in reference 7a.
14. M. E. Hawley, et. al, Phys. Rev. Lett. 57, 629 (1986).
15. Z. Schlesinger, et. al, Phys. Rev. B (1 June 1990).
16. T. Pham, et. al, Phys Rev. B41, 11681 (1990).
17. L. Forro, et. al, submitted to Phys. Rev. Lett. (1990).
18. Z. Schlesinger, et. al, Nature 343, 242 (1990).
19. M. Takigawa, et. al, J. Phys. Soc. Jap. 56, 873 (1987).
20. T. Takahashi, et. al, Syn. Met. 27, A319 (1988).
21. S. E. Barrett, et. al, Phys. Rev. B (April 1990) and references therein.
22. L. Krusin-Elbaum, et. al, Phys. Rev. Lett. 62, 217 (1989); R. L. Greene, et. al, Phys. Rev. Lett. 62, 2886 (1989).
23. D. R. Harshman, et.al, Phys. Rev. Lett. 64, 1293 (1990); also see G. Grüner, this volume.
24. N. E. Phillips, et. al, Phys. Rev. Lett. (submitted May 1990).
25. D. Andraka, et. al, Phys. Rev. B 40 (1989); also see J. E. Graebner et. al, Phys. Rev. B (1990) who find $\Delta C/\gamma T_c = 1.5$. However, Graebner, et. al, use a larger γ than Andraka, et. al. The latter group made a direct measurement of γ so it is more likely to be correct.

THALLIUM-BASED COPPER OXIDE SUPERCONDUCTORS

A.M. Hermann,* H. Duan,* W. Kiehl,* and D. Weeks**

* Department of Physics, University of Colorado, Boulder, CO 80309-0390
** Department of Physics, Hebrew University, Jerusalem 91904, Israel

ABSTRACT

Stable bulk superconductors in the Tl-Ca-Ba-Cu-O system with zero resistance above 120K are discussed. Structural, magnetic, and electronic transport properties, including thermoelectric power data, are presented on a variety of phases in this system. A potential way to increase the critical temperature using site-selective doping is discussed. A novel safe preparation procedure for Tl-Ba-Ca-Cu-O thin film preparation, wherein Tl is added through a vapor-phase reaction to Ca-Ba-Cu-O precursors is discussed. Transport data showing zero resistance temperatures in excess of 100K for smooth, homogeneous thin films prepared by this process are presented.

INTRODUCTION

While the theme of this conference is organic superconductors, it may be possible to bring some knowledge to this field from exciting recent discoveries of inorganic doped copper-oxide compounds that superconduct well above the boiling temperature of liquid nitrogen.

Discoveries of 30K La-Ba-Cu-O superconductors[1] and 90K Y-Ba-Cu-O superconductors[2] have stimulated a worldwide race for new and even higher temperature superconductors. Breakthroughs were made by the discoveries of the 90K Tl-Ba-Cu-O system,[3,4] 110K Bi-Sr-Ca-Cu-O system[5,6] and 120K Tl-Ba-Ca-Cu-O system.[7-9] High temperature superconductivity was also observed in the Tl-Sr-Ca-Cu-O system,[10-12] and in the M-Tl-Sr-Ca-Cu-O with M = Pb[13,14] and rare earths.[15] In this paper, we present structural and electronic characterization of several phases of the Tl-Ba-Ca-Cu-O superconductors. Recent results on vapor-processing of thin film Ba-Ca-Cu-O precursors using unreacted Tl-Ba-Ca-Cu-O as a Tl-source are also presented.

STRUCTURE

The Tl-Ba-Ca-Cu-O system can form a number of superconducting phases. Two phases, $Tl_2Ba_2Ca_2Cu_3O_{10+x}$ (2223) and $Tl_2Ba_2Ca_1Cu_2O_{8+x}$ (2212), were first identified.[16] The 2223 superconductor has a 3.85 x 3.85 x 36.25A tetragonal unit cell. The 2212 superconductor has a 3.85 x 3.85 x 29.55A tetragonal unit cell.[16,17] The 2223 phase is related to 2212 by addition of extra calcium and copper layers. In addition, the superconducting phase in the Ca-free Tl-Ba-Cu-O system is $Tl_2Ba_2CuO_{6+x}$ (2201).[16,18] Fig. 1 shows schematically the arrangements of metal atom planes in these three Tl-based superconducting phases. The 2201 phase has a zero-resistance temperature of about 80K, whereas the 2212 and 2223 phases have zero-resistance temperatures 108K and 125K respectively.[16-21]

Organic Superconductivity, Edited by V.Z. Kresin and W.A. Little, Plenum Press, New York, 1990

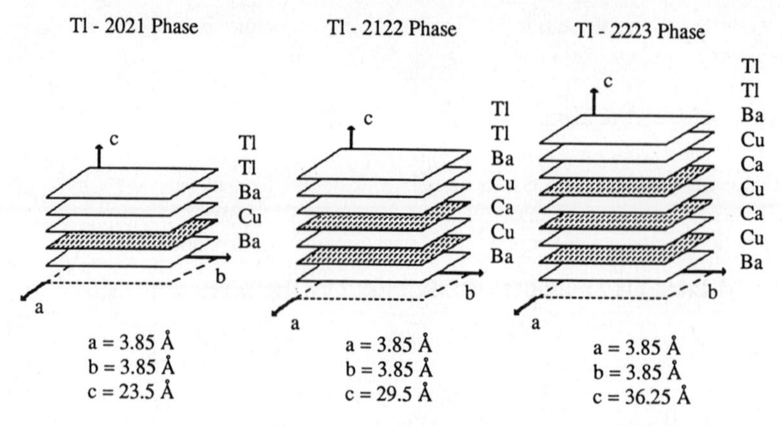

Fig. 1 Arrangements of metal atom planes in the three Tl-based superconducting phases.

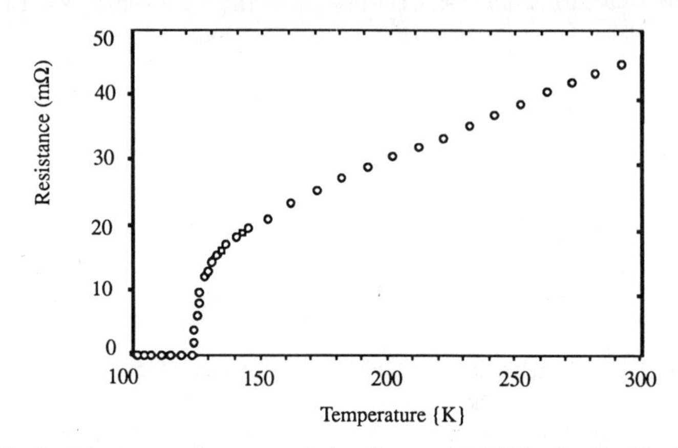

Fig. 2 Resistance-temperature variation for a nominal $Tl_{2.2}Ba_2Ca_2Cu_3O_{10.3+x}$ sample.

A new series of superconducting compounds with a single Tl-O layer, which we denote by $TlBa_2Ca_{n-1}Cu_nO_{2n+2.5}$, were recently also reported.[22,23] The Tl-Ba-Ca-Cu-O superconducting series should be represented using a general formula of $Tl_mBa_2Ca_{n-1}Cu_nO_{1.5m+2n+1}$ with m = 1 and 2, and n = 1, 2, 3, 4, and 5. The T_c of the single Tl-O layer compounds also increases with the number of Cu-Ca layers, and is slightly lower than that of the corresponding double Tl-O layer compounds.

RESISTANCE

Fig. 2 shows resistance-temperature variation for a nominal $Tl_{2.2}Ba_2Ca_2Cu_3O_{10.3+x}$ sample.[9] This sample has an onset temperature near 140K, midpoint of 127K, and zero resistance temperature at 122K.

MAGNETIZATION

Fig. 3 shows[24] dc magnetization (field cooled and zero field cooled) as a function of temperature for an applied field of 1 mT for a nominal $Tl_2Ca_4BaCu_3O_{11+x}$ sample. The insert of Fig. 3 shows data traces for the same sample, and also for a well-prepared $EuBa_2Cu_3O_{7-x}$ sample, where the vertical axis represents the dX"/dH signal of an EPR spectrometer. As is seen from the insert, the onset temperature for the sample A ($Tl_2Ca_4BaCu_3O_{11+x}$) is 118.3K, 23.9K higher than that of the $EuBa_2Cu_3O_{7-x}$ (sample B), whose onset temperature is 94.4K. This onset temperature is consistent with those measured by resistance-temperature variations.

THERMOELECTRIC POWER

The thermoelectric power as a function of temperature for a nominal $Tl_2Ca_2Ba_2Cu_3O_{10+x}$ sample[25] is shown in Fig. 4. The normal-state thermoelectric power is positive, indicating dominant hole conduction. At least three separate ranges of temperature-dependent behavior are apparent.

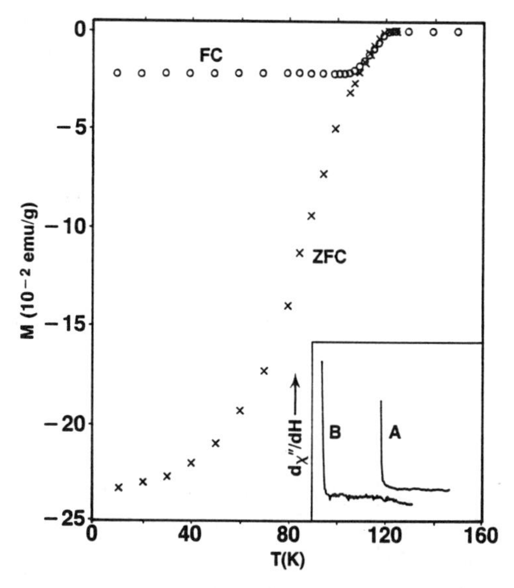

Fig. 3 dc magnetization as a function of temperature for an applied field of 1mT for a nominal $Tl_2Ca_4BaCu_3O_{11+x}$ sample. The insert of Fig. 3 shows data traces for the same sample, and also for a well-prepared $EuBa_2Cu_3O_{7-x}$ sample, where the vertical axis represents the dX"/dH signal of an EPR spectrometer.

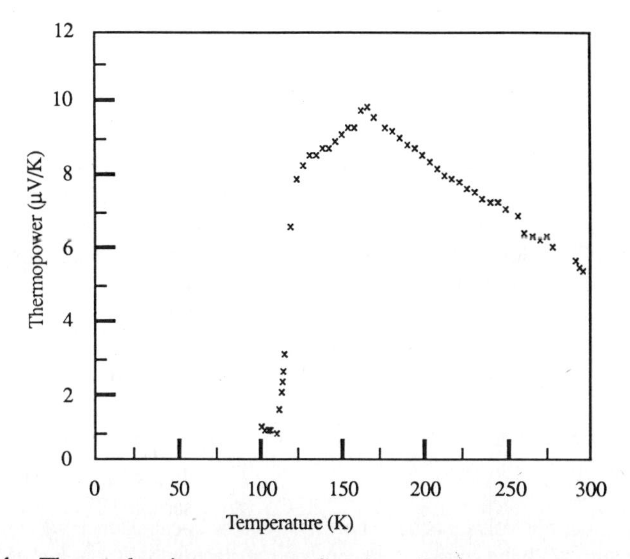

Fig. 4 Thermoelectric power as a function of temperature for a nominal $Tl_2Ca_2Ba_2Cu_3O_{10+x}$ sample.

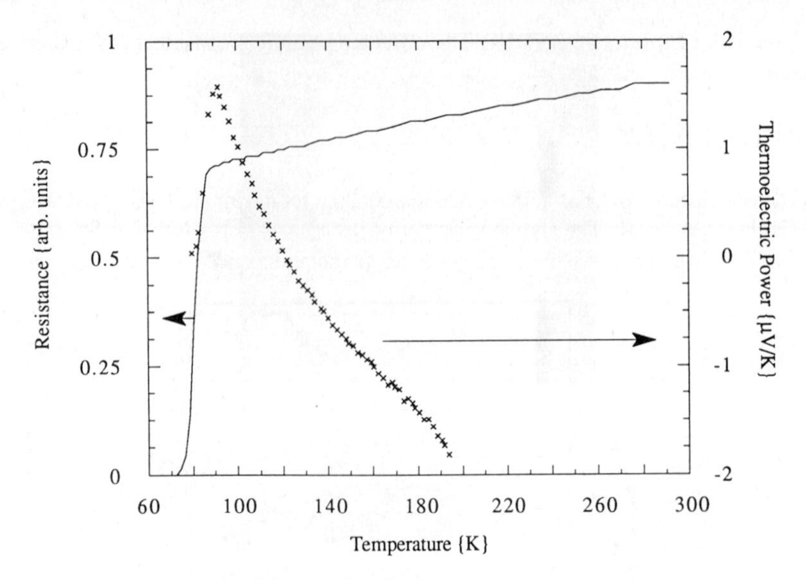

Fig. 5 Resistivity-temperature and thermoelectric power-temperature variations for a ceramic single phase sample of Tl 2201.

Below the transition (the midpoint of the transition was determined to lie at 118K), the thermoelectric power is an increasing function of temperature. Finally, from 175K to room temperature, the thermoelectric power decreases linearly with increasing temperature. The temperature dependence of the Tl-Ca-Ba-Cu-O superconductor is qualitatively similar to that of Y-Ba-Cu-O sample,[26] and is not characteristic of a typical normal metal whose thermoelectric power increases with increasing temperature.

Fig. 5 shows the resistivity-temperature and thermoelectric power-temperature variations for a ceramic sample of the 2201 phase. While this phase is the lowest T_c phase for the double Tl-O compounds, its unusual thermoelectric power-temperature variation with a cross-over from negative to positive thermopower as temperature is lowered (above T_c) warrants further substantial study for the 2201 phase.

JOSEPHSON JUNCTION STUDIES

We have made[27] point contacts between a classical S-wave BCS superconductor (Pb) and $Tl_2Ca_2Ba_2Cu_3O_{10+x}$. A needle shaped piece of Pb was mounted in an apparatus designed to form bulk tunnel junctions.[28] A small coil was mounted on top of the point contact to expose the junction to a dc magnetic field (up to $2x10^{-6}$ T).

At 4.2K we formed several stable Josephson junctions. Fig. 6 shows the the I-V characteristics of a typical junction at 4.2K. A critical current of 1.455mA was found with no magnetic field.

Application of a small external magnetic field gave the periodic behavior predicted by pair tunneling between two superconductors,[29] as shown in Fig. 7.

Above 10K the current-voltage behavior always became linear. Hence we conclude that below the critical temperature of Pb the supercurrent consisted of electron pairs that tunnel from Pb into $Tl_2Ca_2Ba_2Cu_3O_{10+x}$ and that the transport in the $Tl_2Ca_2Ba_2Cu_3O_{10+x}$ is accomplished by electron pairs.

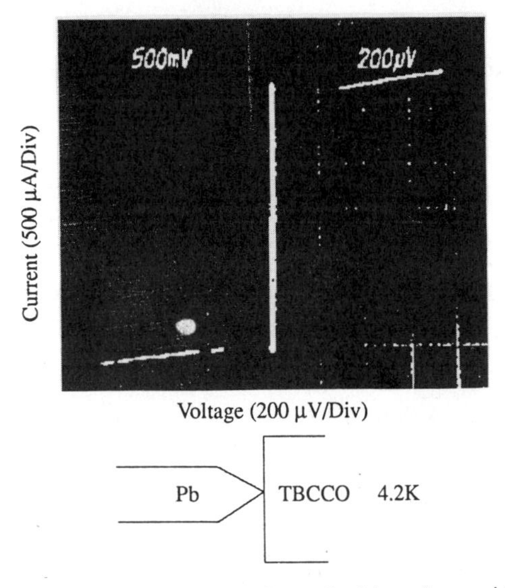

Fig. 6 I-V characteristics of a typical junction at 4.2K.

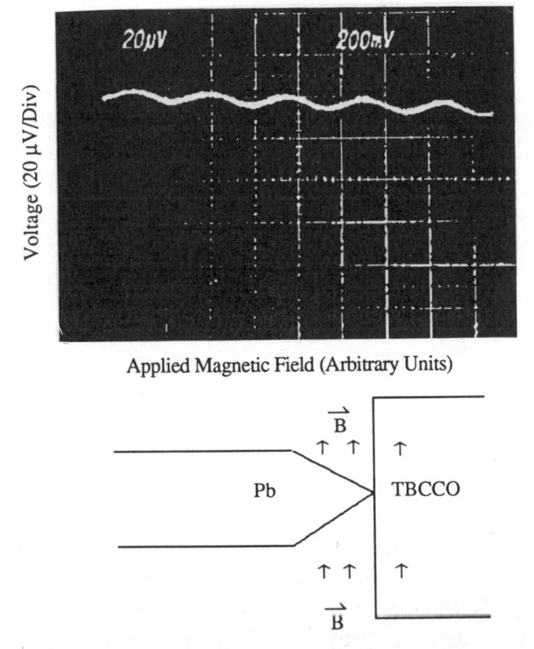

Fig. 7 Characteristics of a Pb-TBCCO tunnel junction in a small external magnetic field.

THEORY AND AN APPROACH TO HIGHER TEMPERATURE COMPOUNDS

Numerous theoretical treatises have been devoted to nearly as many approaches to the understanding of the mechanism of superconductivity in high-temperature superconductors. We list in Table 1 some of the more popular approaches to mechanistic studies. Table 1 also categorizes these approaches as to the origin of the condensation. In this listing, only the Bipolaron mechanism, that of Resonating Valence Bands, and of crystal-field lone-pair condensation, corresponds to a Bose-Einstein condensation.[30] If we assume that superconductivity in the doped copper-oxide superconductors

Table 1

Type of Interaction	BCS Type	Bose-Einstein Condensation
Phonons	•	
Exciton Enhanced Phonons	•	
Excitons – Frenkel (Localized)	•	
Magnetic Interactions		
In-plane	•	
Magnons (Triplet State Pairs)	•	
Bipolaron		•
Spin Bag	•	
Spin Fluctuations	•	
Resonating Valence Bonds		•
2-D Plasmons	•	
Crystal Field – Lone Pairs		•

Groupings of high temperature superconductor theories according to origin of condensation

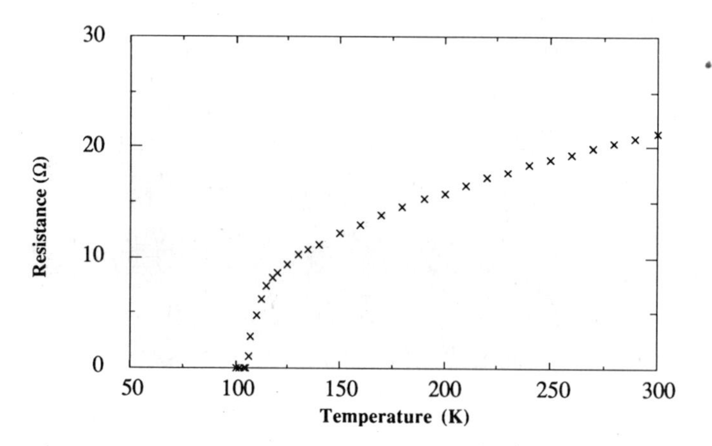

Fig. 8 Resistivity temperature variation for a typical Tl-Ca-Ba-Cu-O film prepared using bulk unreacted Tl-Ca-Ba-Cu-O as a Tl-source

occurs by a BCS mechanism (likely *not* involving phonons directly), we can make some predictions about achieving even higher transition temperatures. The XPS measurements of Meyer et al[31] show an enhanced density of states (DOS) deeper in the valence band (below the Fermi energy, E_F). It may be possible to site-selectively dope the TBCCO compound to access this DOS by lowering of E_F. If successful, BCS theory predicts[32] an enhanced critical temperature with increased DOS.

IMPROVED TL-BA-CA-CU-O THIN FILMS

Many processes to make Tl-Ba-Ca-Cu-O (TBCCO) superconducting thin films have been reported.[33,34,35] After the diffusion process was successfully used to make superconducting bulk samples,[36] it was adapted to make TBCCO superconducting thin films.[37,38] However, the films made by these techniques have been inhomogeneous and rough. We report here the results of studies on a novel safe technique for formation of TBCCO thin films, with significantly reduced surface roughness and greatly improved homogeneity. This technique involves the use of precursor oxide films and Tl-incorporation from unreacted bulk Tl-Ba-Ca-Cu-O sources.

The Ba-Ca-Cu-O (BCCO) precursor thin films used in these experiments were deposited onto MgO, ZrO, or $SiTiO_3$ substrates by magnetron sputtering or laser ablation. The thickness of these films ranged from 5000A to 12000A. The BCCO source targets were pellets that were heated for 24-48 hours at 900°C before the deposition. An annealing in oxygen was carried out midway during the cooling, at 300°C to 500°C for 5-30 minutes. The composition and properties of the precursor films made in this fashion were relatively insensitive to the time and temperature of this anneal.

The Tl-incorporation and superconductor formation were accomplished by a vapor-phase process similar (but not identical) to that reported previously.[36,38] Dense Tl-Ca-Ba-Cu-O pellets were used as Tl-sources. These pellets were made by mixing Tl_2O_3 with bulk precursor material that had been formed by heating (*one* day at 950°) appropriate mixtures of BaO, CaO, and CuO. The nominal bulk stoichiometries reported here for this technique were 2223 (Tl-Ba-Ca-Cu). The Tl_2O_3 oxide/precursor pellets were *not* preheated prior to this use as Tl-sources.

The film precursors deposited onto the substrates were then placed (film downward) on Cu O-rings, which had been placed onto one-half-inch-diameter precursor Tl-Ca-Ba-Cu-O pellets. The pellets + Cu O-rings + films-on-substrates were then placed in an Al_2O_3 boat and placed in a preheated (to 920°C) tube furnace in flowing oxygen for 20-30 minutes. The contents were then cooled to 500°C at a rate of 1-3°C/minute, and then to room temperature by furnace cooling.

All films formed by this technique were superconducting. Fig. 8 is a typical resistance-temperature curve for one of the films on a MgO substrate. The surface roughness is of the order $\pm$ 500A, as compared to surface roughnesses of the order of $\pm$ 3000A for Tl-incorporation from Tl_2O_3 reported previously.[39] Excellent uniformity is also noted from X-ray dot maps except for the Cu map, which does show some non-uniformity, but at a far lower level than that noted previously.[39]

CONCLUSIONS

Layered doped copper oxide superconductors based on Tl have been described. Two-dimensionality appears necessary to the achievement of high critical temperatures (above 120K). Proximity of polarizable sheets adjacent to the Cu-O layers may play an important role in the superconducting properties. It is suggested that the synthesis of organic superconductors with similar dimensionality and structure be attempted.

REFERENCES

1. J.G. Bednorz and K.A. Müller, Z. Phys. B 64, 189 (1986)
2. M.K. Wu, J.R. Ashburn, C.T. Torng, P.H. Hor, R.L. Meng, L. Gao, Z.J. Huang, Y.Z. Wang, and C.W. Chu, Phys. Rev. Lett. 58, 908 (1987)
3. Z.Z. Sheng and A.M. Hermann, Nature 332, 55 (1988)
4. Z.Z. Sheng, A.M. Hermann, A. El Ali, C. Almason, J. Estrada, T. Datta, and R.J. Matson, Phys. Rev. Lett. 60, 937 (1988)
5. H. Maeda, Y. Tanaka, M. Fukutomi, and T. Asano, Jpn. J. Appl. Phys. Lett. 27, L207 (1988)
6. C.W. Chu, J. Bechtold, L. Gao, P.H. Hor, Z.J. Huang, R.L. Meng, Y.Y. Sun, Y.Q. Wang, and Y.Y. Xue, Phys. Rev. Lett. 60, 941 (1988)
7. Z.Z. Sheng and A.M. Hermann, Nature 332, 138 (1988)
8. Z.Z. Sheng, W. Kiehl, J. Bennett, A. El Ali, D. Marsh, G.D. Mooney, F. Arammash, J. Smith, D. Viar, and A.M. Hermann, Appl. Phys. Lett. 52, 1738 (1988)
9. A.M. Hermann, Z.Z. Sheng, D.C. Vier, S. Schultz, and S.B. Oseroff, Phys. Rev. B 37, 9742 (1988)
10. Z.Z. Sheng, A.M. Hermann, D.C. Vier, S. Schultz, S.B. Oseroff, D.J. George, and R.M. Hazen, Phys. Rev. B 38, 7074 (1988)

11. W.L. Lechter, M.S. Osofsky, R.J. Soulen, Jr., V.M. LeTourneau, E.F. Skelton, S.B. Qadri, W.T. Elam, H.A. Hein, L. Humphreys, C. Skowronek, A.K. Singh, J.V. Gilfrich, L.R. Toth, and S.A. Wolf (submitted)
12. S. Matsuda, S. Takeuchi, A. Soeta, T. Suzuki, K. Aihara, and T. Kamo (submitted)
13. M.A. Subramanian, C.C. Torardi, J. Gopalakrishnan, P.L. Gai, J.C. Calabrese, T.R. Askew, R.B. Flippen, and A.M. Sleight (submitted)
14. Z.Z. Sheng and A.M. Hermann (unpublished)
15. Z.Z. Sheng, L. Sheng, X. Fei, and A.M. Hermann, Phys. Rev. B 39, 2918 (1989)
16. R.M. Hazen, L.W. Finger, R.J. Angel, C.T. Prewitt, N.L. Ross, C.G. Hadidiacos, P.J. Heaney, D.R. Veblen, Z. Z. Sheng, A. El Ali, and A.M. Hermann, Phys. Rev. Lett. 60, 1657 (1988)
17. L. Gao, Z.J. Huang, R.L. Meng, P.H. Hor, J. Bechtold, Y.Y. Sun, C.W. Chu, Z.Z. Sheng, and A.M. Hermann, Nature 332, 623 (1988)
18. C.C. Torardi, M.A. Subramanian, J.C. Calabrese, J. Gopalakrishnan, E.M. McCarron, K.J. Morrissey, T.R. Askew, R.B. Flippen, U. Chowdhry, and A.M. Sleight, Phys. Rev. B, 225 (1988)
19. M.A. Subramanian, J.C. Calabrese, C.C. Torardi, J. Gopalakrishnan, T.R. Askew, R.B. Flippen, K.J. Morrissey, U. Chowdhry, A.M. Sleight, Nature 332, 420, (1988)
20. C.C. Torardi, M.A. Subramanian, J.C. Calabrese, J. Gopalakrishnan, K.J. Morrissey, T.R. Askew, R.B. Flippen, U. Chowdhry, and A.M. Sleight, Science 240, 631 (1988)
21. S.S.P. Parkin, V.Y. Lee, E.M. Engler, A.I. Nazzal, T.C. Huang, G. Gorman, R. Savoy, and R. Beyers, Phys. Rev. Lett. 60, 2539 (1988)
22. Y. Luo, Y.L. Zhang, J.K. Liang, and K.K. Fung (submitted)
23. S.S.P. Parkin, V.Y. Lee, A.I. Nazzal, R. Savoy, R. Beyers, and S.J. La Placa, Phys. Rev. Lett. 61, 750 (1988)
24. A.M. Hermann, Z.Z. Sheng, D.C. Vier, S. Schultz, and S.B. Oseroff, Phys. Rev. B 37, 9742 (1988)
25. N. Mitra, J. Trefny, B. Yarar, G. Pine, Z.Z. Sheng, and A.M. Hermann, Phys. Rev. B 38, 7064 (1988)
26. N. Mitra, J. Trefny, M. Young, and B. Yarar, Phys. Rev. B 36, 5581 (1987)
27. W. Eidelloth, F.S. Barnes, Z.Z. Sheng, and A.M. Hermann, Appl. Phys. Comm. 8(4), 191 (1988)
28. W. Eidelloth, F.S. Barnes, S. Geller, L.C. Xu, and E.A. Kraut (in print)
29. A. Barone, G. Paterno, in "Physics and Applications of the Josephson Effect," Wiley, New York (1982)
30. W.A. Little, Science 242, 1390 (1988)
31. H. Meyer, T.J. Wagener, and J.H. Weaver, Phys. Rev. B 39, 7343 (1988)
32. J. Bardeen, L.H. Cooper, and J.R. Schrieffer, Phys. Rev. 108, 1175 (1957)
33. C.X. Qiu and I. Shih, Appl. Phys. Lett. 53, 523 (1988)
34. P. Barboux, J.M. Tarascon, F. Shokoohi, B.J. Wilkens, and C.L. Schwartz, J. Appl. Phys. 64, 6382 (1988)
35. W.L. Olson, M.M. Eddy, T.W. James, R.B. Hammond, G. Gruner and L. Drabeck, Appl. Phys. Lett. 55, 188 (1989)
36. Z.Z. Sheng, L. Sheng, H.M. Su, and A.M. Hermann, Appl. Phys. Lett. 53, 2686 (1988)
37. B. Johs, D. Thompson, N.J. Ianno, J.A. Woollam, A.M. Hermann, Z.Z. Sheng, W. Kiehl, Q. Shams, X. Fei, L. Sheng, and Y.H. Liu, Appl. Phys. Lett. 54 (18), 1810 (1989)
38. S.I. Shah, N. Herron, C.R. Fincher, and W.L. Holstein, Appl. Phys. Lett. 56, 782 (1990)
39. A.M. Hermann and Z.Z. Sheng, in "Science and Technology of Thin Film Superconductors," R.D. McConnell and S.A. Wolf, eds., Plenum Press, New York (1989)

Fundamental Similarities between Organic and Cuprate Superconductors Shown by Muon Spin Relaxation Studies

Y.J. Uemura[1], L.P. Le[1], G.M. Luke[1], B.J. Sternlieb[1],
J.H. Brewer[2], T.M. Riseman[2], G. Saito[3,4], and H. Yamochi[3,4]

1. *Department of Physics, Columbia University, New York, New York 10027, USA*
2. *TRIUMF and Department of Physics, University of British Columbia, Vancouver, B.C., Canada*
3. *Department of Chemistry, Kyoto University, Kitashirakawa, Kyoto 606, Japan*
4. *Institute for Solid State Physics, University of Tokyo, Roppongi, Tokyo 106, Japan*

Muon Spin Relaxation (μSR) measurements on $(BEDT-TTF)_2Cu(NCS)_2$ show that the magnetic field penetration depth λ_{bc} within the conductive b-c plane is about 8000 Å at $T \to 0$. This value corresponds to the superconducting carrier density $n_s = 1.6 \times 10^{20}/cm^3$ for an in-plane effective mass $m^* = 3.5m_e$. Comparisons with μSR results on high-T_c cuprate superconductors demonstrate that both the present organic superconductor and the hole-doped cuprate systems are characterized by a small number of superconducting pairs, i.e. 5 to 10 pairs, in the coherence area ξ^2 within the 2-dimensional conductive plane. Implications of this feature on theories of superconductivity in these systems are discussed.

The magnetic field penetration depth λ of type-II superconductors can be measured by muon spin relaxation (μSR) experiments[1]. Remarkable correlations between T_c and $1/\lambda^2 \propto n_s/m^*$ (superconducting carrier density / effective mass) were found in μSR measurements[2,3] on various high-T_c cuprate superconductors. We have recently extended this μSR study to an organic superconductor $(BEDT-TTF)_2Cu(NCS)_2$ ($T_c \sim 10K$ in zero field). In the T_c versus $1/\lambda^2 \propto n_s/m^*$ plot, the μSR results from this organic system give a point close to the linear relation[2] found for the cuprate high-T_c systems. The extension of these correlations to the organic superconductor is very interesting, since high-T_c cuprate and organic superconductors share various general characteristics, such as, highly 2-dimensional electronic structure, high H_{c2}, short coherence length ξ, and low carrier density. In this paper, we elucidate this aspect with experimental results from μSR and some other measurements, and demonstrate that (1) both the organic and cuprate systems are close to the clean limit situation; (2) the organic superconductor has superconducting carrier density $n_s \sim 1.6 \times 10^{20}/cm^3$; and (3) the mean separation of superconducting carriers within the conductive planes is about 1/3 of the in-plane coherence length $\xi_{in-plane}$ in both the organic and cuprate superconductors. We also discuss the implication of these features for the condensation mechanisms of these superconductors.

Organic Superconductivity, Edited by V.Z. Kresin and W.A. Little, Plenum Press, New York, 1990

$(BEDT-TTF)_2Cu(NCS)_2$ (*bis*-ethylenedithio-tetrathiafulvalene dithiocyano-cuprate)[4] has highly 2-dimensional (2-d) electronic structure with the conductivity $1/\rho$ in the conductive b-c plane 600 times larger than that along the a* direction. The conducting planes, composed of π electrons of the organic molecules, are separated by insulating layers of anions with the interlayer distance $c_{int} = 15.2 = 16.3 \times sin(110.3deg)$ Å along the a* direction. Small plate-like crystals (typically $2 \times 1 \times 0.1$ mm) are obtained using the electrochemical method, with the conductive b-c plane parallel to the plate surface. We gathered many of these crystals (~ 300 mg) with their faces alligned parallel to form a specimen of about 20 mm in diameter and 1 mm thick. The transverse-field μSR measurements were performed at TRIUMF (Vancouver) using an external magnetic field $H_{ext} = 3.1$ kG applied perpendicular to the conductive b-c planes.

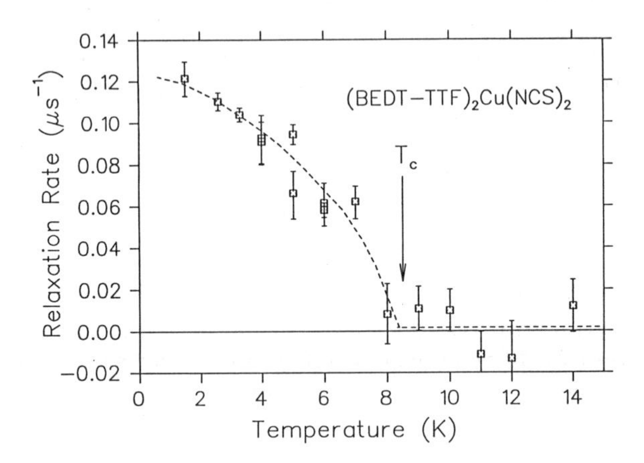

Fig. 1. Temperature dependence of the transverse-field muon spin relaxation rate σ observed in single crystal specimens of the organic superconductor $(BEDT-TTF)_2Cu(NCS)_2$ with the external field $H_{ext} = 3.1kG$ applied perpendicular to the conductive b-c plane. The low temperature relaxation rate ($\sigma = 0.12\mu sec^{-1}$ after correction for background relaxation) correspnds to a penetration depth of $\lambda \sim 8000$ Å. The broken line is guide to the eye.

The muon spin relaxation rate σ was derived by fitting the damping of the precession envelope with a Gaussian function $exp(-0.5\sigma^2 t^2)$. The relaxation rate increased with decreasing temperature below $T \sim 8K$, which corresponds to the critical temperature T_c in $H_{ext} = 3.1kG$. We subtracted the temperature independent background relaxation $\sigma_{bg} \sim 0.19\mu sec^{-1}$ (due to nuclear dipolar fields) from the observed value σ_{obs} to obtain the relaxation σ due to superconductivity as $\sigma^2 = \sigma_{obs}^2 - \sigma_{bg}^2$. Figure 1 shows the temperature dependence $\sigma(T)$. Zero-field μSR studies confirmed the absence of static magnetic order.

In the vortex state of type-II superconductors where H_{ext} is partially screened, the local field within the specimen is inhomogeneous with a width $\Delta H \equiv \sqrt{< (H- < H >)^2 >}$ determined by the penetration depth λ as $\Delta H \propto 1/\lambda^2$. Then the value of λ can be derived from the observed relaxation rate σ as $\sigma = \gamma_\mu \Delta H = \alpha/\lambda^2$, where $\gamma_\mu = 85.1kHz/G$ and α is a proportionality constant which is independent of H_{ext} over a wide range of H_{ext} (see ref. 5). From the low temeprature relaxation rate $\sigma(T \rightarrow 0) = 0.12\mu sec^{-1}$, we obtain the absolute value of the in-plane penetration depth (for $H_{ext} \perp$ conductive plane) to be $\lambda(0) = 8000$ Å. This value is consistend with a recent report by Harshman *et al.*[6]. The lack of data at $T \leq 1.5K$ and insufficient statistics of the data prevent us from determining the symmetry of the superconducting pairs using $\lambda(T)$.

The penetration depth λ is a function of n_s/m^* as

$$\frac{1}{\lambda^2} = \frac{4\pi n_s e^2}{m^* c^2} \times \frac{1}{1 + \xi/l}$$

with a correction term due to the coherence length ξ and the mean free path l. In the present organic superconductor, $H_{c2}(T \rightarrow 0)$ for $H \perp$ b-c plane is about 7 *Tesla* (ref. 4), indicating the in-plane coherence length ξ is about 70 Å. Shubnikov de-Haas (SvdH) experiments by Toyota *et al.*[7] and Oshima *et al.*[8] yield the mean scattering time $\tau \sim 10^{-12} sec$, the Fermi momentum $k_F \sim 1.4 \times 10^7 cm^{-1}$, and the effective mass $m^* \sim 3.5m_e$ for the carriers in the b-c plane, which implies that the mean free path $l = \tau \times \hbar k_F/m^* \sim 500$ Å. A more conservative estimate of l can be made using conductivity results $1/\rho = 7500\Omega^{-1}cm^{-1}$ just above T_c (ref. 9) which provides a lower limit $l \geq 200$ Å. Thus, we conclude that the present organic superconductor is close to the clean limit $\xi/l \ll 1$.

Adopting the approximation that $\xi/l \rightarrow 0$, we obtain $n_s/(m^*/m_e) = 4.6 \times 10^{19}/cm^3$ from the relaxation rate $\sigma(T \rightarrow 0)$. For $m^* = 3.5m_e$, the superconducting carrier density in this system becomes $n_s = 1.6 \times 10^{20}/cm^3$. We note that this carrier density is significantly smaller than the number $n \sim 1.2 \times 10^{21}/cm^3$ expected if one carrier exists per formula unit. The SvdH experiment[7], however, found that the Fermi surface of the hole carriers consists only of 18 % of the first Brillouin zone, giving a hole carrier density of $2.1 \times 10^{20}/cm^3$. Therefore, the carrier density measured in the present μSR study can be explained if one assumes that these holes are responsible for superconductivity in $(BEDT - TTF)_2Cu(NCS)_2$.

Now it is interesting to compare this case with the cuprate high-T_c systems. In $YBa_2Cu_3O_7$ (YBCO), the coherence length of carriers within the conductive CuO_2 planes (a-b plane) is about $\xi_{ab} = 15 \sim 25$ Å, as inferred[10] by the very high H_{c2}. The in-plane resistivity $\rho_{ab} \sim 50\mu\Omega cm$ of a single crystal specimen[11] just above T_c gives a lower limit of the mean free path l as $l \geq 100$ Å. An optical study[12] on a film specimen suggests $l \sim 160$ Å. Thus, the cuprate system also falls in the clean limit situation.

We have performed μSR experiments on more than 30 different specimens of various cuprate high-T_c superconductors. As reported in ref. 2, we found remarkable correlations in a plot of T_c versus $\sigma(T \rightarrow 0) \propto 1/\lambda^2 \propto n_s/m^*$. Figure 2 shows the results on the cuprates. In this study, we used non-oriented sintered ceramic specimens, which are expected to have better homogeneities of the chemical concentrations compared to single crystal specimens[13]. In non-oriented ceramic specimens of highly anisotropic cuprate systems with $\lambda_{in-plane}$ ($H_{ext} \perp$ a-b plane) much shorter than λ_{soft} for $H_{ext} \parallel$ a or b-axis, the observed penetration depth λ_{obs} is determined predominantly by $\lambda_{in-plane}$, as shown by a model calculation[14] which finds $\lambda_{obs} = 1.23 \times \lambda_{in-plane}$ for $\lambda_{soft}/\lambda_{in-plane} \geq 5$. Therefore, the results in Fig. 2 represent the relation between T_c and n_s/m^* for the carriers in the conductive CuO_2 planes.

As the hole-carrier concentration is increased, T_c initially increases, then saturates and is even suppressed in the heavily doped region. The initial increase of T_c follows a linear relation $T_c \propto n_s/m^*$ shared by compounds with single, double, and triple CuO_2 layers. We called the overall features of Fig. 2 "universal correlations". Recently, the results on $YBa_2Cu_3O_y$ in Fig. 2 have been confirmed by independent works performed at PSI (Zurich) on $YBa_2Cu_3O_y$ (ref. 15) and $YBa_2Cu_3O_7$ doped with H (ref. 16), as well as by our work on $(Y_{1-x}Pr_x)Ba_2Cu_3O_7$ (ref. 3) shown in Fig. 2 by solid circles. The good quantitative agreement of these results indicate that T_c is determined by the carrier density in the CuO_2 plane regardless of the chemical substitutions used to control n_s. The good reproducibility of the results in Fig. 2 provides further support to our picture of the clean limit, since it is hardly expected if the system were in the dirty limit and the values of λ depended heavily on the mean free path l which is a highly material-sensitive parameter. We also note that the average separation c_{int} of the conductive CuO_2 planes is 6 ± 1 Å for all the cuprate systems plotted in Fig. 2. Therefore, the correlations in fig. 2 can be regarded to exist between T_c and the areal carrier density on the plane n^s_{area} divided by m^*, where $n^s_{area} = n_s \times c_{int}$.

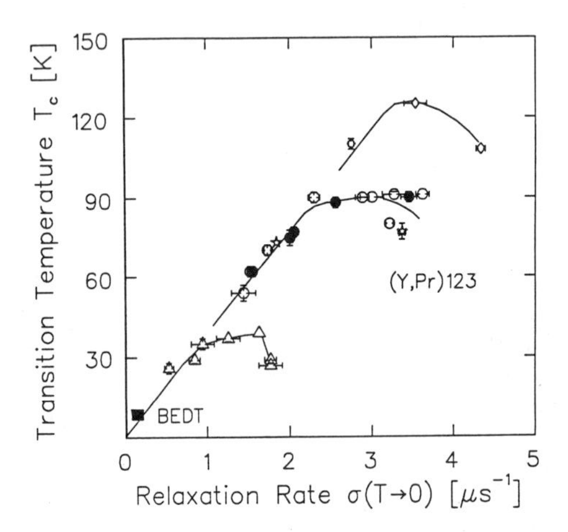

Fig. 2. The superconducting transition temperature T_c of various high-T_c superconductors plotted versus the low temperature values of the muon spin relaxation rate $\sigma(T \rightarrow 0)$. Both T_c and σ have been determined by μSR measurements. In the "clean limit" $\xi/l \ll 1$, $\sigma \propto 1/\lambda^2 \propto n_s/m^*$. The open symbols represent points from Fig. 2 of ref. 2 (see the figure caption in ref. 2 for details), where the points for the 123 system are obtained with the oxygen depleted specimens $YBa_2Cu_3O_y$. The closed circles represent $(Y_{1-x}Pr_x)Ba_2Cu_3O_7$ with x = 0.3 to 0.05 (ref. 3), and the closed square near the origin represents $(BEDT-TTF)_2Cu(NCS)_2$.

When we plot the present results from the organic BEDT system in Fig. 2, the point comes close to the linear relation found for the cuprates. This is very interesting since both systems share some basic features such as 2-d electronic structure and low carrier density. To be more precise, here we compare the μSR results of the organic superconductor with those we obtained on c-axis oriented ceramic specimen of $YBa_2Cu_3O_7$ (ref. 17) by applying H_{ext} perpendicular to the conductive a-b plane. Table 1 compares the results from the BEDT and YBCO systems. The relaxation rate $\sigma(T \rightarrow 0)$ is about 30 times different between the two systems, as is the volume carrier density n_s divided by m^*. Using the interplane distances c_{int}, and assuming the effective mass to be $m^* = 3.5m_e$ for BEDT and $m^* \sim 2.3m_e$ (ref. 18) for YBCO, respectively,

Table I. Comparisons of parameters in BEDT and YBCO systems.

Parameter	Symbol	$(BEDT\text{-}TTF)_2Cu(NCS)_2$	$YBa_2Cu_3O_7$
conductive plane		b-c	a-b (CuO_2)
average interplane distance	c_{int}	15.2 [Å]	5.8 [Å]
H_{c2} for H perpendicular to conductive plane	H_{c2}	7 [T] [4,8]	≧ 68 [T] [10]
in-plane coherence length	$\xi_{in\ plane}$	70 [Å]	16-22 [Å]
mean free path just above T_c	l	≧ 200 [Å] [7]	≧ 100 [Å] [12]
effective mass	m^*	3.5 m_e [7]	2 - 2.5 m_e [18]
RESULTS OBTAINED FROM THE μSR MEASURMENTS			
sample shape		single crystals	c-axis oriented ceramic
T_c determined by μSR	T_c	8 [K]	90 [K]
low temperature relaxation rate	$\sigma(T\to 0)$	0.12 [μsec^{-1}]	3.5 [μsec^{-1}] [17]
penetration depth (in plane value)	$\lambda(T\to 0)$	8000 [Å]	1450 [Å]
carrier density divided by m^*	$n_s / (m^*/m_e)$	4.6×10^{19} [$/cm^3$]	1.3×10^{21} [$/cm^3$]
superconducting carrier density	n_s	1.6×10^{20} [$/cm^3$]	3.1×10^{21} [$/cm^3$]
areal carrier density on conductive plane	n^s_{area}	2.5×10^{13} [$/cm^2$]	1.8×10^{14} [$/cm^2$]
mean carrier separation	d_{carr}	20 [Å]	7.5 [Å]
coherence length divided by carier separation	ξ / d_{carr}	3.5	2.7

we can estimate the areal carrier density n^s_{area} in the conductive planes. Then we calculate the mean separation d_{carr} of superconducting carriers within the 2-d planes to be 20 Å for BEDT and 7.5 Å for YBCO. It is important to note that the in-plane coherence length ξ is about 3 to 4 times longer than the mean carrier separation d_{carr} in both of the BEDT and YBCO systems. This implies that there are about 10 to 15 superconducting carriers (*i.e.,* 5 to 8 pairs) in the 2-d coherence area of ξ^2.

The point for BEDT in Fig. 2 lies on left hand side of the linear line for the cuprates. From the above argument, we realize that this can be understood in terms of the difference in the interplane distance c_{int} (in other words, the 3-d to 2-d conversion factor) between the organic and cuprate systems. In addition, we note here that not only the oriented ceramic specimen of YBCO (with $\sigma \sim 3.5 \mu sec^{-1}$ corresponding to an angular averaged value $\sigma \sim 2.5 \mu sec^{-1}$ in the case of non-oriented ceramic specimen) but all the specimens on the linear line of $T_c \propto \sigma$ are expected to safisfy the relation $\xi / d_{carr} \sim 3$ for the following reasons. The Fermi energy ϵ_F of 2-d electron gas is proportional to n/m^*, leading to the Fermi velocity $v_F \propto \sqrt{n}/m^*$. Combining this with $k_B T_c \propto \Delta \propto v_F/\xi$, where Δ denotes the energy gap, the coherence length is given as $\xi \propto (T_c)^{-1} \times \sqrt{n}/m^*$. Since the systems on the linear line of Fig. 2 satisfy $T_c \propto n_s/m^*$, these systems have $\xi \propto \sqrt{1/n_s}$. The mean carrier separation d_{carr}, calculated from $d^2_{carr} \times c_{int} = 1/n_s$, is also proportional to $\sqrt{1/n_s}$. Therefore, the linear relation in Fig. 2 implies that all those cuprate systems on the line have $\xi / d_{carr} \sim 3$.

This fundamental feature $\xi / d_{carr} \sim 3$, common to the organic and many cuprate systems, is very important when we consider the condensation mechanisms of these superconductors. The BCS theory[19] for ordinary superconductors is developed for the case of $\xi / d_{carr} \geq 100$, where the Cooper pairs are overlapping with many other pairs in real space. This leads to the condensation in momentum space. The opposite extremum is the Bose-Einstein (B-E) condensation for $\xi / d_{carr} \sim 1$ where bosons exist independently in real space without overlapping with each other (for example, consider the case of ^4He). The present work has shown that the cuprate and organic superconductors belong to neither the conventional BCS nor the standard B-E cases but lie in the middle of them with several pairs overlapping. This is the most unexplored region.

It is possible to come to the relation $T_c \propto n_s/m^*$ by starting from either of the two extremum cases. With decreasing carrier density of fermions, the Fermi enrgy ϵ_F is reduced, and one can expect the situation where ϵ_F becomes smaller than the energy scale $\hbar\omega_B$ of the bosons which mediate the pairing. Then the T_c formula $T_c \propto \hbar\omega_B exp(-1/N(\epsilon_F)\Lambda)$ in the BCS weak coupling limit ($\epsilon_F \gg \hbar\omega_B$) is replaced by the one which has ϵ_F as the pre-exponential factor, leading to $T_c \propto \epsilon_F \propto n_s/m^*$. This situation correponds to "no retardation" of the pairing interaction[20]. Starting from the B-E situation, it stems in the following way. The B-E condensation does not occur in purely 2-dimensional systems. When one calculates the 3-d B-E condensation temperature using the values of n_s and m^* appropriate for YBCO, for example, T_c becomes too high. In quasi 2-dimensional systems, however, one can expect the condensation temperature to be of the order of 100 K and to scale as $T_c \propto n_s/m^*$ (see ref. 21). It is also possible to expect $T_c \propto n_s$ in Anyon theories[22].

In summary, we demonstrated that organic and cuprate high-T_c superconductors share fundamental characteristics, *i.e.,* $\xi / d_{carr} \sim 3$, in addition to various common general features such as 2-d electronic structure, high H_{c2}, short ξ, long λ, and low carrier density. Although the microscopic pairing mechanism of the carriers may be different between the organic and cuprate systems, the present results suggest that there is a good chance that these two types of superconductors have the same condensation mechanism, which could be different from conventional BCS and B-E condensations.

Acknowledgement

We would like to thank R. Kadono, R.F. Kiefl, S.R. Kreitzman, C.L. Seaman, M.B. Maple, C.E. Stronach, and many other scientists and co-authors of ref. 2 for collaboration on the μSR experiments; V.J. Emery, R. Friedberg, T.D. Lee, T. Ishiguro for stimulating discussions; NSF (DMR-89-13784), David and Lucile Packard Foundation (YJU), and NSERC of Canada, and the Japanese Ministry of Education, Culture and Science (Grant-in-Aid) for financial support.

References

1. See, for example, Y.J. Uemura *et al.*, Physica *C*162-164, 857 (1989); H. Keller, IBM J. Res. Develo, 33, 314 (1989); and references in these two review papers.
2. Y.J. Uemura *et al.*, Phys. Rev. Lett. 62, 2317 (1989).
3. C.L. Seaman*et al.*, Phys. Rev. *B*, in press.; Y.J. Uemura *et al.*, *Proceedings for the NATO Advance Research Workshop on "Dynamics of Magnetic Fluctuations in High Temperature Superconductors"*, (Crete, 1989), Plenum, 1990, in press.
4. G. Saito, Physica *C*162 – 164, 577 (1989), and references therein.
5. P. Pincus *et al.*, Phys. Lett. 13, 31 (1964); E.H. Brandt, Phys. Rev. *B*37, 2349 (1989). These two papers give about 30 % different values for the constant α, resulting in 16 % difference in the absolute value of λ. For consistency with our earlier papers, here we use the constant α derived by Pincus *et al.* after modifying it for the triangular lattice.
6. D.R. Harshman *et al.*, Phys. Rev. Lett. 64, 1293 (1990). The definition of σ in this paper is different from the one in the present paper by a factor of $\sqrt{2}$. Harshman *et al.* argued the BEDT system to be close to the dirty limit situation, which we disagree.
7. N. Toyota *et al.*, J. Phys. Soc. Japan 57, 2616 (1988).
8. K. Oshima *et al.*, Physica *C*153-155, 1148 (1988).
9. Typical resistivity results can be found in ref. 4 or ref. 8. According to T. Ishiguro (private communication), better single crystals show the conductivity just above T_c about 1000 times larger than the room temperature values.
10. U. Welp *et al.*, Phys. Rev. Lett. 62, 1908 (1989) obtained $\xi_{inplane}$ = 16 Å.: Y. Iye *et al.*, Physica *C*153-155, 26 (1988) estimated $H_{c2} \sim 680kG$ for $H \parallel$ c-axis, which corresponds to $\xi_{inplane}$ = 22 Å.
11. H. Takagi and S. Uchida, unpublished data.
12. K. Kamaras *et al.*, Phys. Rev. Lett. 64, 84 (1990).
13. The best single crystal of the $(La_{2-x}Sr_x)CuO_4$ system with $x \sim 0.15$ (nominal) was made by Kojima and Tanaka using the zone melting method (the so-called Koshu crystals). Our μSR measurements on this Koshu crystal indicated that there is a considerable spread of the doped carrier concentration within the specimen, and the average carrier concentration was significantly lower than that of the optimum ceramic specimen. This example shows the difficulty in producing high-quality single crystals with sufficient number of doped hole carriers.
14. W. Barford and J.M.F. Gunn, Physica *C*156, 515 (1988).
15. B. Pumpin *et al.*, presented at the International Conference on Muon Spin Rotation, Oxford, 1990, to be published in Hyperfine Interactions.
16. A. Weidinger *et al.*, *ibid.*.
17. Y.J. Uemura *et al.*, J. de Physique (Paris) 49, C8-2087 (1988).
18. $m^*/m_e = 2 \sim 2.5$ in YBCO is inferred from optical measurements. See, for example, B. Batlogg, in *High Temperature Superconductivity*, ed. by K.S. Bedell *et al.*, Addison-Wesley, California 1990, p.p. 60; In $Bi_2Sr_2CaCu_2O_8$ systems, photo emission studies have found the mass $m^* \sim 2m_e$ from the dispersion relation: see P.A. Lee, *ibid*, p. 113; In La_2CuO_{4+y} system, C.Y. Chen *et al.* found $m^* = (2\pm1)m_e$ from the measurement of dielectric constant. All these results suggest that the effective mass in the cuprate systems is about 2-3m_e.
19. J. Bardeen, L.N. Cooper and J.R. Schrieffer, Phys. Rev. 108, 1175 (1957).
20. V.J. Emery and G. Reiter, Phys. Rev. *B*38, 4547 (1988).
21. R. Friedberg and T.D. Lee, Phys. Rev. *B*40, 6745 (1989).
22. R. Laughlin, Phys. Rev. Lett 60, 2677 (1988); Science 242, 525 (1988); B.I. Halperin *et al.*, Phys. Rev. *B*40, 8726 (1989).

ORGANICS VS CUPRATES: WHY IS T_c STILL SO LOW IN THE ORGANICS

S.A. Wolf* and V.Z. Kresin**

*Naval Research Lab., Washington DC 20375-5000
** Lawrence Berkeley Lab., Berkeley CA 94720

ABSTRACT

Based on our previous evaluation of the major parameters of the cuprate superconductors, the analysis - organic superconductors vs cuprate superconductors is carried out. The organics are found to have a short coherence length and, as a result, display two-gap structure. Many of the normal and superconducting parameters of the two classes of compounds are quite similar. However, the presence of nesting states in the organic conductors is an important factor; this is a major difference between the two classes of materials and may be partly responsible for the differences in the transition temperatures.

INTRODUCTION

Remarkable progress in the field of organic superconductivity made during the last 10 years has demonstrated the great potential of these materials for future development. Therefore, it is of definite interest to understand the origin of the present limitations of T_c. We think that a comparison of the organic superconductors and the high T_c oxides [1] is a fruitful approach to this problem.

The organic superconductors which belong to $(ET)_2X$ family, (e.g. $(BEDT\text{-}TTF)_2\ Cu(SCN)_2$, which has one of the highest values of T_c amongst the organics) and high T_c cuprates, have a lot of similarities and some differences. The present analysis is based on our approach to the physics of high T_c [1] and recent experimental data on organic superconductivity.

Organic Superconductivity, Edited by V.Z. Kresin and W.A. Little, Plenum Press, New York, 1990

Normal Properties

As was noted above, high T_c cuprates and organic superconductors have a number of similarities. Both classes of materials are characterized by a large anisotropy. $(ET)_2X$ materials were ET is the abbreviation for BEDT-TTF have a quasi layered structure[2]; the same property turns out to be a common feature of cuprates and it is one decisive factor determining their high T_c [1]. In addition, both classes have relatively small and quite similar carrier concentration. In our recent papers [1,3] we described in detail the evaluation of the major normal and superconducting parameters of cuprates. The values of these parameters are presented in Tables I and II. Our evaluation is based on Fermiology. For La-Sr-Cu-O compounds, the Fermi surface is cylindrically shaped. The Fermi surfaces of $(ET)_2Cu(SCN)_2$ ($T_c \cong$ 10.4K) [4] and $(ET)_2KHg(SCN)_4$ ($T_c <$ 0.5K) [5] are also nearly cylindrical. Their parameters have been obtained recently [4,5] by using conventional methods (Shubnikov - de Haas oscillations, and magnetoresistance); the ability to use these methods is due to a relatively low value of T_c and upper critical field (one can study the normal state in the low temperature region). The comparison of the normal state parameters of the cuprates, organic superconductors and conventional metals can be seen from Table I. The Fermi velocity, v_F, the Fermi energy, E_F, and the Fermi wave vector, k_F, for $(ET)_2Cu(SCN)_2$ were estimated from the upper critical field H_{c2} by the following equations:

$$v_F = 4.4\, \Delta(0)/\hbar[\phi_o/2\pi H_{c2}]^{1/2}$$

$$E_F = m^* v_F^2/2 \text{ and } k_F = m^* v_F/\hbar$$

where $\Delta(0)$ is the zero temperature energy gap, and ϕ_o is the flux quantum and m* is the measured effective mass. We used a value for $\Delta(0)$ of 4.0 $k_B T_c$ [6].

According to [4], the $(ET)_2KHg(SCN)_4$ material has a large in-plane anisotropy of the normal conductivity (1:2). This anisotropy is probably present for the $(ET)_2Cu(SCN)_2$ compound as well. The Cu-O sheet in the cuprates is not anisotropic although Y-Ba-Cu-O has an anisotropy in the a-b plane due to the contribution of the chains [7] . The presence of this anisotropy in the organics means that the cross-sections of the cylindrical Fermi surfaces of $(ET)_2X$ and La-Sr-Cu-O cut by the plane Pz=const have a different geometry. Namely, the cross-section for $(ET)_2X$ is a stretched ellipse whereas for La-Sr-Cu-O, the deviation from circular shape is not large. The presence of the stretched ellipse means the presence of nesting states and this is an important factor which affects the value of T_c (see below).

The value of the effective mass, m*, in $(ET)_2KHg(SCN)_4$ is 1.4 and is 3.5 for $(ET)_2Cu(SCN)_2$ whereas for La-Sr-Cu-O, m*≅5. This difference is related to the strengths of the electron-phonon coupling (see next section).

TABLE I. Normal state parameters of conventional metals the cuprates and organics.

Parameters	Conventional	LaSrCuO	YBaCuO	$(ET)_2KHg(SCN)_4$	$(ET)_2Cu(SCN)_2$
m*	$1\text{-}15m_e$	$5m_e$	$5m_e$|$25m_e$	$1.4m_e$	$3.5m_e$
E_F (eV)	5-10 eV	0.1 eV	0.3 eV	0.05 eV	0.07 eV
k_F $(cm)^{-1}$	$\sim 10^8$	3.5×10^7	--------	$\leq 2 \times 10^7$	$\leq 2 \times 10^7$
v_F (cm-sec)	$1\text{-}2 \times 10^8$	8×10^6	10^7| 2×10^6	$\approx 10^7$	$\approx 10^7$

Superconducting Properties

Either using values of the Fermi velocities or the upper critical field H_{c2}, one can estimate the values of the coherence lengths in these materials. According to [1,3] ξ_o for the cuprates is short (~15-25 Å). It turns out that the value of ξ_o for $(ET)_2Cu(SCN)_2$ is also short (≅70Å) relative to conventional superconductors. At the same time, the mean free path in the organic superconductor is large (≅340Å, see Ref. [4]). Therefore, the criterion $\xi_o < l$ is satisfied; it means (see [3]), that in the presence of the overlapping energy bands, one can observe two-gaps. A very similar situation occurs in the Y-Ba-CuO compound (in this case $\xi_o << l$,see [3]).

Note that the analysis of the band structure in $(ET)_2X$ [8] shows that two bands cross the Fermi level. Moreover, tunneling data [9] indicates the presence of two gaps in a similar organic compound. As a result, we think that the presence of the two-gap structure along with the anisotropy of each gap is perfectly realistic.

The presence of a two-gap structure is an important factor which is directly related to the origin of organic superconductivity. First of all, one should stress (see also [3]) that the one-to-one correspondence between the ratio $\Delta(o)/T_c$ and the strength of the coupling is valid only for the one-gap model. The two-gap superconductivity is characterized by two gaps and three coupling constants, and one problem should be studied with a considerable care.

Secondly, a number of experimental data (see e.g. [10]), reveal temperature dependences of several quantities (e.g. spin susceptibility) that are different from the usual BCS exponential dependence. This has been interpreted as a manifestation of an unconventional non-phononic mechanism (probably, magnetic) of superconductivity. We would like to emphasize that the presence of multigap structure allows us the opportunity to give a different interpretation. In this case the temperature dependence is a sum of exponentials, and such a sum can match the observed power law.

A large value of the effective mass in the La-Sr-Cu-O compound (see Table I) is connected with strong coupling between the carriers and lattice [3]; indeed $m^*(o)=m^b (1+\lambda)$, where m^b is the band value, and λ is the coupling constant. For La-Sr-Cu-O, $\lambda \cong 2.5$, and, therefore, $m^*(o) \cong 3.5\ m^b$. Even though this value of λ represents very strong coupling, it cannot alone account for the measured T_c of 40 K.

By comparing the measured effective masses of the two organic compounds listed in Table I in which one has a high T_c (10.4 K) and the other is not superconducting above 0.5 K, we can make the simple assumption that the enhanced effective mass of the superconducting compound is due to a much larger electron-phonon coupling constant. By setting the m^* of $(ET)_2KHg(SCN)_4$ equal to the m^b of the high T_c compound and using the equation that $m^* = m^b(1=\lambda)$ we estimate that λ for $(ET)_2Cu(SCN)_2$ is approximately 1.5. This is indicative that the organics are strong coupled but not very strong coupled as we found for La-Sr-Cu-O. Indeed, a value of 1.5 for λ can adequately account for the measured transition temperature of 10.4 K. (see Table II)

We noted above, that the Fermi surface of the organic superconductor contains a large number of nesting states. This is very favorable for the occurance of a charge density wave(CDW) transition in this material. A CDW has the effect of opening up an energy gap over the nested parts of the Fermi surface, removing this phase space from forming Cooper pairs. Qualitatively, this means that the effective dimensionality of the high T_c organic superconductor is intermediate between quasi-2D and quasi-1D, in contrast to the much more quasi-2D character of the Cu-O planes in

the cuprates. Again, the presence of the nesting states in this case is a negative factor (cf., Ref. [11]) because of the transition at $T>T_c$ to the charge density wave (CDW) state. We think that this factor is important and its elimination might lead to further increases in T_c of organic low-dimensional materials.

SUMMARY

We have shown that the cuprates and organics are very similar in their normal state properties differing mainly in their smaller

TABLE II. Superconducting parameters of conventional metals the cuprates and organics.

Parameters	Conventional	LaSrCuO	YBaCuO	$(ET)_2KHg(SCN)_4$	$(ET)_2Cu(SCN)_2$
T_c	<23 K	40 K	95K	<0.5 K	10.4 K
Δ /E_F	10^{-4}	10^{-1}	$2x10^{-1} \| 10^{-1}$	--------	$5x10^{-2}$
λ_{el-ph}	< 2	2.5	--------	--------	~1.5
ξ_o (Å)	10^3-10^4	20	15 | 7	--------	70

values for the Fermi energy and k-vector. This reduced phase space for pairing is due to their intermediate dimensionality (between quasi-1D and quasi-2D) and nested Fermi surface.

ACKNOWLEDGEMENT

We both thank ONR for its support of this research. In particular one of us (VZK) is supported by the Office of Naval Research under Contract No. N00014-89-F0006 for work carried out at the Lawrence Berkeley Laboratory under Contract No. DE-AC03-76S-F00098.

REFERENCES

1. V. Z. Kresin and S. A. Wolf, Phys. Rev. B41,4278 (1990)

2. F. L. Pratt et. al., Phys. Rev. Lett. 61, 2721 (1988)

3. V. Z. Kresin and S. A. Wolf, preprint

4. K. Oshima et. al. Phys. Rev. B38, 938 (1988)

5. T. Osada et. al.,Technical Report of ISSP, Serial A #2211, (1989)

6. See Y. Uemura and other articles, this proceedings.

7. T. A. Friedman et. al, preprint.

8. T. Mori et. al., Bull. Chem. Soc. Japan 57, 627 (1986)

9. K. E. Gray, M. E. Hawley and E. R. Moog, *Novel Superconductivity* ed. by S. A. Wolf and V. Z. Kresin, Plenum, NY (1987) pgs. 611-627.

10. D. Jerome and F. Creuzet, *ibid.* pgs. 103-135.

11. M. Ichimura et. al., Phys. Rev. B41, 6387 (1990)

NEW ORGANIC SUPERCONDUCTORS

STRUCTURE-PROPERTY RELATIONSHIPS FOR ß- AND κ-PHASE BEDT-TTF SALTS AND THEIR USE IN THE SYNTHESIS OF κ-(BEDT-TTF)$_2$Cu[N(CN)$_2$]Br: A SALT HAVING THE HIGHEST-T_c (INDUCTIVE ONSET = 11.6 K, RESISTIVE ONSET = 12.5 K) YET OBSERVED IN AN ORGANIC SUPERCONDUCTOR

Jack M. Williams, Aravinda M. Kini, Urs Geiser, Hau H. Wang, K. Douglas Carlson, W. K. Kwok, K. G. Vandervoort, James E. Thompson, Daniel L. Stupka, D. Jung† and M.-H. Whangbo†

Chemistry and Materials Science Divisions, Argonne National Laboratory, Argonne, Illinois, 60439; †Department of Chemistry, North Carolina State University, Raleigh, North Carolina, 27695

INTRODUCTION

The electron-donor molecule BEDT-TTF (or ET, **1**) and its salts have yielded the highest number of ambient pressure organic superconductors and also the highest T_c's reported to date compared to any other electron-donor molecule. The most highly developed structure-property correlations for ET salts are for the isostructural ß-(ET)$_2$X,

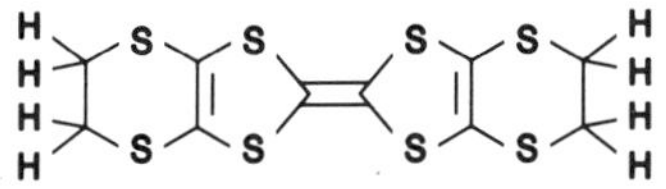

BEDT-TTF (ET)

Figure 1

X = mono-valent linear anion, superconductors.[1,2] In these salts both the S···S distances in the corrugated-sheet[3] networks, and T_c, increase concomitantly with the linear anion length, i.e., T_c of I_3^- (0.5 kbar), 8 K > AuI_2^-, 4.98 K > IBr_2^-, 2.8 K.[2] The synthesis of these materials has been summarized in detail in the literature.[2]

ß-Phase (ET)$_2$X, X = Anion, Superconductors

In the isostructural ß-phase materials, the anions reside within a cavity of ethylene-group hydrogen atoms formed by the corrugated-sheet[3] layers of ET molecules as shown in the stereodiagram, **2**.

Organic Superconductivity, Edited by V.Z. Kresin and W.A. Little, Plenum Press, New York, 1990

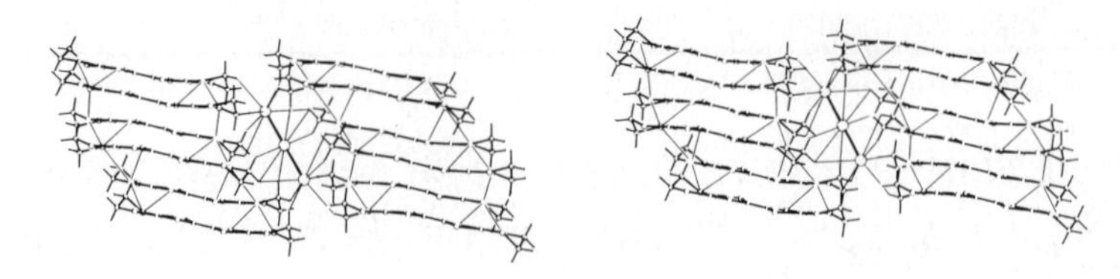

Figure 2

ß-$(ET)_2X$

X		V_c (Å^3/298 K)	T_c (K)
I_3^-	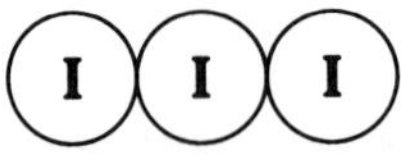 I I I	855.9	1.5
AuI_2^-	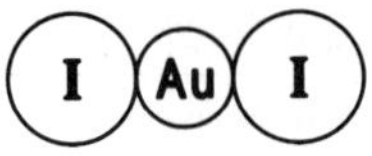 I Au I	845.2	5
I_2Br^-	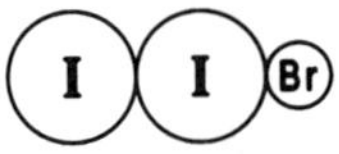 I I Br	842.3	none
IBr_2^-	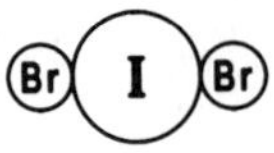 Br I Br	828.7	2.80
$ClIBr^-$	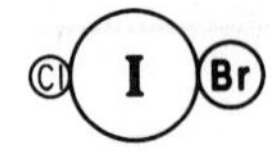 Cl I Br	821.3	?
ICl_2^-	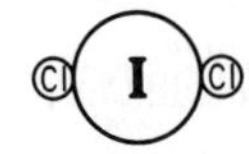 Cl I Cl	814.3	T_{MI}=22.1

Figure 3

The effect of varying the anion length is to change the unit cell volume in a like fashion and at the same time alter the system of intra- and interstack S···S contact distances between ET molecues. The linear anions that yield ß or ß-like structures are shown in **3** where anion length, unit cell volume, and T_c (where appropriate) are given. Thus, over the range of anions given in **3**, the electrical properties range from that of metals with metal-insulator transitions in the shortest anions to ambient pressure superconductivity in the longer anions (IBr_2^-, AuI_2^- and I_3^-). ß-$(ET)_2I_3$ is unique because a modest pressure of only 0.5 kbar is sufficient to convert it to a new structure[4] [ß*-$(ET)_2I_3$] with T_c = 8 K, which is the maximum found to date in ß-$(ET)_2X$ type superconductors. These observations (unit cell volume and T_c) formed the basis of a useful structure-property correlation[1,2] for the isostructural salts as shown in **4**. From **4** we can see that anions longer than I_3^- might give new ß-$(ET)_2X$ superconductors with $T_c > 8$ K!

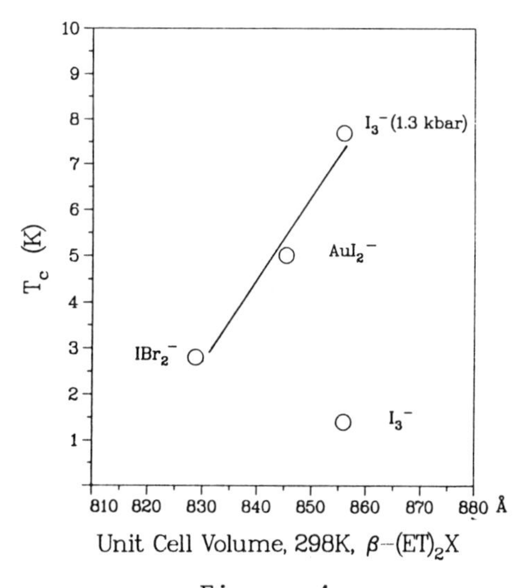

Figure 4

Since triatomic anions containing terminal halide atoms that are longer than I_3^- are presently unknown, new anionic species[2] such as $(NCS\text{-}M\text{-}SCN)^-$ and $(NC\text{-}M\text{-}CN)^-$, M = metal, have been investigated as the source of potential ß-$(ET)_2X$ superconductors. A conjectured correlation similar to that vein in **4**, but based on anion length, is shown in **5**.

An inspection of **5** reveals that T_c's as high as ~40 K might be achieved in ß-$(ET)_2X$ phase materials if the linear anions depicted in **5** could be prepared and incorporated in the ß-$(ET)_2X$ structure. In an attempt to prepare longer anions that I_3^-, candidate species such as $(NC\text{-}Ag\text{-}CN)^-$, $(I\text{-}Ag\text{-}I)^-$and $(NCS\text{-}Au\text{-}SCN)^-$ were suggested,[2] which, coupled with the introduction of the in situ crown ether-anion preparatory route,[5] led to the synthesis of the first polymeric anion ET conductors[5,6] viz., $(ET)Ag_4(CN)_5$ and $(ET)_3Ag_{6.4}I_8$, respectively.

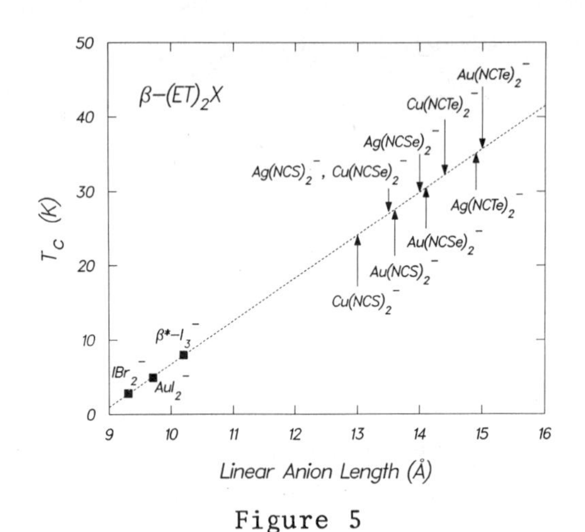

Figure 5

κ-Phase $(ET)_2X$, X = Anion, Superconductors

These findings vide supra were necessary precursors to the discovery[7] of ambient pressure superconductivity in the first polymeric anion superconductor, κ-$(ET)_2Cu(NCS)_2$, with the highest T_c (10.4 K, resistive midpoint) in an organic superconductor, at that time.[7-9] However, this salt belongs to an entirely different structural class than the ß-phase salts, viz., the so-called "κ-type" containing orthogonally arranged molecular dimers rather than a corrugated-sheet network.[8,9b] Thus, complex polymeric anions offered a new route to the synthesis of ET superconductors. However, our previous structure-property correlations[2] for ß-phase systems do not apply to κ-phase salts because of their different structural features.

The negative pressure derivative of T_c for κ-$(ET)_2Cu(NCS)_2$, is larger than that for ß-$(ET)_2X$ salts, and, in fact, is the largest known (–3 to –3.6 K/kbar) for any superconductor.[10,11] These findings, and the recent analyses of structural features of other known κ-phase salts,[12-14] again suggest that the expansion of the ET donor-molecule network, perhaps attainable through the incorporation of even larger polymeric anions, is a viable approach to further raise T_c's in κ-phase materials. Similar arguments based on "effective volume" of ET have also been put forward by Saito et al.[15] Although no precise structure-property correlations of the type determined for ß-phase materials exist for κ-phase systems, the main predictors of superconductivity[12-14] in the latter materials are (1) "bond over ring" intradimer molecular overlap and (2) relatively short (~3.35 Å) intradimer separations.

κ-$(ET)_2Cu[N(CN)_2]Br$, A New Organic Superconductor

Our investigations based on the above guidelines led us to consider bis(pseudo-halide)cuprate(I) type of anions larger than $Cu(NCS)_2^-$, in particular, the substitution of dicyanamide ion, $(NC\text{-}N\text{-}CN)^-$, for thiocyanate. We also considered substitution reactions on the trigonal copper atom site that eventually resulted in the synthesis of a new superconducting cation-radical salt, κ-$(ET)_2Cu[N(CN)_2]Br$. The dicyanamide ion, $[N(CN)_2]^-$, although bent, is larger than the thiocyanate, $(SCN)^-$, anion. The new salt is an ambient pressure superconductor with an inductive onset $T_c = 11.6$ K (resistive onset = 12.5 K), which represents a new high for an organic superconductor.[20]

Preliminary ESR measurements on a single crystal of κ-$(ET)_2Cu[N(CN)_2]Br$ indicated a single absorption line spectrum with a peak-to-peak linewidth at room temperature in the range of 60-85 G, depending on crystal orientation. Low temperature studies revealed an estimated relative spin susceptibility that was approximately constant (300 K-100 K) indicative of a metallic sample. Both, the magnitude of the peak-to-peak linewidth, and its temperature dependence are similar to those of previously known κ-phase compounds, e.g., κ-$(ET)_2Cu(NCS)_2$[16,17] and κ-$(ET)_4Hg_{2.89}Br_8$.[18]

The crystal structure of κ-$(ET)_2Cu[N(CN)_2]Br$ was determined by use of single crystal X-ray diffraction data. Unit cell data (space group *Pnma*, $Z = 4$) are as follows (298 K): a = 12.942(3) Å, b = 30.016(4) Å, c = 8.539(3) Å, V = 3317(1) Å^3. The structure contains alternating layers (perpendicular to the b axis) of ET donor-molecules and polymeric $\{Cu[N(CN)_2]Br\}_\infty^-$ anions. The overall arrangement with respect to symmetry elements and unit cell directions is the same as that in κ-$(MDT\text{-}TTF)_2AuI_2$.[19] The ET donor-molecule network (see stereodiagram **6**) contains centrosymmetric face-to-face dimers arranged in a typical κ-fashion,[12] i.e., adjacent dimers are approximately perpendicular to each other (74.5°). The central C=C bond of one molecule, if projected along the normal to the molecular plane, is on top of one of the 5-membered rings of the other molecule within the dimer ("bond-over-ring") arrangement. This feature is common to all known κ-phase superconductors.[12,13] All S···S intramolecular contacts within a dimer exceed the sum of the van der Waals radii (3.6 Å), but shorter contacts exist between dimers, as shown in **6**. The topology of the S···S contacts in this salt is different from that found in κ-$(ET)_2Cu(NCS)_2$[9b]

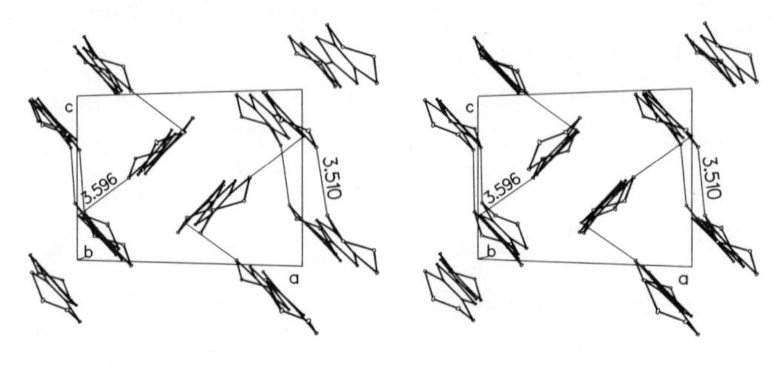

Figure 6

The polymeric anion, which resides on a mirror plane, is shown in **7**. It contains infinite zig-zag chains of ···Cu–dicyanamide–Cu··· units, and the copper atoms complete their coordination spheres with the bromine atom. Trigonal coordination around Cu^+ is occasionally observed, most notably in κ-$(ET)_2Cu(NCS)_2$, where similar chains exist. One notable difference between the two salts is that in κ-$(ET)_2Cu[N(CN)_2]Br$ the chains run along the longer of the two unit cell axes within the layer (i.e., *a*), whereas in κ-$(ET)_2Cu(NCS)_2$ they run along the shorter direction.

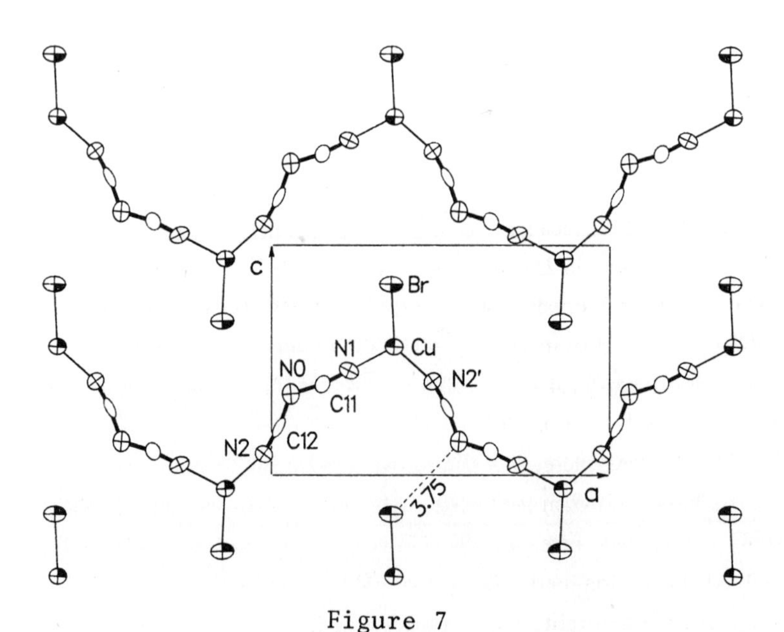

Figure 7

Superconductivity in κ-$(ET)_2Cu[N(CN)_2]Br$ was established by the use of rf penetration depth measurements, low-field dc magnetization measurements, and resistivity measurements. Superconducting transition curves determined by the rf method are illustrated in **8** for two single-crystal specimens of ~200 µg mass each. This figure

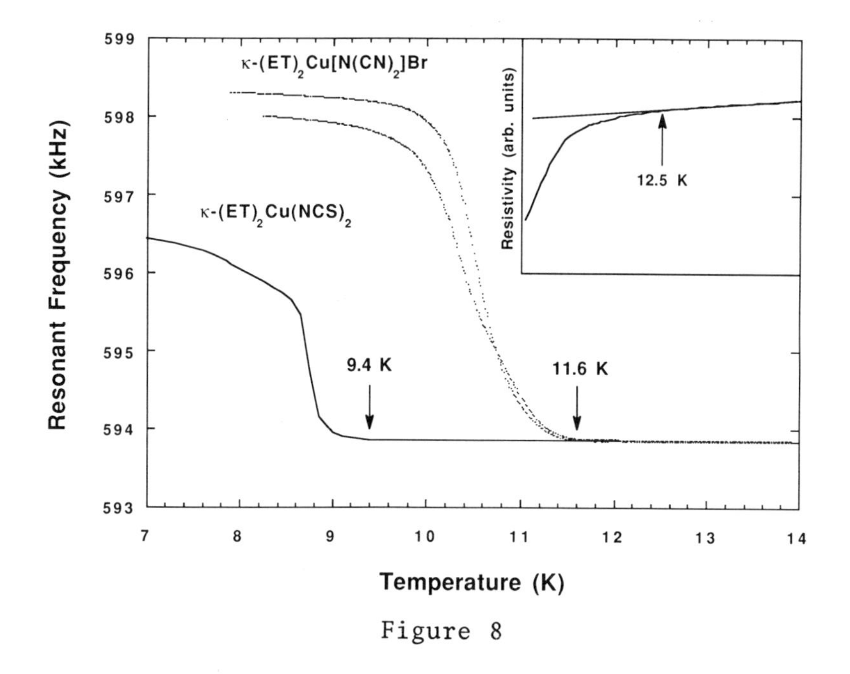

Figure 8

includes, for comparison, the transition curve previously determined[9b] for κ-$(ET)_2Cu(NCS)_2$ by the same technique, apparatus, and thermometry. In these measurements, the crystals were cooled overnight from room temperature to 60 K and further cooled to ~25 K at a rate of ~0.5 K/min to avoid possible problems with frozen-in crystallographic disorder that often lowers T_c in organic superconductors. The transition curves were determined by measurements of the frequency at intervals of 0.01 K with comparably slow cooling from 25 K to ~8 K, a temperature below signal saturation.

The curves for both crystal specimens of κ-$(ET)_2Cu[N(CN)_2]Br$ exhibit a weak superconducting onset temperature of 11.60 ± 0.05 K, a strong increase in the resonant frequency beginning near 11.4 K, and saturation of the increased resonant frequency beginning near 9.6 K. In contrast, the transition curve for κ-$(ET)_2Cu(NCS)_2$ exhibits[9b] an onset of 9.4 K, a strong increase in the resonant frequency beginning near 8.9 K, and saturation near 6 K (not shown in the Figure). Magnetization experiments were carried out by use of a SQUID magnetometer, and for cooling in a magnetic field H of 1 Oe, with $H \parallel b$, the onset of the diamagnetic susceptibility occurred at the same temperature, 11.6 ± 0.1 K. The susceptibility curve had essentially the same shape and saturation temperature as that of the rf transition curve. We estimate ~30% bulk superconductivity at 4 K from our measurements of both dc-field cooled and zero-field cooled magnetization. Clearly the salt is a volume superconductor with a T_c substantially higher (~1-2 K) than that of κ-$(ET)_2Cu(NCS)_2$. Four-probe resistivity measurements of a single crystal specimen of κ-$(ET)_2Cu[N(CN)_2]Br$ gave a resistive onset T_c of 12.5 ± 0.1 K (see inset in **8**) in two different experiments (with the same crystal specimen) in which a different apparatus, with different thermometry, was utilized. A complete 4-probe resistivity curve is given in **9** indicating (see inset) again, an onset of 12.5 ± 0.1 K, a midpoint of 11.2 ±0.1 K, and completion at 10.5 ± 0.1 K. The linearly extrapolated resistive onset, i.e., the intersection of lines extrapolated from the normal state resistivity and from the linear drop of the resistance in the superconducting state, is 11.6 ± 0.1 K, in perfect agreement with the inductive onset of 11.60 ± 0.05 K. All thermometers used in the several measurements reported here have been calibrated by Lake Shore Cryotronics, Inc., against standards calibrated by the U.S. National Institute of Standards and Technology. It is also noteworthy that κ-$(ET)_2Cu[N(CN)_2]Br$ displays a broad resisitivity maximum at ~90-100 K, a feature previously observed in other κ-phase salts, κ-$(ET)_2Cu(NCS)_2$[7,9,11] and κ-$(DMET)_2AuBr_2$.[23]

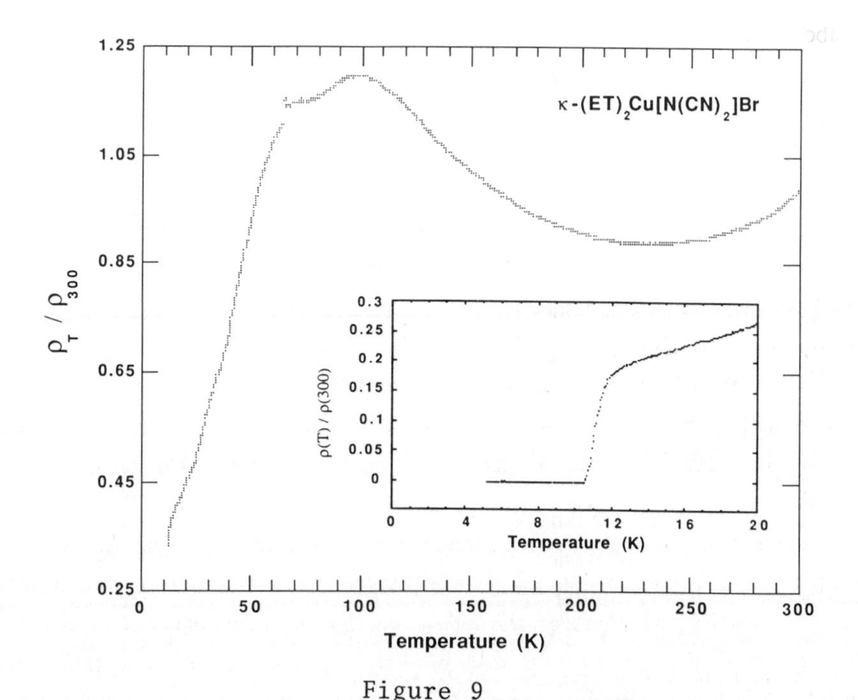

Figure 9

In order to put the above cited new developments into proper perspective, the T_c increment (~1-2 K) of κ-$(ET)_2Cu[N(CN)_2]Br$ over κ-$(ET)_2Cu(NCS)_2$ is approximately the same as the increment observed of κ-$(ET)_2Cu(NCS)_2$ over the previous highest-T_c organic superconductor ß*-$(ET)_2I_3$ (T_c = 8 K).

The calculated dispersion relations and Fermi surface (two highest-occupied bands based on a tight-binding band electronic structure calculation)[21] reveals that with the formal oxidation of $(ET)_2^+$, the highest occupied band of κ-$(ET)_2Cu[N(CN)_2]Br$ is half-filled, and its Fermi surface consists of overlapping distorted circles centered at Γ and its equivalent points in reciprocal space. Therefore, κ-$(ET)_2Cu[N(CN)_2]Br$ is expected to be a two-dimensional metal. The band dispersion relations and the Fermi surface of κ-$(ET)_2Cu[N(CN)_2]Br$ are very similar to those of other κ-phase salts.[9b,12,14,15,22]

CONCLUSION

A new ambient pressure organic superconductor, κ-$(ET)_2Cu[N(CN)_2]Br$, has been discovered with an inductive onset $T_c = 11.6 \pm 0.1$ K (resistive onset = 12.5 ± 0.1 K). The anion in the new salt consists of a polymeric mixed (halide)(pseudohalide)cuprate(I) species. Because of the large number of metal/halide/pseudohalide substitutions that can be made in this anionic species, it opens up an entirely new conceptual approach to the design of additional conducting and superconducting (with even higher T_c's) cation-radical salts by use of planar polymeric anions. Studies along theses lines are currently under way in our laboratory.

Finally, some perspective on the present status and future prospects of organic superconductors in relation to the high-T_c oxide superconductors can be gained from an inspection of **10**. It is also becoming increasingly clear that in terms of structure (both are layered superconductors) and the physics (both are type II-superconductors with anisotropic coherence lengths, anisotropic H_{c2}, and in which superconductivity is competing with magnetic ground states, etc.), the organic superconductors are more similar to the high-T_c oxide superconductors than to other classes of superconductors. Thus, the prospects for discovering high-T_c in organic systems appear very bright!

ACKNOWLEDGMENT

Work at Argonne National Laboratory and at North Carolina State University is supported by the Office of Basic Energy Sciences, Division of Materials Sciences, U.S. Department of Energy, under Contract W-31-109-ENG-38 and Grant DE-FG05-86ER45259, respectively. K.G.V. is a Laboratory Graduate Student participant from the University of Illinois (Chicago, IL), D.L.S. and J.E.T. are student research participants

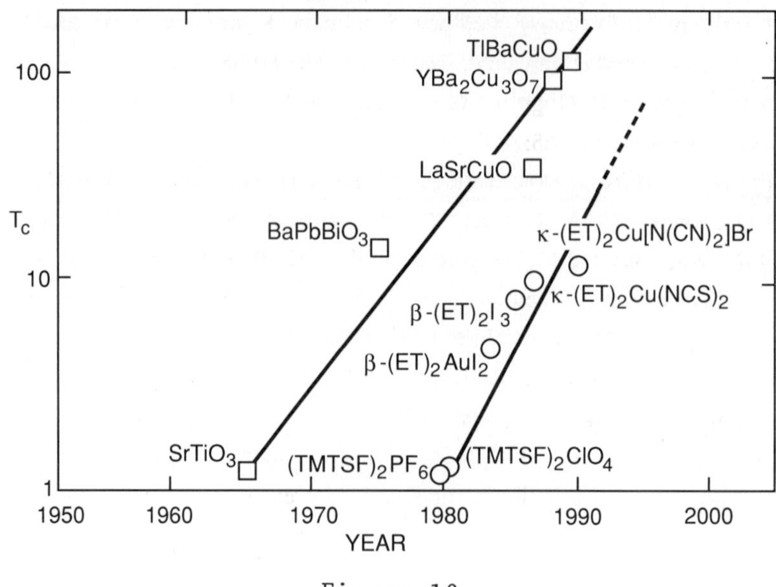

Figure 10

from the University of Illinois (Urbana, IL) and the University of Michigan (Ann Arbor, MI), respectively, all sponsored by the Argonne Division of Educational Programs..

REFERENCES

1. J. M. Williams, A. J. Schultz, H. H. Wang, K. D. Carlson, M. A. Beno, T. J. Emge, U. Geiser, M. E. Hawley, K. E. Gray, E. L. Venturini, J. F. Kwak, L. J. Azevedo, J. E. Schirber and M.-H. Whangbo in Proceedings of the Yamada Conference XV on Physics and Chemistry of Quasi One-Dimensional Conductors, Lake Kawaguchi, Yamanashi, Japan, May 26-30, 1986, S. Tanaka, K. Uchinokura, Eds., Physica B 143:346 (1986).
2. J. M. Williams, H. H. Wang, T. J. Emge, U. Geiser, M. A. Beno, K. D. Carlson, R. J. Thorn, A. J. Schultz and M.-H. Whangbo, Prog. Inorg. Chem. 35:51 (1987).
3. J. M. Williams, M. A. Beno, H. H. Wang, P. E. Reed, L. J. Azevedo and J. E. Schirber, Inorg. Chem. 23:1790 (1984).
4. A. J. Schultz, H. H. Wang, J. M. Williams and A. Filhol, J. Am. Chem. Soc. 108:7853 (1986).
5. U. Geiser, H. H. Wang, K. M. Donega, B. A. Anderson, J. M. Williams and J. F. Kwak, Inorg. Chem. 25:401 (1986).
6. U. Geiser, H. H. Wang, L. E. Gerdom, M. A. Firestone, L. M. Sowa, J. M. Williams and M.-H. Whangbo, J. Am. Chem. Soc. 107:8305 (1985).
7. H. Urayama, H. Yamochi, G. Saito, K. Nozawa, T. Sugano, M. Kinoshita, S. Sato, K. Oshima, A. Kawamoto and J. Tanaka, Chem. Lett. 55 (1988).

8. H. Urayama, H. Yamochi, G. Saito, S. Sato, A. Kawamoto, A. Tanaka, T. Mori, Y. Maruyama and H. Inokuchi, Chem. Lett. 463 (1988).
9. (a) S. Gärtner, E. Gogu, I. Heinen, H. J. Keller, T. Klutz and D. Schweitzer, Solid State Commun. 65:1531 (1989).
(b) K. D. Carlson, U. Geiser, A. M. Kini, H. H. Wang, L. K. Montgomery, W. K. Kwok, M. A. Beno, J. M. Williams, C. S. Cariss, G. W. Crabtree, M.-H. Whangbo and M. Evain, Inorg. Chem. 27:965, 2904 (1988).
10. J. E. Schirber, E. L. Venturini, A. M. Kini, H. H. Wang, J. R. Whitworth and J. M. Williams, Physica C 152:157 (1988).
11. W. Kang, D. Jérome, C. Lenoir and P. Batail, J. Phys.: Condens. Matter 2:1665 (1990).
12. D. Jung, M. Evain, J. J. Novoa, M.-H. Whangbo, M. A. Beno, A. M. Kini, A. J. Schultz, J. M. Williams and P. J. Nigrey, Inorg. Chem. 28:4516 (1989) and 29:1594 (1990).
13. (a) A. M. Kini, M. A. Beno, K. D. Carlson, J. R. Ferraro, U. Geiser, A. J. Schultz, H. H. Wang, J. M. Williams and M.-H. Whangbo in Proceedings of the First ISSP International Conference on the Physics and Chemistry of Organic Superconductors, Tokyo, Japan, August 17-30, 1989, G. Saito, S. Kagoshima, Eds. Springer-Verlag: Heidelberg, in press.
(b) J. M. Williams, H. H. Wang, A. M. Kini, K. D. Carlson, M. A. Beno, U. Geiser, M.-H. Whangbo, D. Jung, M. Evain and J. J. Novoa, Mol. Cryst. Liq. Cryst. 181:59 (1990).
14. M.-H. Whangbo. D. Jung, H. H. Wang, M. A. Beno, J. M. Williams and K. Kikuchi, Mol. Cryst. Liq. Cryst. 181:1 (1990).
15. G. Saito, H. Urayama, H. Yamochi and K. Oshima, Synth. Met. 27:A331 (1988).
16. H. Urayama, H. Yamochi, G. Saito, T. Sugano, M. Kinoshita, T. Inabe, T. Mori, Y. Maruyama and H. Inokuchi, Chem. Lett. 1057 (1988).
17. H. H. Wang, L. K. Montgomery, A. M. Kini, K. D. Carlson, M. A. Beno, U. Geiser, C. S. Cariss, J. M. Williams and E. L. Venturini, Physica C 156:173 (1988).
18. H. H. Wang, B. A. Vogt, U. Geiser, M. A. Beno, K. D. Carlson, S. Kleinjan, N. Thorup and J. M. Williams, Mol. Cryst. Liq. Cryst. 181:135 (1990).
19. (a) G. C. Papavassiliou, G. A. Mousdis, J. S. Zambounis, A. Terzis, S. Hountas, B. Hilti, C. W. Mayer, J. Pfeiffer in Proceedings of International Conference on Science and Technology of Synthetic Metals, Santa Fe, NM, June 26-July 2, 1988, M. Aldissi, Ed., Synth. Met. 27:B379 (1988).
(b) A. M. Kini, M. A. Beno, D. Son, H. H. Wang, K. D. Carlson, L. C. Porter, U. Welp, B. A. Vogt, J. M. Williams, D. Jung, M. Evain, M.-H. Whangbo, D. L. Overmeyer and J. E. Schirber, Solid State Commun. 69:503 (1989).

20. A. M. Kini, U. Geiser, H. H. Wang, K. D. Carlson, J. M. Williams, W. K. Kwok, K. G. Vandervoort, J. E. Thompson, D L. Stupka, D. Jung, M.-H. Whangbo, Inorg. Chem. 29:xxx (1990).
21. (a) M.-H. Whangbo, J. M. Williams, P. C. W. Leung, M. A. Beno, T. J. Emge, H. H. Wang, K. D. Carlson and G. W. Crabtree, J. Am. Chem. Soc. 107:5815 (1985).
    (b) M.-H. Whangbo and R. Hoffmann, J. Am. Chem. Soc. 100:6093 (1978).
22. (a) K. Oshima, T. Mori, H. Inokuchi, H. Urayama, H. Yamochi and G. Saito Phys. Rev. B: Condens. Matter 38:938 (1988).
    (b) K. Oshima, T. Mori, H. Inokuchi, H. Urayama, H. Yamochi and G. Saito, Synth. Met. 27:A165 (1988).
23. K. Kukuchi, Y. Honda, Y. Ishikawa, K. Saito, I. Ikemoto, K. Murata, H. Anzai, T. Ishiguro and K. Kobayashi, Solid State Commun. 66:405 (1988).

NEW AMBIENT PRESSURE ORGANIC SUPERCONDUCTORS: α-(BEDT-TTF)$_2$(NH$_4$)Hg(SCN)$_4$, βm-(BEDO-TTF)$_3$Cu$_2$(NCS)$_3$, AND κ-(BEDT-TTF)$_2$Cu[N(CN)$_2$]Br

Hau H. Wang, Mark A. Beno, K. Douglas Carlson, Urs Geiser, Aravinda M. Kini, Lawrence K. Montgomery, James E. Thompson and Jack M. Williams

Chemistry and Materials Science Divisions, Argonne National Laboratory
Argonne, Illinois, 60439, U. S. A.

INTRODUCTION

More than one hundred and twenty conducting salts based on the organic donor-molecule BEDT-TTF are known, where BEDT-TTF is bis(ethylenedithio)tetrathiafulvalene (abbreviated herein as ET). Several of the early salts possessed tetrahedral and octahedral anions, such as $(ET)_2ClO_4(TCE)$,[1] $(ET)_2PF_6$,[2] $(ET)_2ReO_4$,[3] and $(ET)_2BrO_4$.[4] The perchlorate salt is metallic to 1.4 K,[1] and the perrenate derivative was the first ET based organic superconductor (T_c 2 K, 4.5 kbar).[3] Since the discovery of ambient pressure superconductivity in β-$(ET)_2I_3$ (T_c 1.4 K),[5] other isostructural β-$(ET)_2X$ salts have been prepared with higher T_c's. These salts are β-$(ET)_2IBr_2$ (T_c 2.8 K),[6] β-$(ET)_2AuI_2$ (T_c 4.98 K),[7] and β^*-$(ET)_2I_3$ (T_c 8 K).[8] A structure-property correlation for the β-type salts has been reviewed in this volume;[9] it predicts that T_c's higher than 8 K are possible if β-salts with linear anions longer than I_3^- can be synthesized. During the search for new linear anions, a variety of compounds was discovered with polymeric anions. For example, with the $Ag(CN)_2^-$ anion, in addition to the expected $(ET)_2Ag(CN)_2$ salt,[10] afforded the first ET salt with a polymeric anion, $(ET)Ag_4(CN)_5$.[11] The linear AgI_2^-[12] or cubic $Ag_4I_6^{2-}$[13] anions, yielded $(ET)_3Ag_{6.4}I_8$.[13] In a similar fashion the CuI_2^- anion leads to $(ET)_2Cu_5I_6$.[14,15] The latter two compounds are good metals to low temperatures and consist of layers of donor molecules and polymeric anions. The report of superconductivity in κ-$(ET)_4Hg_3X_8$ (X = Cl, T_c 5.3 K 29 kbar[16] and X= Br, T_c 4.3 K

Organic Superconductivity, Edited by V.Z. Kresin and W.A. Little, Plenum Press, New York, 1990

ambient pressure[17] and 6.7 K 3.5 kbar[18]) and κ-$(ET)_2Cu(NCS)_2$ (Tc 10.4 K)[19] further stimulated the search for novel polymeric anions. A general synthetic strategy for preparing new salts containing polymeric anions is to couple a coordinatively unsaturated neutral transition metal halide/pseudohalide with a simple halide or pseudohalide during an electrocrystallization synthesis. A number of new compounds have been made in this way, for example, $(ET)BiI_4$,[20] $(ET)Ag_{1.6}(SCN)_2$,[21] and $(ET)Ag_{2.4}Br_3$.[22] In an attempt to replace the halide in κ-$(ET)_4Hg_3X_8$ with a pseudo halide, a β-like $(ET)_4Hg_3(SCN)_8$[23,24] compound was obtained. Using slightly different experimental conditions during electrocrystallization, two entirely different salts, α-$(ET)_2(M)Hg(SCN)_4$ (M^+ = K^+ and NH_4^+), have been reported by Oshima *et al.*[25,26] In this article, we discuss three new ambient pressure organic superconductors with novel polymeric anions, α-$(ET)_2(NH_4)Hg(SCN)_4$,[27] βm-$(BO)_3Cu_2(NCS)_3$,[28] and κ-$(ET)_2Cu[N(CN)_2]Br$.[29]

α-$(ET)_2(M)Hg(SCN)_4$, M^+ = K^+, NH_4^+, and Rb^+

The α-phase salts are prepared by electrocrystallization of ET (one eq), $Hg(SCN)_2$ (10 eq), MSCN (20 eq M^+ = K^+, NH_4^+, and Rb^+), and 18-crown-6 (10 eq) in 1,1,2-trichloroethane with 10% (vol) ethanol. The applied current density is maintained in the range of 0.2 to 0.5 $\mu A/cm^2$. Crystal growth is carried out at room temperature for approximately two to six weeks. The resulting crystals are shiny black thick platelets. The room temperature ESR peak-to-peak linewidth falls in the range of 55 to 85 G for all three compounds (*vide infra*). Screening of crystals employing ESR spectroscopy reveals a second minor unidentified phase with a linewidth of less than 10 G. The typical morphology for the minor phase is thin needle-like platelets. It is worth noting in this regard that a 3:1 salt, $(ET)_3[Hg(SCN)_3]$, has been reported.[25]

The α-phase potassium and ammonium salts were identified by their unit cell parameters,[25-27] and a single-crystal structural determination was carried out on the rubidium salt. Detailed structural results of the rubidium salt will be published elsewhere. The unit cell parameters of all three compounds are listed in Table 1.

The unit cell volume increases by approximately 0.5% in going from potassium to ammonium and another 0.5% in continuing from ammonium to rubidium. The two-dimensional polymeric anion layer is shown in Figure 1. The mercury atoms are located in tetrahedral sites with four sulfur bonded SCN^- ligands. The rubidium atoms on the other hand are located in eight coordination sites with four sulfur bonded SCN^- ligands on one side and four nitrogen bonded NCS^- ligands on the other side in an overall square antiprism configuration. The donor layer belongs to a typical α-type packing motif.

Table 1. Unit Cell Parameters for α-$(ET)_2(M)Hg(SCN)_4$

	K^+ [25]	NH_4^+ [26,27]	Rb^+
a (Å)	10.082	10.089	10.087 (b)
b (Å)	20.565	20.613	20.642 (c)
c (Å)	9.933	9.968	9.998 (a)
α (°)	103.70	103.67	103.54 (β)
β (°)	90.91	90.47	90.53 (γ)
γ (°)	93.06	93.36	93.23 (α)
V (Å^3)	1997.0	2010	2020.1

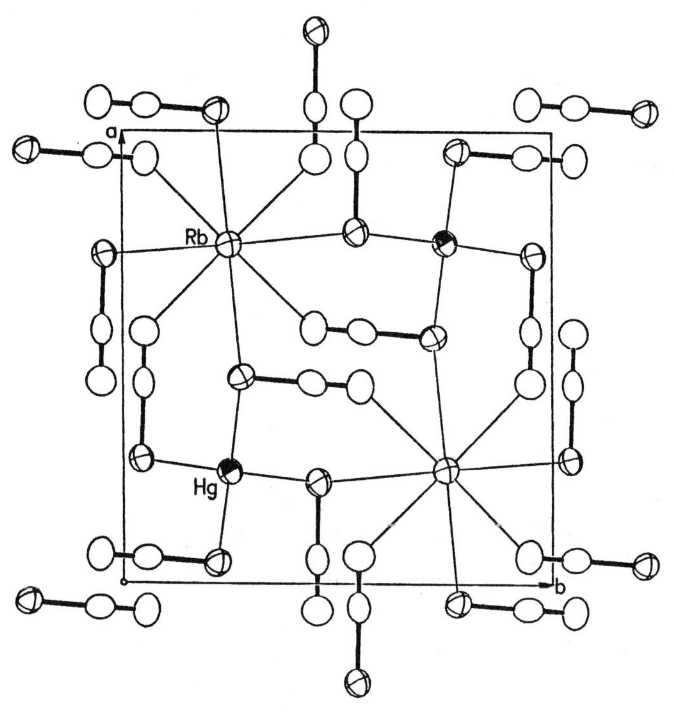

Figure 1. The anion layer of α-$(ET)_2RbHg(SCN)_4$ viewing along c^* axis

An oriented single-crystal ESR study has been carried out on the rubidium salt. A square platelet crystal of α-$(ET)_2RbHg(SCN)_4$ was mounted on a quartz rod with the crystal plane oriented vertically in the microwave cavity. The *b* axis (see Table 1, Rb salt) pointed upward. The zero and ninety degree orientations correspond to the static magnetic field parallel and perpendicular to the crystal plane, respectively. The measured g-values (circles) and peak-to-peak linewidths (triangles) are plotted in Figure 2. The corresponding solid curves are the least squares fit to the measured data with the following equations:

$$g_{obs}^2 = \sum_{i,j=1}^{3} g_{ij}^2\, l_i\, l_j \quad \text{and} \quad \Delta H = \sum_{i,j=1}^{3} \Delta H_{ij}\, l_i\, l_j$$

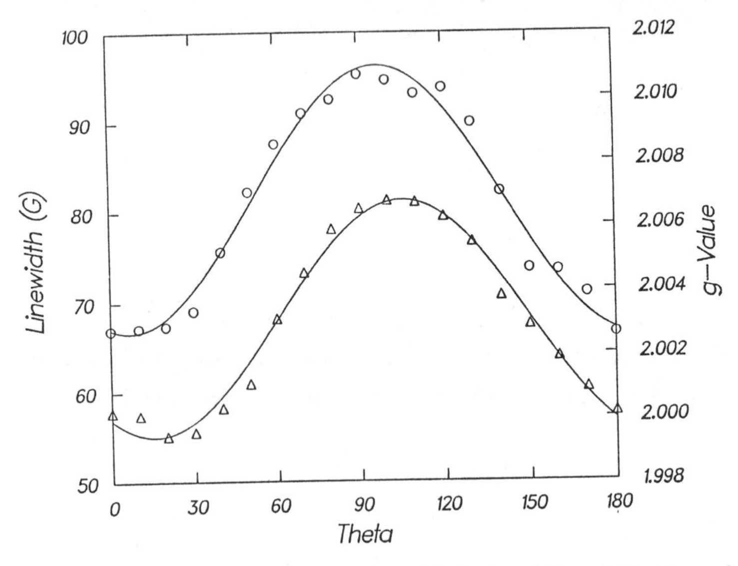

Figure 2. Orientation dependence of g-values (circles) and linewidths (triangles) of α-$(ET)_2RbHg(SCN)_4$ at room temperature.

The l_i and l_j are the direction cosines of the principal axes. The g-values and linewidths for the rubidium salt range from 2.002 to 2.011 and 55 G to 81 G, respectively. The results are very similar to that of the ammonium salt.[27] As shown in Figure 2, the minimum and maximum g-values correspond to the static magnetic field parallel to the crystallographic *a* and *c** axes. The g-value maximum near 90°, where the static field is approximately parallel to the central C=C double bond of each ET molecule, is commonly observed in other ET salts, such as α-$(ET)_2(NH_4)Hg(SCN)_4$[27] and α-$(ET)_2I_3$.[30]

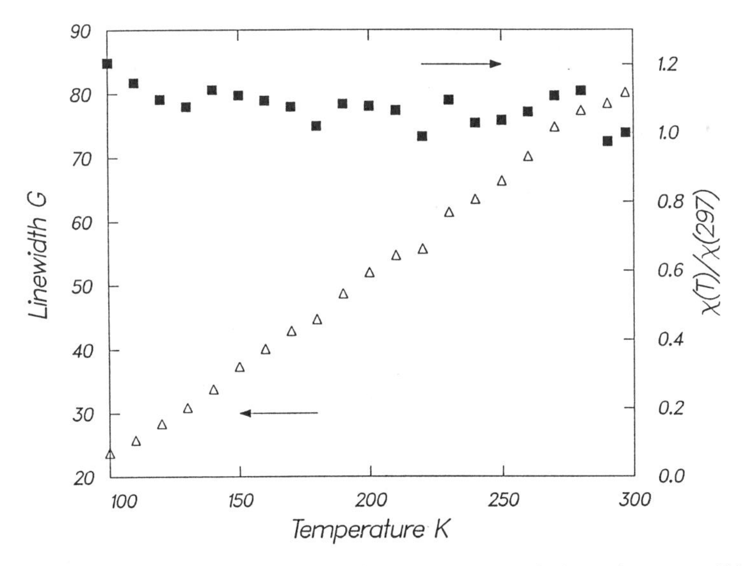

Figure 3. The temperature dependence of linewidth and relative spin susceptibility of α-$(ET)_2RbHg(SCN)_4$.

The temperature dependences of the ESR linewidth (ΔH) and the relative spin susceptibility (χ) of α-$(ET)_2RbHg(SCN)_4$ from 300 to 100 K are plotted in Figure 3. The peak-to-peak linewidth decreases monotonically with decreasing temperature from 81 G at 300 K to 24 G at 100 K. The spin susceptibility remains constant in the temperature range measured, which is consistent with the Pauli paramagnetism of a metallic sample. This behavior is a common feature of the α,[27] β,[31,32] and θ-phase[33] salts of ET. The decrease in linewidth with decreasing temperature can be qualitatively understood by use of the Elliott formula:[34,35]

$$\Delta H \sim (\Delta g)^2/\tau \qquad \text{and} \qquad \sigma = ne^2\tau/m_{eff}$$

where $\Delta g = g_{obs} - 2.0023$ (a constant within experimental error) and τ is the conduction electron relaxation time. Since the conductivity (σ) increases with decreasing temperature, the linewidth is expected to become sharper as the temperature is lowered. Low temperature ESR measurements have been carried out on α-$(ET)_2(NH_4)Hg(SCN)_4$ from 150 to 4.2 K. The behavior between 150 K and 20 K is identical to that of the 300 K to 100 K, namely, the linewidth decreases with decreasing temperature and the spin susceptibility remains constant. Below 20 K, the relative spin susceptibility starts to drop, which is very likely due to a microwave penetration depth effect.[36] The penetration depth becomes smaller when the conductivity increases significantly, and only outer portion of the sample is measured. The linewidth decreases at a much slower rate below 20 K and is 0.33G at 4.2 K. This linewidth is even sharper than some of the β-phase crystals at a similar temperature.[37] One possible reason for the extremely sharp linewidth of α-$(ET)_2(NH_4)Hg(SCN)_4$ is that the contribution from the conduction electrons becomes negligible at 4 K (*vide supra*) and only the residue linewidth (similar to the residue resistivity) remains. The residue linewidth could be caused by chemical impurities or crystal defects. Further experiments are needed to clarify this point.

The possible presence of superconductivity in the three α-phase salts was explored with RF penetration depth measurements at 0.5 K. In this method, superconductivity is exhibited by an increase in resonant frequency over that of the empty coil caused by the exclusion of the RF field from the sample by the persistent shielding currents. Only the ammonium salt displayed a strong superconducting signal (onset 1.15 K).

All known α-phase ET salts are listed in Table 2 along with their conductive properties. The *a* x *b* cross section area of the unit cell is also included so that the packing of the donor layers of different salts can be compared directly. A cross section area near 100 $Å^2$ seems to favor conductivity. One remaining question is that why does not the structure-property correlation for the β-phase salts, i.e., larger unit cell volume give rise to higher T_c's apply to the α-$(ET)_2(M)Hg(SCN)_4$ salts. No apparent structural disorder is observed in these materials. One difference among the K^+, NH_4^+ and Rb^+ salts is that the NH_4^+ cation is a molecular species and might provide a unique phonon mode that leads to superconductivity.

Table 2. Summary of α-Phase ET Compounds

Compounds	Conductance	Unit Cell Volume ($Å^3$)	Cross Section a x b ($Å^2$)	Reference
α-$(ET)_2[RbHg(SCN)_4]$	metal	2020.1	100.85	this work
α-$(ET)_2(NH_4)Hg(SCN)_4$	$T_c = 1.15$	2010	100.54	27
α-$(ET)_2[KHg(SCN)_4]$	metal at 1.5K	1997	100.14	25
θ-$(ET)_2I_3$	metal	3386	100.03	38
α-$(ET)_2I_3$	T_{MI}= 135K	1698	99.21	39
α-$(ET)_2I_2Br$	T_{MI}= 245K	1688	98.9	40
α-$(ET)_2Cu(NCS)_2$	T_{MI}= 200K	1672	98.25	41
α-$(ET)_2IBr_2$	semiconductor	1652	107.14	42
α-$(ET)_2BrICl$	semiconductor	1647	106.86	43

β_m-(BEDO-TTF)$_3$Cu$_2$(NCS)$_3$

The new donor molecule BEDO-TTF, bis(ethylenedioxy)tetrathiafulvalene (abbreviated herein as BO) contains four outer oxygen atoms and four inner sulfur atoms[44]. The BO donor molecule is more susceptible to oxidation than ET. Electrocrystallization syntheses of BO salts are preferably carried out under inert atmosphere. Three BO salts have been characterized structurally , i.e., $(BO)_2AuBr_2$,[45] $(BO)_{2.4}(I_3)$,[46] and $(BO)_3Cu_2(NCS)_3$.[28] Several other BO salts suffer from inferior crystal quality, e.g., $(BO)_2AuI_2$, $(BO)_2PF_6$. In the case of $(BO)_2ClO_4$[45] only the structure of the donor layer can be solved. The β_m-$(BO)_3Cu_2(NCS)_3$ crystals are prepared by electrocrystallization of BO (one eq), CuSCN (10 eq), KSCN (10 eq), and 18-crown-6 (10 eq) in 1,1,2-trichloroethane and 10% (vol) ethanol. The current density employed is ~0.1μA/cm^2, and the crystal growth is carried out in a nitrogen-purged drybox. Typical crystals are rectangular boxes. Moreover the crystal size is generally very small due to facile nucleation. During sample preparation, the electrode polarity was inadvertently reversed for one of the cells for a period of two days. The polarity was subsequently switched back. The polarity switch substantially reduced the rate of nucleation and larger crystals were obtained. It is interesting to note that this observation is reproducible. The larger crystals were used for structural determinations.

The detailed crystal structure of βm-$(BO)_3Cu_2(NCS)_3$ has been reported elsewhere.[28] The novel polymeric anion layer is shown in Figure 4. Both Cu1 and Cu2 are three coordinate with a trigonal planar configuration. The Cu1 atom coordinates to two sulfur and one nitrogen atoms, while the Cu2 atom coordinates to two nitrogen and one sulfur atoms. All SCN^- ligands are bridged. The resulting anion network is an infinite two-dimensional layer. This is in constrast to the anion layer in κ-$(ET)_2Cu(NCS)_2$, where $Cu(NCS)_2^-$ forms a zigzag chain with one bridging and one terminal SCN^- group. Two different anion layers ($Cu_2(NCS)_3^-$ and $Cu(NCS)_2^-$) are both "self-assembled" considering the almost identical synthetic conditions for both salts. Apparently the anion cavity created by the donor packing predetermines the appropriate polymeric anion to be accommodated.

An oriented single-crystal ESR study was carried out on β_m-$(BO)_3Cu_2(NCS)_3$. The measured g-values (circles) and peak-to-peak linewidths (triangles) are plotted in Figure 5. The platelet crystal of β_m-$(BO)_3Cu_2(NCS)_3$ was mounted horizontally in the microwave cavity. The 0° and 90° positions correspond to orientations where the static magnetic field is approximately parallel and perpendicular to the *b* axis, respectively.[28] The correlation between crystallographic axes and the crystal morphology in this BO salt is different from the majority of the $(ET)_2X$ derivatives. The two solid lines in Figure 5 are the least square fits employing the same equations as discussed above The linewidths

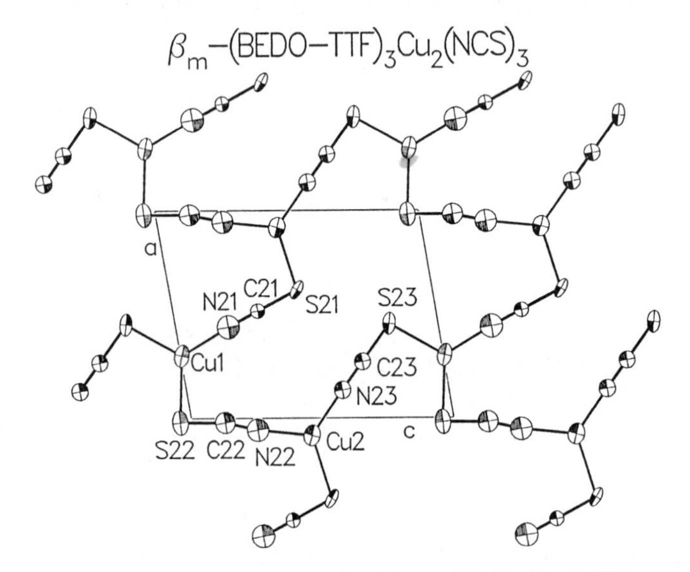

Figure 4. The polymeric anion layer of $(BO)_3Cu_2(NCS)_3$.

range from 17.5 to 26.6 G and the g-values from 2.002 to 2.012. It is worth noting that the ESR linewidth for β_m-$(BO)_3Cu_2(NCS)_3$ is different from all other BO salts.[45] This is reasonable, since the linewidth is sensitive to the crystal symmetry and crystal structure. Accordingly, a linewidth near 20 G might well serve as an indication of new BO-based superconductors.

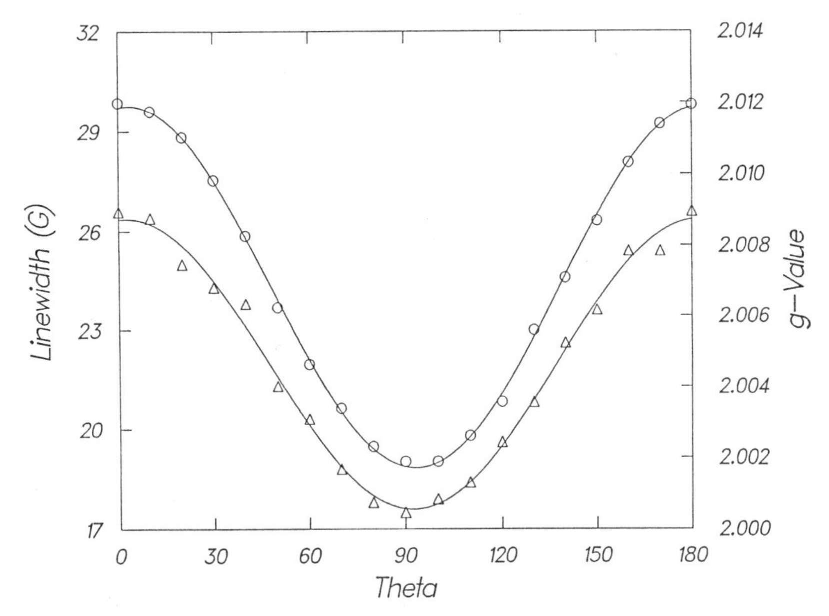

Figure 5. The orientation dependence of g-values (circles) and linewidths (triangles) of $(BO)_3Cu_2(NCS)_3$ at room temperature.

κ-$(ET)_2Cu[N(CN)_2]Br$

κ-phase compounds similar to the 10.4 K superconductor κ-$(ET)_2Cu(NCS)_2$ have long been actively sought since 1988. The $Cu(NCS)_2^-$ polymeric network consists of tri-coordinated Cu(I) with two bridging SCN^- ligands that form a zigzag chain and a terminal SCN^- that completes the third coordination site. Many attempts to replace the terminal SCN^- group with other halides or pseudohalides such as I^- or CN^- and maintain the zigzag chain, have been in vain. The resulting products are usually a mixture of the κ-$(ET)_2Cu(NCS)_2$ and intractable materials. During attempts to prepare $(ET)_2Cu[N(CN)_2]_2$, κ-$(ET)_2Cu[N(CN)_2]Br$ was isolated.[29] The synthesis will be published elsewhere.

The crystal structure has been determined[29] and yielded the following unit cell parameters: orthorhombic space group *Pnma*, $a = 12.942$ Å, $b = 30.016$ Å, $c = 8.539$ Å, $V = 3317$ Å^3. The donor packing motif is typical of κ-phase salts. The polymeric anion layer is shown in Figure 6. The $Cu[N(CN)_2]Br^-$ layer consists of tri-coordinated Cu(I) with two bridging $(NC)N(CN)^-$ ligands forming a zigzag chain and a terminal bromide to complete the coordination around Cu atom. The terminal bromine atom has a close contact with a nitrogen atom from an adjacent chain so that each zigzag chain is not totally isolated.

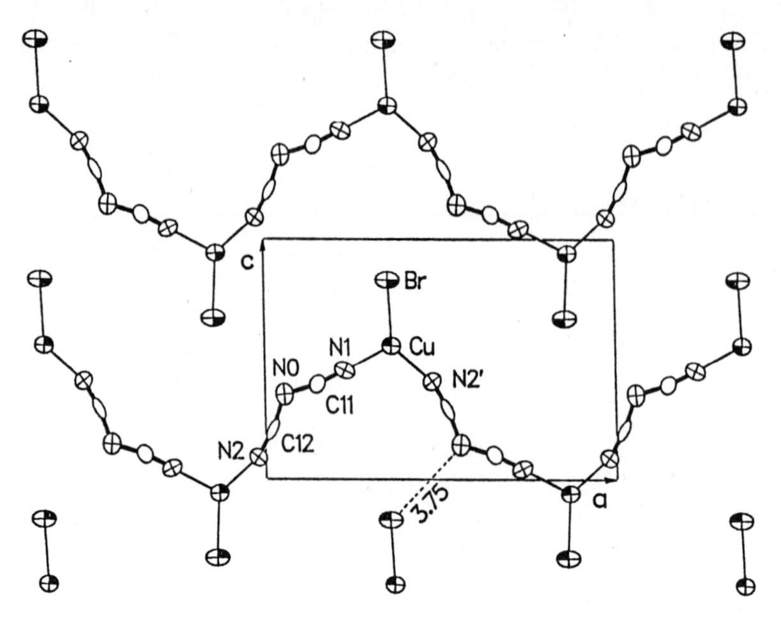

Figure 6. The polymeric anion layer of κ-$(ET)_2Cu[N(CN)_2]Br$ viewing along *b* axis.

An oriented single-crystal ESR study was carried out on κ-$(ET)_2Cu[N(CN)_2]Br$. The typical crystal morphology is a thick diamond-shaped platelet. The *b* axis is normal to the crystal plane and the *a* axis is coincident with one of the diagonals of the platelet. A platelet crystal was mounted vertically in the microwave cavity (*a* axis pointed upward). The measured g-values (circles) and peak-to-peak linewidths (triangles) are plotted in Figure 7. The 0° and 90° angles correspond to the static magnetic field parallel and perpendicular to the crystal plane, respectively. The two solid lines are the least squares fit to the experimental values. The maximum g-value occurs at 90°, where the static magnetic field is parallel to the *b* axis and the central C=C double bond of the ET molecules. This observation parallels what is observed in other ET salts. The maximum g-value at 90° corresponds to the minimum linewidth. This behavior appears to be unique to the κ-phase and has also been observed in κ-$(ET)_2Cu(NCS)_2$.[42] The linewidth varies from 60 G to 80 G, which is characteristic of κ-phase salts.

Four probe conductivity measurements on a single crystal of κ-$(ET)_2Cu[N(CN)_2]Br$ reveal metallic character between 300 and ~220 K, a large resistive hump near 100 K, and strongly metallic behavior below 60 K.[9] The superconductive onset temperature is 12.5 K, with midpoint at 11.2 K and zero resistance at 10.5 K. Superconductivity was also detected with RF penetration depth measurements and low field dc magnetization experiments employing single crystals weighing ~200μg.[29] The onset temperatures for these two measurements were both 11.6 K. The RF signal reaches saturation near 9.8 K. All the measurements for superconductivity indicate a very sharp transition.

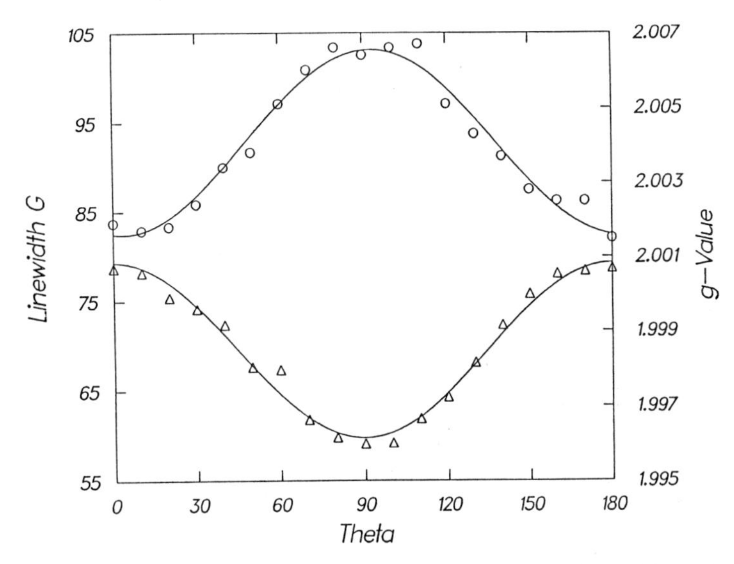

Figure 7. The orientation dependence of g-values (circles) and linewidths (triangles) of κ-$(ET)_2Cu[N(CN)_2]Br$ at room temperature.

The single crystal that was used for the RF penetration depth measurements was also used for low temperature ESR studies. It was oriented with the *b* axis (90° in the aforementioned orientation study) parallel to the static magnetic field. The linewidth increases slowly with decreasing temperature from 60 G at 300 K to 79 G at 60 K with a rate of increase of about 0.08 G/K. The linewidth and relative spin susceptibility are plotted against temperature in Figure 8. The ESR behavior of κ-$(ET)_2Cu[N(CN)_2]Br$ between 300 and 60 K is similar to that of κ-$(ET)_2Cu(NCS)_2$[47,48] and κ-$(ET)_4Hg_{2.89}Br_8$.[33] Below 60 K, the linewidth starts to decrease with decreasing temperature. At 40 K, in addition to the main absorption peak, a second peak with a sharper linewidth and ~3% intensity starts appearing. The second peak behaves in a paramagnetic fashion between 40 and 15 K. The origin of the new peak could be either a small amount of chemical impurities or crystal defects. It should be pointed out that a sharp uncharacterized peak has also been observed in κ-$(ET)_2Cu(NCS)_2$ below 20 K.[47,48] Further experiments are in progress. The linewidth of the main absorption peak decreases with decreasing temperature from 79 G at 60 K to 40 G at 10 K with a rate of decrease at 0.8 G/K, which is one order of magnitude faster than the aforementioned linewidth increase rate between 300 and 60 K. The observed linewidth decrease in κ-$(ET)_2Cu[N(CN)_2]Br$ below 60 K contrasts with the behavior of κ-$(ET)_2Cu(NCS)_2$ and κ-$(ET)_4Hg_{2.89}Br_8$. Nevertheless, the linewidth behavior of the title compound is totally consistent with the conductivity measurements, i.e. the apparent semiconductive to metallic behavior.[9] The spin susceptibility (squares) shows a slight increase between 300 and 100 K. Below 100 K, it starts to drop. At 15 K, it reaches 80% of the room temperature value. Below 10 K the main ESR absorption disappears,which is in agreement with the onset of superconductivity.

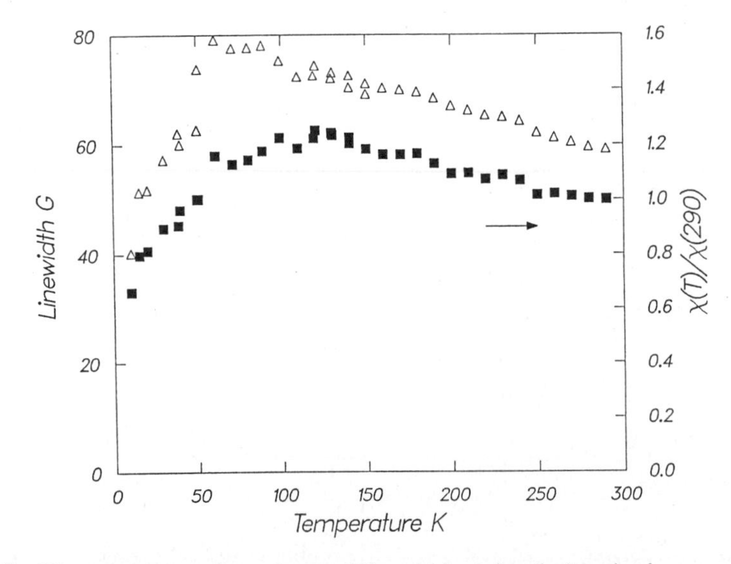

Figure 8. The temperature dependence of linewidths (triangles) and spin susceptibility (square) of κ-$(ET)_2Cu[N(CN)_2]Br$.

CONCLUSIONS

The use of polymeric anions continues to generate interesting new conducting materials, including the two highest T_c organic superconductors, κ-$(ET)_2Cu(NCS)_2$ and κ-$(ET)_2Cu[N(CN)_2]Br$. In contrast to the limited number of known linear monovalent anions, the possibilities for synthesizing new and previously unknown polymeric anions are considerable. Further research along these lines should prove to be rewarding.

Over the past decade, the T_c's of organic superconductors have increased by an order of magnitude from the 1.2 K value of $(TMTSF)_2ClO_4$ to 12.5 K for κ-$(ET)_2Cu[N(CN)_2]Br$. The oxide superconductors have made a greater jump from 11 K for $BaBi_{0.75}Pb_{0.25}O_3$ to 125 K of the Tl based high T_c materials. Considering the similarities among these two systems, it is not unreasonable to expect the organic superconductors to make major improvements in the future.

ACKNOWLEDGMENT

Work at Argonne National Laboratory is supported by the Office of Basic Energy Sciences, Division of Materials Sciences, U.S. Department of Energy, under Contract W-31-109-ENG-38. L.K.M. is a scientist in residence, sponsored by the Argonne Division of Educational Programs, on leave from the Department of Chemistry, Indiana University, Bloomington, IN. J.E.T. is a student research participant sponsored by the Argonne Division of Educational Programs, from the University of Michigan, Ann Arbor, MI.

REFERENCES

1. H. Kobayashi, A. Kobayashi, Y. Sasaki, G. Saito, T. Enoki and H. Inokuchi, J. Am. Chem. Soc. 105:297 (1983).
2. (a) H. Kobayashi, T. Mori, R. Kato, A. Kobayashi, Y. Sasaki, G. Saito and H. Inokuchi, Chem. Lett. 581 (1983).
   (b) H. Kobayashi, R. Kato, T. Mori, A. Kobayashi, Y. Sasaki, G. Saito and H. Inokuchi, Chem. Lett. 759 (1984).
3. S. S. P. Parkin, E. M. Engler, R. R. Schumaker, R. Lagier, V. Y. Lee, J. C. Scott and R. L. Greene, Phys. Rev. Lett. 50:270 (1983).
4. J. M. Williams, M. A. Beno, H. H. Wang, P. E. Reed, L. J. Azevedo and J. E. Schirber, Inorg. Chem. 23:1790 (1984).
5. E. B. Yagubskii, I. F. Shchegolev, V. N. Laukhin, P. A. Kononovich, M. V. Kartsovnik, A. V. Zvarykina and L. I. Buravov, J.E.T.P. Lett. 39:12 (1984).
6. J. M. Williams, H. H. Wang, M. A. Beno, T. J. Emge, L. M. Sowa, P. T. Copps, F. Behroozi, L. N. Hall, K. D. Carlson and G. W Crabtree, Inorg. Chem. 23:3839 (1984).
7. H. H. Wang, M. A. Beno, U. Geiser, M. A. Firestone, K. S. Webb, L. Nuñez, G. W. Crabtree, K. D. Carlson, J. M. Williams, L. J. Azevedo, J. F. Kwak and J. E. Schirber, Inorg. Chem. 24:2465 (1985).

8. (a) V. N. Laukhin, E. E. Kostyuchenko, Yu. V. Sushko, I. F. Shchegolev and E. B. Yagubskii, J.E.T.P. Lett. 41:81 (1985).
(b) K. Murata, M. Tokumoto, H. Anzai, H. Bando, G. Saito, K. Kajimura and T. Ishiguro, J. Phys Soc. Jpn. 54:1236 (1985).
9. J. M. Williams, A. M. Kini, U. Geiser, H. H. Wang, K. D. Carlson, W. K. Kwok, K. G. Vandervoort, J. E. Thompson, D. L. Stupka, D. Jung and M.-H. Whangbo, Plenary lecture, International Conference on Organic Superconductors, May 20-24, 1990, South Lake Tahoe, CA.
10. M. A. Beno, M. A. Firestone, P. C. W. Leung, L. M. Sowa, H. H. Wang, J. M. Williams and M.-H. Whangbo, Solid State Commun. 57:735 (1986).
11. U. Geiser, H. H. Wang, L. E. Gerdom, M. A. Firestone, L. M. Sowa, J. M. Williams and M.-H. Whangbo, J. Amer. Chem. Soc. 107:8305 (1985).
12. H. H. Wang, U. Geiser and J. M. Williams, unpublished results, unit cell parameters for K(18-crown-6)AgI_2: monoclinic, $P2_{1/n}$ $a = 8.790(4)$ Å, $b = 8.447(5)$ Å, $c = 13.864(7)$ Å, $\beta = 103.14(4)°$, $V = 1002.3(9)$ Å^3. Z = 4.
13. U. Geiser, H. H. Wang, K. M. Donega, B. A. Anderson, J. M. Williams and J. F. Kwak, Inorg. Chem. 25:401 (1986). Unit cell parameters for K(18-crown-6)$Ag_2I_3 \cdot 2H_2O$, cubic F_{d3m} $a = 22.124(4)$ Å, V = 10,829(b) Å^3, Z = 16.
14. L. I. Buravov, A. V. Zvarykina, M. V. Kartsovnik, N. D. Kushch, V. N. Laukhin, R. M. Lobkovskaya, V. A. Merzhanov, L. N. Fedutin, R. P. Shibaeva and E. B. Yagubskii, Sov. Phys. JETP 65(2):336 (1987).
15. J. M. Williams, H. H. Wang, T. J. Emge, U. Geiser, M. A. Beno, P. C. W. Leung, K. D. Carlson, R. J. Thorn, A. J. Schultz and M.-H. Whangbo, Prog. in Inorg. Chem. 35:51 (1987).
16. R. N. Lyubovskaya, R. B. Lyubovskii, R. P. Shibaeva, M. A. Aldoshina, L. M. Gol'denberg, L. P. Rozenberg, M. L. Khidekel'and Yu. F. Shul'pyakov, JETP Lett. 42:468 (1985).
17. (a) R. N. Lyubovskaya, E. A. Zhilyaeva, A. V. Zvarykina, V. N. Laukhin, R. B. Lyubovskii and S. I. Pesotskii, JETP Lett 45:530 (1987).
(b) R. N. Lyubovskaya, E. I. Zhilyaeva, S. I. Pesotskii, R. B. Lyubovskii, L. O. Atovmyan, O. A. D'yachenko and T. G. Takhirov, JETP Lett. 46:188 (1987).
18. J. E. Schirber, D. L. Overmyer, E. L. Venturini, H. H. Wang, K. D. Carlson, W. K. Kwok, S. Kleinjan and J. M. Williams, Physica C 161:412 (1989).
19. H. Urayama, H. Yamochi, G. Saito, K. Nozawa, T. Sugano, M. Kinoshita, S. Sato, K. Oshima, A. Kawamoto and J. Tanaka, Chem. Lett. 55 (1988).
20. U. Geiser, H. H. Wang, S. M. Budz, M. J. Lowry, J. M. Williams, J. Ren and M.-H. Whangbo, Inorg. Chem. 29:1611 (1990).

21. U. Geiser, M. A. Beno, A. M. Kini, H. H. Wang, A. J. Schultz, B. D. Gates, C. S. Cariss, K. D. Carlson and J. M. Williams, Synth. Met. 27:A235 (1988).
22. U. Geiser, H. H. Wang, P. R. Rust, L. M. Tonge and J. M. Williams, Mol. Cryst. Liq. Cryst. 181:117 (1990).
23. H. H. Wang, M. A. Beno, K. D. Carlson, N. Thorup, A. Murray, L. C. Porter, J. M. Williams, K. Maly, P. Coppens, and M.-H. Whangbo, to be published.
24. P. Coppens, K. Maly and V. Petricek, Mol. Cryst. Liq. Cryst. 181:81 (1990).
25. M. Oshima, H. Mori, G. Saito and K. Oshima, Chem. Lett. 1159 (1989).
26. M. Oshima, H. Mori, G. Saito and K. Oshima, The Physics and Chemistry of Organic Superconductors, eds. G. Saito and S. Kagoshima, Springer-Verlag.
27. H. H. Wang, K. D. Carlson, U. Geiser, W. K. Kwok, M. D. Vashon, J. E. Thompson, N. F. Larsen, G. D. McCabe, R. S. Hulscher and J. M. Williams Physica C 166:57 (1990).
28. M. A. Beno, H. H. Wang, A. M. Kini, K. D. Carlson, U. Geiser, W. K. Kwok, J. E. Thompson, J. M. Williams, J. Ren and M.-H. Whangbo Inorg. Chem. 29:1599 (1990).
29. A. M. Kini, U. Geiser, H. H. Wang, K. D. Carlson, J. M. Williams, W. K. Kwok, K. G. Vandervoort, J. E. Thompson, D. L. Stupka, D. Jung and M.-H. Whangbo, Inorg. Chem., accepted for publication (1990).
30. T. Sugano, G. Saito and M. Kinoshita, Phys. Rev. B 34:117 (1986).
31. E. L. Venturini, L. J. Azevedo, J. E. Schirber, J. M. Williams and H. H. Wang, Phys. Rev. B 32:2819 (1985).
32. L. Forró, G. Sekretarczyk, M. Krupski, D. Schweitzer and H. Keller, Phys. Rev. B. 35:2501 (1987).
33. H. H. Wang, B. A. Vogt, U. Geiser, M. A. Beno, K. D. Carlson, S. Kleinjan, N. Thorup and J. M. Williams, Mol. Crystl. Liq. Crystl. 181:135 (1990).
34. R. J. Elliott, Phys. Rev. 96:266 (1954).
35. T. Enoki, K. Imaeda, M. Kobayashi, H. Inokuchi and G. Saito, Phys. Rev. B 33:1553 (1986).
36. H. Hurdequint, F. Creuzet, D. Jeróme, Synth. Metals 27:A183 (1988).
37. E. L. Venturini, J. E. Schirber, H. H. Wang and J. M. Williams, Synth. Metals 27:981 (1988).
38. H. Kobayashi, R. Kato, A. Kobayashi, S. Moriyama, Y. Nishio, K. Kajita and W. Sasaki, Synth. Metals 27:A283 (1988).
39. K. Bender, I. Hennig, D. Schweitzer, K. Dietz, H. Endres and H. J. Keller, Mol. Cryst. Liq. Cryst. 108:359 (1984).
40. Z. Daoben, W. Ping, W. Meixiang, Y. Zhaolou and Z. Naijue, Solid State Commun. 57:843 (1986).

41. M. Tokumoto, H. Anzai, K. Takahashi, N. Kinoshita, K. Murata, T. Ishiguro, Y. Tanaka, Y. Hayakawa, H. Nagamori and K. Nagasaka, Synth. Metals 27:A171 (1988).
42. R. P. Shibaeva, R. M. Lobkovskaya, M. A. Simonov, E. B. Yagubskii and A. A. Ignat'ev, Kristallografiya 31:1105 (1986) (Sov. Phys. Crystallogr. 31:654 (1986)).
43. H. Kobayashi, R. Kato, A. Kobayashi, G. Saito, M. Tokumoto, H. Anzai and T. Ishiguro, Chem. Lett. 93 (1986).
44. T. Suzuki, H. Yamochi, G. Srdanov, K. Hinkelmann and F. Wudl, J. Am. Chem. Soc. 111:3108 (1989).
45. M. A. Beno, H. H. Wang, K. D. Carlson, A. M. Kini, G. M. Frankenbach, J. R. Ferraro, N. Larson, G. D. McCabe, J. E. Thompson, C. Purnama, M. Vashon, J. M. Williams, D. Jung, and M.-H. Whangbo, Mol. Cryst. Liq. Cryst. 181:145 (1990).
46. F. Wudl, H. Yamochi, T. Suzuki, H. Isotalo, C. Fite, H. Kasmai, K. Liou, G. Srdanov, P. Coppens, K. Maly and A. Frost-Jensen, J. Am. Chem. Soc. 112:2461 (1990).
47. H. Urayama, H. Yamochi, G. Saito, S. Sato, T. Sugano, M. Kinoshita, A. Kawamoto, J. Tanaka, T. Inabe, T. Mori, Y. Maruyama, H. Inokuchi and K. Oshima, Synth. Metals 27:A393 (1988).
48. H. H. Wang, L. K. Montgomery, A. M. Kini, K. D. Carlson, M. A. Beno, U. Geiser, C. S. Cariss, J. M. Williams and E. L. Venturini, Physica C 156:173 (1988).

ELECTROMAGNETIC PROPERTIES OF EXOTIC SYSTEMS

THE INFRARED RESPONSE OF THE EXOTIC SUPERCONDUCTORS

T. Timusk

Department of Physics
McMaster University
Hamilton, Ontario
Canada, L8S 4M1

INTRODUCTION

Far-infrared spectroscopy has been widely used to probe the low-lying excitations in the new superconductors. The organic charge transfer salts, the copper oxides and most recently the heavy Fermion materials have all been examined in the 10 cm^{-1} (1 meV) to 1000 cm^{-1} (100 meV) region by infrared reflectance techniques. Kramers Kronig analysis of the reflectance magnitude is a fairly reliable technique that yields the frequency dependent optical conductivity which can be related, in many cases, directly to fundamental processes.

Since the materials, in their normal state, are metals showing a temperature dependent resistivity with a positive temperature coefficient one would also expect to see, in analogy with ordinary metals, a corresponding temperature dependent Drude response in the infrared. Added to this free carrier absorption, at higher frequency usually in ordinary metals, one encounters particle-hole or interband transitions.

Deviations from this simple behavior would signal interactions of the free carriers with low lying excitations. In the case of ordinary superconductors where the electron-phonon interaction is the main source of temperature dependent scattering the processes are easily recognized, particularly in the superconducting state. Predicted by Holstein[1] the inelastic scattering of electrons by phonons gives rise to absorption thresholds at frequencies of the transverse and longitudinal branches in the phonon density of states. This effect was first demonstrated by Joyce and Richards[2] in lead. In the superconducting state these Holstein bands are almost perfect replicas of the phonon densities of states and rival in sharpness and detail the spectra obtained by electron tunnelling spectroscopy.[3] A characteristic feature of the Holstein phonon bands in a superconductor with a gap Δ is the shift to higher frequency by 2Δ in the superconducting state. If the area under the Drude band, centered on zero frequency, is taken to be unity then the area of the Holstein side-band is λ the dimensionless Eliashberg coupling constant.

Another source of deviation from Drude behavior is the direct coupling of low lying excitations to the electromagnetic field. The most obvious example is the coupling of phonons to the field in polar insulators with more than one atom per unit cell, the reststrahlen band. Because of the high conductivity in ordinary metals the electromagnetic field shields the ions from this direct interaction and phonon bands are not easily seen in ordinary metals.

In this review we summarize some of the recent experimental data in the far infrared range on the optical conductivity of the exotic superconductors. We will see that while to a theorist the systems are very disparate: ranging from the quasi-one dimensional organics through the 2D oxides to the clearly three dimensional heavy Fermion systems, to an experimentalist they are in fact very similar. The infrared reflectance is decidedly non-Drude in all the systems, dominated by strong inelastic processes in the 0.1 to 0.5 eV range and anomalous phonon features with unusual line shape and strength at lower frequencies.

Accompanying the non-Drude infrared conductivity is an unusual dc transport. A power law temperature dependence at low temperatures dominates and there is no sign of residual resistance.

THE CONTINUOUS SPECTRUM

In the Drude model the frequency dependent conductivity is given by:

$$4\pi\sigma(\omega) = \omega_p^2\tau/(1+(\omega\tau)^2), \tag{1}$$

where $\omega_p = 4\pi Ne^2/m$ is the plasma frequency. The number of electrons per unit volume is N, m is their effective mass and $1/\tau$ is the scattering rate. The scattering rate and the plasma frequency can be determined independently from an analysis of the frequency dependent infrared conductivity given by eq. 1. In this model in the low frequency limit, the conductivity should agree with the dc conductivity as measured by a four-probe resistance technique.

Applied to the organic superconductors in their normal state this analysis predicts a conductivity peak of the order of 10000 Ω^{-1} cm^{-1} in amplitude, centered at zero frequency and a width of the order of 10 cm^{-1}. This strong mode would dominate the optical properties. Figure 1 shows the optical conductivity of TTF-TCNQ based on recent reflectance measurements.[4] At 85 K the material is an excellent conductor with a dc conductivity of the order of 5000 Ω^{-1} cm^{-1} and one would expect to see metallic conductivity. However the infrared conductivity consists of a broad featureless band with a gap-like region below 400 cm^{-1}. There is no sign of the expected low frequency Drude peak.

This discrepancy between the dc and far infrared conductivities has led to the suggestion by Lee *et al.*[5] that the transport is by means of sliding charge density wave (CDW) fluctuations. A fully developed CDW would have a gap that would open at the transition temperature. In a 1D system fluctuations suppress the opening of the gap and a pseudogap develops. In this fluctuation regime the charge transport is by sliding charge density waves of very large effective mass $m^* = m(1+\lambda)$. The Drude damping rate would then be reduced by a factor of $(1+\lambda)$ and the peak would be too narrow to be observable by far infrared techniques. This picture has applied to the infrared properties of TTF-TCNQ.[6,7] Recent microwave measurements seem to confirm the presence of this low-lying mode.[8]

In this picture the midinfrared band would be the result of transitions across the charge density wave gap. The spectral weight of this band would be λ times the weight of the zero frequency mode.[5] The total spectral weight, obtained from the oscillator strength sum rule is $\int \sigma d\omega = \omega_p^2/8$ and is of the order of 1 eV^2.

The heavy Fermion materials also show a non-Drude infrared conductivity. As in the organics there is a peak in the midinfrared. In the same range of temperatures the dc resistivity shows "Kondo" behavior, rising as the temperature is lowered. Naively interpreted we have a system with an extremely high scattering rate, $1/\tau = 0.35$ eV[9] corresponding to an unphysical mean free path that is less than a lattice spacing.

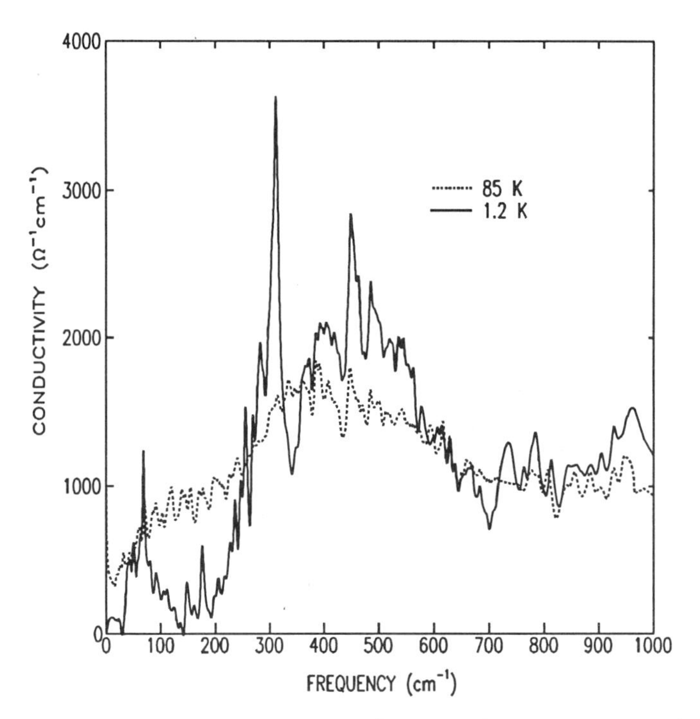

FIGURE 1. The optical conductivity of TTF-TCNQ at 85 K in the conducting state and at 1.2 K in the CDW state. The high temperature spectrum is dominated by a gap like depression at low frequency that develops into a full gap in the low temperature insulating state. At the same time as the CDW order appears strong phonon lines are electronically activated.

We note that, as in the organics, the midinfrared absorption of the heavy Fermion materials is of electronic strength.

In the high temperature region there is no discrepancy between the midinfrared conductivity and the dc value in the heavy Fermion materials. The two values agree to within a few percent. Unlike the organic conductors, there is no pseudogap, and no evidence of a collective mode at high temperature.

When the temperature is lowered below the coherence temperature, T_{co}, the dc conductivity begins to rise and interesting changes occur in the far infrared conductivity. Figure 2a shows the conductivity of URu_2Si_2 at three temperatures.[9] A region of depressed conductivity develops in the 50 - 300 cm^{-1} region and at the same time the narrow peak grows at low frequency. It is reasonable to assume that this narrow mode is the high frequency tail of the peak that is responsible for the rising dc conductivity in this temperature range.

One way of analyzing this non-Drude conductivity is in terms of a frequency dependent scattering rate and mass renormalization.[10] Figure 2b shows the scattering rate at three temperatures as a function of frequency. Above the coherence temperature the scattering rate is nearly constant *i.e.* Drude-like. As coherence develops the scattering is dominated by a peak at $\hbar\omega \approx 2T_{co}$. Another result of the onset of coherence is a severe depression of the scattering rate at low frequency.

A test of the validity of the frequency dependent scattering rate analysis is a comparison of $\Gamma(\omega)$ with $\rho(T)$. In the case of URu_2Si_2 the agreement is excellent; the

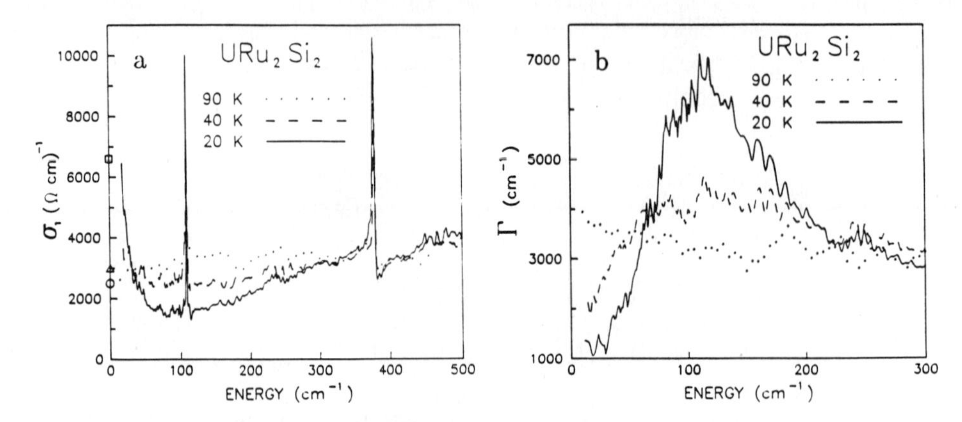

FIGURE 2. (a) The optical conductivity of URu_2Si_2 a heavy Fermion metal. At 90 K the material has a Drude like conductivity with a very large relaxation rate. As the temperature is lowered below the coherence temperature $\approx$ 90K the low frequency conductivity is depressed and a sharp peak builds up centered at zero frequency. This non-Drude behavior is interpreted in terms of the buildup of mass of the carriers as a result of strong interactions. (b) The frequency dependent scattering rate corresponding to the conductivities in a). At 90 K $\Gamma(\omega)$ is monotonically decreasing (Drude model has $\Gamma(\omega)$ constant) while at low temperature the scattering rate is suppressed. A sign of the validity of the analysis is the similarity of low the temperature scattering rate curve to the temperature dependence of the dc conductivity.

dc resistivity rises to peak at $kT \approx 2\hbar\omega_{co}$ where $\omega_{co} \approx 100$ cm^{-1} and then drops off slowly.

Turning finally to the oxide materials we again find an unconventional conductivity. As in the two systems described above the midinfrared is dominated by a broad temperature independent band while the low frequency region has a very conventional Drude-like absorption. The width of the Drude band, $\Gamma = 1.0 - 1.5kT$, has the same linear temperature dependence as the dc resistivity.[11,12] The plasma frequency is in good agreement with the hole concentration obtained from the chemical doping level provided a value of 2–3 m_e is taken for the effective mass of the charge carriers.

Further evidence of conventional electrodynamic properties of these materials is the behavior in the superconducting state. Most of the variation of the optical constants can be explained in terms of a clean limit London model where in the superconducting state all the spectral weight in the Drude band shifts to a delta function at the origin. The strength of this mode can be measured by a variety of techniques: μSR,[13] dc magnetization[14] or by comparing the area under the Drude curve in the normal state with the contribution to the real part of ϵ_1 from the condensate in the superconducting state. In all cases a value between 1.0 – 1.4 eV is found for ω_p.

It should be emphasized that the Drude plasma frequency determined from low frequency conductivity fits is much lower than the plasma frequency found from the observation of the near infrared plasma edge which is of the order of 2.5 to 3.0 eV. There are several possible reasons for this difference, a mass renormalization with m^* of the order of 2 – 3 is one, additional particle-hole transitions that contribute to the sum rule that makes up the oscillator strength is another. Recent work of Orenstein *et al.*[15] attempts to separate the various contributions by a careful analysis of the conductivity sum rule in a series of samples with various oxygen contents.

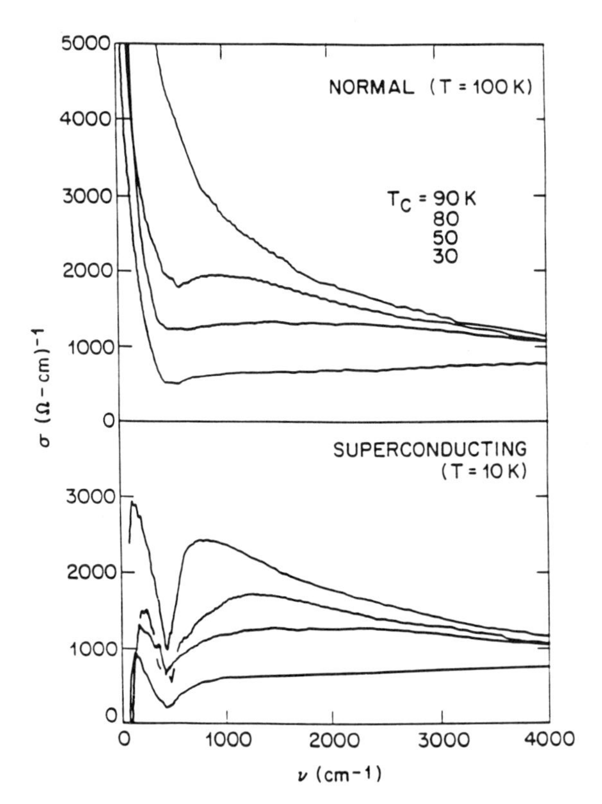

FIGURE 3. The optical conductivity of $YBa_2Cu_3O_7$ single crystals for light polarized in the highly conducting ab plane. The curves refer to various doping levels giving T_c's from 30K to 90 K. The spectrum is dominated by a strong midinfrared band that grows with doping and the center of the band shifts to lower frequency with increasing doping level. The curves in the top panel are for the normal state and the bottom in the superconducting state. The Drude tail can be seen in the normal state. We suggest the depression in the conductivity at 450 cm^{-1} is due to phonons that interact with the electronic continuum. Its position is independent of doping level and temperature.

Infrared measurements on high quality laser ablated films[16] show that the scattering rate varies linearly with temperature and the plasma frequency is temperature independent. In agreement with the linear temperature dependence of the dc resistivity this behavior should be contrasted with the Hall effect results that show a temperature dependent Hall constant, which, simply interpreted leads to a temperature dependent carrier density.

In addition to the low frequency Drude band a common feature of all the oxide superconductors is a broad midinfrared band. This decidedly non-Drude contribution to $\sigma(\omega)$ starts with a sharp onset at 150 cm^{-1} in $YBa_2Cu_3O_7$ and at 300 cm^{-1} in $Bi_2Sr_2CaCu_2O_8$. With an oscillator strength of the order of one electron per formula unit it accounts for the spectral weight expected for the carriers introduced by chemical doping.

Recent measurements on samples with varying carrier concentration obtained by varying the doping level show that the shape of the midinfrared band is quite variable.[17,18] In very low T_c materials the band is weak and centered at 5000 cm^{-1}.

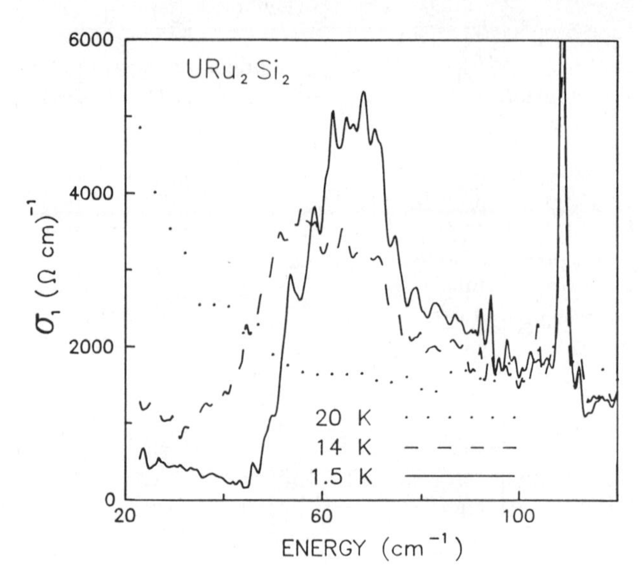

FIGURE 4. The optical conductivity of URu_2Si_2 at low temperature. Above the Neél temperature the behavior at low frequency is Drude-like. Below the SDW transition a gap develops in the conductivity in the 50 cm^{-1} region. It is clear that as the SDW order develops spectral weight is transferred from the Drude absorption to the SDW peak. Unlike the case of the organic conductors there is very little change in the phonon spectrum as the SDW transition occurs.

As the doping is increased the peak shifts to lower frequency and becomes more dominant, reaching a conductivity of 2500 (Ω^{-1} cm^{-1}) in the best crystals. Figure 3 shows the midinfrared band in $YBa_2Cu_3O_7$for a series of oxygen concentrations.[17]

The most viable explanation of the midinfrared band at this time is that it is largely an electronic particle-hole band. Its strength is of electronic magnitude and the fact that it does not change on entry to the superconducting state means that carriers other than the ones that are responsible for charge transport are involved. The reason for this is that one would expect an inelastic, Holstein type, side band of the free carrier conductivity to shift by 2Δ to higher frequency on entry to the superconducting state.[19] Since there is evidence that the gap in the oxides is large[16] the shift should be observable. No such shift has been reported.

GAPS AND PSEUDO GAPS

One of the manifestations of strong coupling of charge carriers to low lying excitations is the development of gaps and pseudo gaps in the electronic density of states near the Fermi energy. The optical conductivity will often exhibit a gap at low frequency as well. In an ordinary BCS superconductor a gap develops in the conductivity from zero frequency up to $\hbar\omega = 2\Delta$ at which point the conductivity approaches the normal state value for that frequency. The missing spectral weight in the gap region is transferred to the collective mode at zero frequency, a delta function for a superconductor at zero kelvin. The fraction of spectral weight that appears above the gap depends on the scattering rate of the electrons. In the clean limit where the gap is much larger than the scattering rate the conductivity above the gap vanishes. This appears to be the case in the high temperature superconductors.[16]

Charge and spin density wave systems develop gaps too. Here the spectral weight in the collective mode and in the transitions across the gap are divided in the ratio of λ which is independent of the scattering rate in the normal state. In the clean limit the calculation of Lee *et al.* shows a sharp band with an onset at 2Δ with a width that is of the order of Δ. The calculation has not been done in the case of strong scattering.

Figure 4 shows the optical conductivity of URu_2Si_2 below $T_N = 17$ K where the material displays SDW order[20,21]. A clear gap develops at a frequency $2\Delta/kT_N \approx 3.5 - 4$, slightly above the mean field value of 3.5. Unlike a superconducting gap the spectral weight removed from the gap region has not gone to the collective mode at zero frequency but has piled up in the region just above the gap as suggested by the the theory of Lee *et al.*

In figure 5 we show the optical conductivity of $(TMTSF)_2SbF_6$ from the work of Ng *et al.*[22] This organic conductor undergoes a SDW transition at 12 K. The contrast with the URu_2Si_2 spectrum is striking. There is no evidence of depressed conductivity or a peak in the frequency region expected from mean field theory. Evidence that a transition has taken place comes from the appearance of strong electronically activated phase phonon lines. The frequencies and amplitudes of the lines follow mean field theory and are in good agreement with a calculation of Fenton *et al.*[23] that assumes the coupling to the phonons is via a higher harmonic CDW of the SDW.

We turn next to the CDW systems where gaps are well documented. In organic conductors such as TTF-TCNQ shown in figure 1 the development of the gap is not always complete. At 85 K, shown as the dotted line there is a depression of conductivity below 300 cm^{-1}. This pseudogap develops into a more complete gap as the temperature is lowered and the transition is completed. In the gap region one generally sees sharp lines of relatively modest oscillator strength. For example the peak in the 40 cm^{-1} region of TTF-TCNQ has been attributed to the pinned charge density wave mode. We argue against this explanation[4] and suggest with Volkov[24] that this mode may be the antiphase oscillation of the CDW's on the two chains. There is also the possibility that the mode is a low lying phonon activated by the CDW.

PHONONS

The role of phonons in the exotic superconductors is subtle. There is little evidence of a phonon contribution to the scattering of the carriers responsible for dc transport. For example none of the systems show the characteristic Bloch-Grüneisen linear temperature dependence. Materials that remain conducting to the lowest temperature show temperature dependent resistivity variations below 10 K when all the phonons have frozen out.

Nevertheless the infrared spectra show dramatic manifestations of the presence of phonons. One example is the the spectrum of $(TMTSF)_2SbF_6$ shown in figure 5. Very strong phonon peaks appear as the material enters the SDW state. The oscillator strength is too large to be due to direct absorption by phonons and detailed comparisons with Raman spectra show that totally symmetric molecular vibrations are involved.[25] The theory of Rice[26] accounts for all the observations. In this picture totally symmetric vibrations of the molecules pump charge over large distances giving rise to electronic magnitude oscillator strengths at phonon frequencies. A condition for this effect to occur is that the molecule has to be dimerized, typically as a result of a charge density wave.

The phenomenon of the electronic activation of lattice vibrations is a case of strong electron-phonon interaction but it is not clear that it is equivalent to the electron-phonon scattering that limits the electrical resistance in ordinary metals.

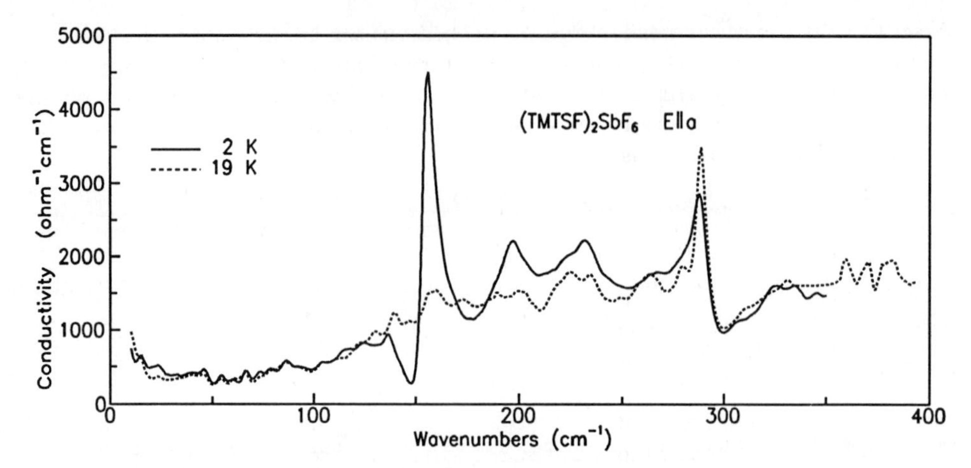

FIGURE 5. The optical conductivity of $(TMTSF)_2SbF_6$. Like the TTF-TCNQ charge density wave material this SDW system has a region of low frequency depressed conductivity that is present above the phase transition. However there is very little change to the continuous spectrum that can be associated with the SDW order. There is no gap expected in the 0 – 100 cm^{-1} region. The strong phonons develop at 150 cm^{-1} at the SDW transition and their frequency and strength follow the order parameter closely.

In optical spectra the electronically activated processes are easily recognized by the appearance of asymmetric lines at phonon frequencies.[27] The exact line shape depends on the nature of the background electronic absorption varying from dispersive shapes to positive and negative going symmetric lines superimposed on the continuum.

The heavy Fermion materials also show sharp optical phonon lines[28,9] but these are not the result of electronic activation. Since the materials are very poor metals the conduction electrons are unable to screen out the field and a direct phonon absorption is possible. The reflectance and absorption curves can be fit with simple addition of a phonon oscillator with strength factor $(\omega_p/\omega_0)^2 \approx 1$, typical of phonon strength in insulators.

Where Kramers Kronig analysis has been done the phonon line shapes in conductivity are fairly symmetrical, a sign of lack of interaction between the electron and phonon systems. In figure 2a the 108 cm^{-1} phonon has a perfectly symmetrical shape whereas the slight asymmetry of the 377 cm^{-1} phonon might signal a certain amount of interaction with the electronic background. In summary, the appearance of phonon lines in the spectra of heavy Fermion compounds is a combination of a complex crystal structure with symmetry-allowed optical modes and a poor metallic conductivity.

The oxide superconductors, in contrast have much higher conductivity and phonons are not expected to appear for light polarized in the highly conducting ab plane. (In the poorly conducting c direction strong phonons dominate the spectrum[29]) However there is evidence from neutron scattering[30,31] that certain phonons in the 500 cm^{-1} region in $YBa_2Cu_3O_7$ interact strongly with electrons. This interaction would induce structure in the this spectral region that would not vary with temperature and doping level.

It has been suggested[19] that the reproducible feature seen in the spectra of $YBa_2Cu_3O_7$ in the 500 cm^{-1} region[17] which has been widely attributed to electronic structure is due to phonons activated by the interaction with an electronic background.

A linear coupling of electronic states to phonons of the type that Rice introduced is proposed. In the organic conductors the phonons are activated by charge density waves. The source of symmetry breaking in the oxides is not clear. A similar conclusion has been reached by Foster et al.[32] who have analysed the reflectance of high quality $Tl_2Ba_2CaCu_2O_8$ films.

SUMMARY

In this review we have discussed the three systems of exotic superconductors: the organics, the heavy Fermions and the high T_c oxides. In putting them together we do not want to imply that a common mechanism explains the puzzling infrared properties of all three systems. What they have in common is that they are all unconventional metals. Analogies and concepts that work in the simple metals fail here. Nevertheless there are several unifying factors that suggest that the three systems should be studied together.

They are all poor metals. This makes infrared reflectance measurements of the type described here possible. Kramers Kronig analysis gives reliable spectral information since the signals resulting from 10 % absorption typical in these materials are easily analysed yielding high resolution spectra that relate directly to fundamental processes.

Phenomena that can be ignored in the study of simple metals, charge and spin density waves, activated phonons, low-lying electronic states manifest themselves in all the spectra here. On the other hand there are other phenomena such as the superconducting gap that seem difficult to establish with certainty in the exotic systems.

ACKNOWLEDGEMENTS

The author thanks H. Basista, A.J. Berlinsky, D.A. Bonn, K. Bechgaard, J.P. Carbotte, M.F. Collins, B. Gaulin, J.E. Greedan, D. Jérome, C. Kallin, A.J. Millis, H.K. Ng, J. Orenstein, J. Preston, M. Reedyk, T.M. Rice, K. Renk, C.V. Stager, D.B. Tanner, G.A. Thomas, J. Voit, for many illuminating discussions and insights. The materials made by K. Bechgaard, J.D. Garrett, A.H. O'Reilly, and L.F. Schneemeyer have made much of this work possible. The work at McMaster is supported in part by the National Science and Engineering Research Council (NSERC), the Canadian Institute of Advanced Research (CIAR) and the Ontario Center for Materials Research (OCMR).

REFERENCES

1. T. D. Holstein, Phys. Rev. **96**, 535, (1954); P. B. Allen, Phys. Rev. B **3**, 305, (1971).

2. R.R. Joyce and P.L. Richards, *Phys. Rev. Lett.* **24**, 1007 (1970).

3. B. Farnworth and T. Timusk, *Phys. Rev. B* **14**, 5119 (1976).

4. H. Basista, D.A. Bonn, T. Timusk, J. Voit, D. Jérome, K. Bechgaard, (to be published).

5. P. A. Lee, T. M. Rice, P. W. Anderson, Phys. Rev. Lett **31**, 462, (1973); Solid State Comm. **14**, 703, (1974).

6. C.S. Jacobsen, D.B. Tanner, A.F. Garito and A.J. Heeger, *Phys. Rev. Lett.* **33**, 1559 (1974).

7. D.B. Tanner et al. *Phys. Rev. Lett.* **47**, 597 (1981).

8. D. Quinlivan, Y. Kim, K. Holczer, G. Grüner, and F. Wudl (unpublished)

9. D.A. Bonn, J.D. Garrett, and T. Timusk, *Phys. Rev. Lett.* **61**, 1305 (1988).

10. J.W. Allen and J.C. Mikkelsen, *Phys. Rev. B* **15**, 2952 (1977).

11. D.B. Tanner (private communication)

12. L. Forro, G.L. Carr, G.P. Williams, D. Mandrus, L. Mihaly (unpublished)

13. D.R. Harshman, G. Aeppli, E.J. Ansoldo, B. Batlogg, J.H. Brewer, J.F. Carolan, R.J. Cava, M. Celio, A.C.D. Chaklader, W.N. Hardy, S.R. Kreitzman, G.M. Luke, D.R. Noakes, and M. Sheba, *Phys. Rev. B* **36**, 2368 (1987).

14. L. Krusin-Elbaum, R.L. Greene, F. Holtzberg, A.P. Malozemoff,and Y. Yeshurun, *Phys. Rev. Lett.* **62**, 217 (1989).

15. J. Orenstein *et al.* (unpublished).

16. K. Kamarás, S.L. Herr, C.D. Porter, N. Tache D.B. Tanner, S. Etemad, T. Venkatesan, E. Chase, A. Inham, X.D. Wu, M.S. Hegde, and B. Dutta, *Phys. Rev. Lett.* **64**, 84 (1990).

17. S.L. Cooper, G.A. Thomas, J. Orenstein, D.H. Rapkine, M. Capizzi, T. Timusk, A.J. Millis, L.F. Schneemeyer, and J.V. Waszczak, *Phys. Rev. B* **40**, 11358 (1989).

18. S. Tajima, S. Tanaka, T. Ido, S. Uchida, (to be published) 2nd int. symposium in Tskuba.

19. T. Timusk, D.B. Tanner,(unpublished).

20. M.B. Maple,J.W. Chen, Y. Dalichauoch, T. Kohara, C. Rossel, M.W. McElfresh, and J.D. Thompson, *Phys. Rev. Lett.* **56**, 185 (1986).

21. C. Broholm,J.K. Kjems, W.J.L. Buyers, C.D. Matthews, T.T.M. Palstra, A.A. Menkovsky, and J.A. Mydosh, *Phys. Rev. Lett.* **58**, 1467 (1987).

22. H. K. Ng, T. Timusk, and K. Bechgaard, *Phys. Rev. B* **30**, 5842 (1984).

23. E.W. Fenton and G.C. Psaltakis, *Solid State Comm.* **47**, 767 (1983).

24. A.A. Volkov, B.P. Gorshunov, G.V. Kozlov, *JETP* **65**, 856 (1987).

25. R. Bozio and C. Pecile, *Solid State Comm.* **37**, 193 (1981).

26. M.J. Rice, *Phys. Rev. Lett.* **37**, 36 (1976).

27. M.J. Rice, V.M. Yartsev, and C.S. Jacobsen, *Phys. Rev. B* **21**, 3437 (1980).

28. F.E. Pinkerton, A.J. Sievers, M.B. Maple,and B.C. Sales, *Phys. Rev. B* **29**, 609 (1984).

29. R.T. Collins, Z. Schlesinger, F. Holtzberg, and C. Field, *Phys. Rev. Lett.* **63**, 422 (1989).

30. B. Renker, F. Gompf, E. Gering, D. Ewert, H. Rietschek, and A. Dianoux, *Z. Phys. B – Condensed Matter* **73**, 309, (1988).

31. W. Reichard, N. Pyka, L. Pintschovius, B. Hennion, and G. Collin, Int. Conf. on Superconductivity, Stanford, 1989, *Physica C.* **162 – 164**, 464, (1989).

32. C.M. Foster, K.F. Voss, T.W. Hagler, D. Mihailović, A.J. Heeger, M.M. Eddy, W.L. Olsen, and E.J. Smith, (unpublished).

THE ELECTRODYNAMICS OF THE SUPERCONDUCTING STATE OF κ-(BEDT-TTF)$_2$Cu(MCS)$_2$

K. Holczer,* O. Klein, and G. Grüner
Department of Physics and Solid State Science Center
University of California, Los Angeles, CA 90024

H. Yamochi and F. Wudl
Department of Physics
University of California, Santa Barbara, CA 93106

INTRODUCTION

We have measured the temperature dependence of the penetration depth λ and surface resistance R_s using a resonant microwave cavity technique for different orientations of the ac magnetic field with respect to the conducting layers. The behavior of $\lambda(T)$ and $R_s(T)$ strongly suggests singlet pairing, and the anisotropy of the penetration depth and surface resistance are suggestive weakly coupled superconducting layers.

REVIEW

The organic metal κ-(BEDT-TFF)$_2$Cu(NCS)$_2$ is a member of the organic linear chain compounds which become superconducting at relatively high temperatures.[1] Various experiments in the superconducting state of this material indicate deviations from what is expected for conventional singlet pairing. The temperature dependence of the critical field,[2] NMR relaxation time, and the anomalous behavior of the specific heat[4] have been interpreted in terms of triplet pairing,[5] spin density wave transition below T_c and vortex glass transition[3] as expected for a ground state with higher momentum pairing.[6] While both NMR and specific heat measurements are sensitive also to excitations other than those of the superconducting state, the parameters which characterize the electrodynamics, the penetration depth λ and surface resistance R_s are free from such complications. Consequently, the magnitude and temperature dependence of these parameters may in principle, distinguish between the various symmetry superconducting states. Recent experiments,[6,7] employing muon spin rotation are controversial, one[6] suggests s wave pairing, the other[7] shows important deviations from the BCS behavior. Low frequency ac magnetization measurements[8] with H_{ac} parallel to the layers led to an unusually large penetration depth along the a-b plane, and a temperature dependence suggesting higher momentum pairing.

---

*On leave from Central Research Institute for Physics, H1525 Budapest, POB 49, Hungary.

We have conducted experiments on the low frequency ($\hbar\omega < \Delta$) electrodynamical response in the normal and superconducting state of κ-$(BEDT\text{-}TTF)_2Cu(NCS)_2$, by measuring the parameter called the surface impedance Z_s at millimeter wave frequencies.

The surface impedance is defined as[9]

$$Z_s = \frac{E_o}{\int^{\infty} j dx} , \qquad (1)$$

where E_o is the electric field at the surface, j is the ac current in the sample and the x direction is perpendicular to the surface. Z_s is, in terms of the complex conductivity $\sigma = \sigma_1 - i\sigma_2$, given by

$$Z_s = \left(\frac{\mu_o \omega}{\sigma_1 - i\sigma_2} \right)^{1/2} = R_s + iX_s , \qquad (2)$$

where μ_o is the permeability of free space, R_s and X_s are the surface resistance and surface reactance, and ω the measuring frequency. In the normal state, at frequencies $\omega\tau < 1$, $\sigma_1 \gg \sigma_2$ and consequently

$$R_s = X_s = \left(\frac{\mu_o \omega}{2\sigma} \right)^{1/2} , \qquad (3)$$

i.e., the surface resistance is equal to the surface reactance in the so-called Hagen-Rubens limit. Consequently, at low frequencies the resistivity is given by $\rho(T) = 2R_s^2/\mu_o\omega$.

In the superconducting state well below T_c the surface reactance is given by

$$X_s(T) = \mu_o \omega \lambda(T) , \qquad (4)$$

where λ is the penetration depth, and the measured $X_s(T)$ can be directly compared with various models of the superconducting state. The surface resistance has been calculated in the local limit by Mattis and Bardeen,[10] and $R_s(T)$ is in good agreement with experiments conducted on classical superconductors.

The two parameters are also related through a sum rule, given by

$$\lambda = \frac{c}{(8A)^{1/2}} , \qquad A = \int_o^{\infty} (\sigma_n - \sigma_s) d\omega , \qquad (5)$$

where σ_n and σ_s are the conductivities in the normal and in the superconducting state as a function of ω. Equation (4) leads to $\lambda = c/\omega_p$, with $\omega_p = (4\pi ne^2/m_b)^{1/2}$ in the clean limit $h/\tau \ll \Delta$. For the dirty limit $h/\tau \gg \Delta$, an equation for λ, equivalent to the Pippard expression $\lambda(T{=}0) = \lambda_L(T{=}0) \times [1+(\xi_o/\ell)]^{1/2}$ is recovered in this limit. $\sigma_n = \mu_o\omega/2R_{SN}^2$ is independent of the frequency, and $\sigma_s(\omega)$ can be evaluated from the known expressions, given by Mattis and Bardeen.

$$\sigma_s(\omega) = \left[1 - \left(\frac{2\Delta}{\hbar\omega} \right)^{1.65} \right] \qquad (6)$$

for $\omega > 2\Delta/\hbar$ at T = 0. Using Eq. (4), Eq. (6) leads to[11]

$$\frac{X_s}{R_s} = A \left(\frac{\hbar\omega}{\Delta} \right)^{1/2} , \qquad (7)$$

with A = 1.27.

EXPERIMENTAL TECHNIQUES AND RESULTS

The experiments were conducted using a resonant cavity. The superconducting materials were placed inside the cavity, and the change in the resonance frequency f_o and quality factor Q measured as a function of the temperature. The changes normally represent a small perturbation on the resonance, and can be linearly related to the changes in the surface impedance of the sample:

$$\Delta\left(\frac{1}{Q}\right) - 2j\left(\frac{\Delta\omega_o}{\omega_o}\right) = \gamma\Delta Z_s = \gamma\left(\Delta R_s + j\Delta X_s\right), \tag{8}$$

where γ is a geometrical factor reflecting the dimensions of the sample and cavity. Depending on the position of the specimen in the cavity, the currents flow along different directions. With the ac magnetic field perpendicular to the conducting plane, all the currents flow within the plane and consequently the parameters λ and R_s refer to this current flow. We denote these parameters as $\lambda_{||}$ and $R_{s||}$. Alternatively, H_{ac} can be parallel to the conducting layers, with the current flow partially parallel and partially perpendicular to the layers. In this case a combination of $\lambda_{||}$ and $\lambda_{\perp}$ - or $R_{s||}$ and $R_{s||}$ - are measured. With A and B being the dimensions of the specimen for the current flow parallel (A) and perpendicular (B) to the layers

$$\begin{aligned} \lambda_{mean} &= \left(A\lambda_{||} + B\lambda_{\perp}\right)\left(A+B\right)^{-1} \\ R_{s_{|}}\ mean &= \left(AR_{s_{||}} + BR_{s\perp}\right)\left(A+B\right)^{-1} \end{aligned} \tag{9}$$

The quality factor Q and ω_o were measured by employing a novel feedback configuration similar to those used in conventional ESR spectrometers.[12] This allows precise evaluation of both parameters R_s and X_s.

EXPERIMENTAL RESULTS

Penetration depth (surface reactance and surface resistance measurements have been conducted by placing the specimen on the bottom of a TE_{011} cavity operating at $f = \omega/2\pi = 102$ GHz. Configurations with either the in-plane properties and a combination of in-plane and out-of-plane parameters were examined.

For a configuration where H_{ac} is perpendicular to the conducting layers, our measurement is not sensitive enough to allow a precise evaluation of $\lambda_{||}(T)$. We find, however, a substantial penetration depth for the configuration indicated in Fig. 1, with the ac magnetic field is parallel to the c direction, i.e., with the highly conducting b-c plane. In the configuration shown in Fig. 1 both the penetration depth along the layers $\lambda_{||}$ and perpendicular to the layers $\lambda_{\perp}$ will contribute to the total frequency shift. The frequency shift due to the specimen is given by

$$\frac{\Delta f}{f_o} = \frac{1}{2}\ \frac{H_o^2}{\langle H_o^2\rangle}\ \frac{(S_{||}\lambda_{||} + S_{\perp}\lambda_{\perp})}{V}, \tag{10}$$

where S is the surface area of the specimen, V the volume of the cavity, H_o the magnetic field at the surface and $\langle H_o^2\rangle$ the magnetic field strength square averaged over the area of the bottom plate of the cavity. For the TE_{011} configuration we have employed, $H_o^2/\langle H_o^2\rangle = 2.09$. Due to the pronounced anisotropy of the electric properties, and because of the strong anisotropy of the coherence length ξ_o, it is expected that $\lambda_{\perp} \gg \lambda_{||}$ and $\lambda_{\perp}$ dominates the frequency shift. The T = 0 limit of Δf has been established by using the fact that, just above T_c, in the metallic state $X_s = R_s$, and consequently the frequency shift Δf is equal to $\frac{1}{2}\ \Delta\ (1/Q)$. This leads to $\lambda_{\perp}(0) = 36\mu m$, a value significantly larger than that of $\lambda_{||}\ (T=0) = 9800Å$ found by μSR experiment.[6] The temperature dependence $\lambda_{\perp}(T)$ as evaluated from $\Delta f(T)$ is shown in Fig. 1, and the dashed line is the prediction of the two fluid model

$$\lambda(T) = \lambda(0)\left[1 - \left(\frac{T}{T_c}\right)^4\right]^{-1/2}. \tag{11}$$

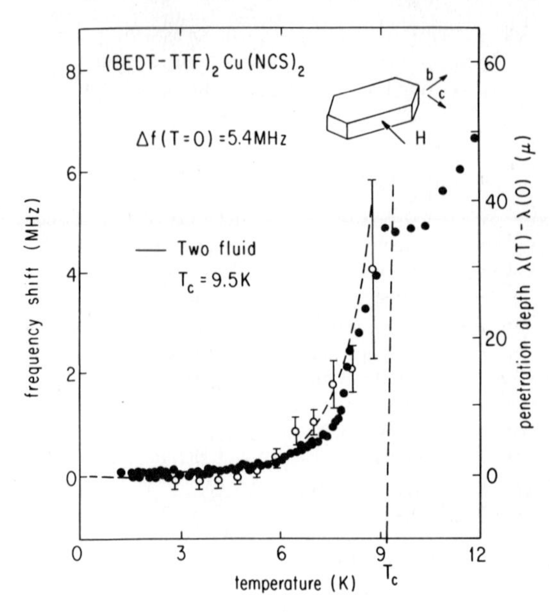

Fig. 1. Temperature dependence of the penetration depth both perpendicular ($\lambda_\perp$) and parallel ($40 \times \lambda_\parallel$ to the layers. $\lambda_\parallel$ is taken from Ref. 6. The full line is the prediction of the two-fluid model for $\lambda(T)$.

Also displayed in the figure is $40 \times [\lambda_\parallel(T) - \lambda_\parallel(T=0)]$ measured by μSR method[6] for comparison. Both experiments strongly suggest singlet pairing and a single particle gap which develops along the entire Fermi surface.

Both parameters can be understood using conventional electrodynamics to the normal and superconducting state. The dc conductivity measured above the transition $\sigma_{dc} = 2.0\times10^3\ \Omega^{-1}\ cm^{-1}$ (T=12K) can be combined with the measured plasma frequency[14] $\omega_p = 0.6$eV to obtain the relaxation time τ, using the relation $1/\tau = \omega_p^2\ (4\pi\sigma)^{-1}$. This leads to $1/\tau = 4 \times 10^{13}\ sec^{-1}$, larger than the gap frequency $\omega_g = 3.5k_BT_c = 5 \times 10^{12}\ sec^{-1}$ expected for the weak coupling BCS limit. With these parameters, the T = 0 penetration depth, using the mean free path corrections,[13] is given by

$$\lambda(o) = \lambda_L(o) \left[1 + \frac{\xi_o}{\ell}\right]^{1/2} = c/\omega_p \left[1 + \frac{2/\tau}{\pi\omega_g}\right]^{1/2}, \tag{12}$$

where ξ_o is the coherence length, and ℓ the mean free path. With the previous values of $1/\tau$, Δ and ω_p one obtains $\lambda(T=0) = 7800$Å, in excellent agreement with the measured value.[6] The same values give $\lambda_L(0) = 3300$Å for the London penetration depth.[13]

Perpendicular to the planes, the coherence length $\xi_\perp < d$, where d is the interplanar separation. Under such circumstances, the situation is close to that of Josephson coupled planes, with the resulting penetration depth[15]

$$\lambda_\perp = \left(\frac{hc^2\rho_\perp}{8\pi^3\Delta}\right)^{1/2}, \tag{13}$$

where $\rho_\perp$ is the conductivity perpendicular to the planes. Our surface resistance measurements (see below) give $\rho_\perp = 1\Omega$cm an approximate value just above the transition, this, together with $\Delta = 1.76k_BT_c$ leads to $\lambda_\perp = 32\mu$m in excellent agreement with the experimentally obtained value.

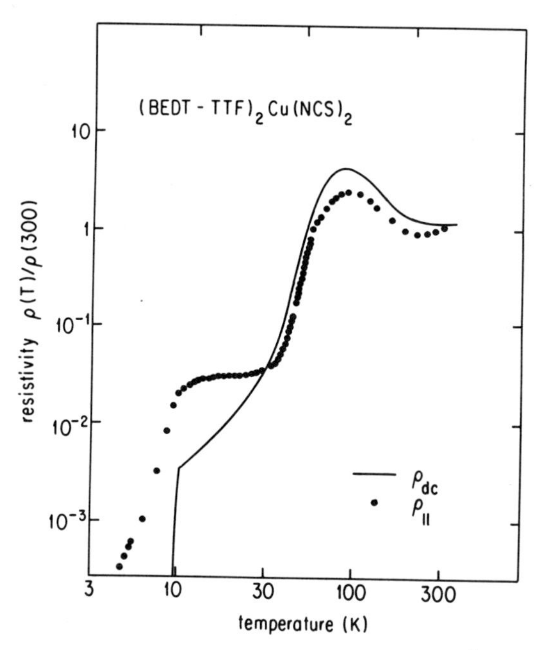

Fig. 2. Temperature dependence of the parameter R_s^2/ω for the configuration with the ac magnetic field perpendicular to the conducting plane. Note, that for a Drude metal, $R_s^2/\omega = \rho_{dc}$, with ρ_{dc} also plotted in the figure.[16]

Figure 2 shows $R_s{}^2/\omega$ in the configuration where the current flows in the b-c plane, and consequently the surface resistance reflects temperature dependence of the in-plane resistivity of $\rho_{\parallel}$. An important feature of Figure 2 is that $\rho_{\parallel}$ measured just above T_c exceeds ρ_{dc}.[16] Whether this is due to a strongly frequency dependent conductivity in the metallic state remains to be seen.

The temperature dependence of the normalized surface resistance R_s/R_N, with R_N the value measured just above T_c is shown in Fig. 3 together with Rs/R_N measured in Nb by employing the same experimental technique. Note, that for the experiments shown in this figure, all the ac currents flow in the plane. We have, however, found a behavior rather similar for configurations where the current flows perpendicular to the planes. The full line is a calculation based on the Mattis- Bardeen model which is appropriate for the situation where the mean free path ℓ is much smaller than the coherence length ξ_o. For κ-(BEDT- TTF)$_2$Cu(NCS)$_2$, ξ_o = 180Å,[1] and the measured relaxation time, together with $v_F = 10^7$ cm/sec gives ℓ = 50Å, consequently $\ell/\pi\xi_o \approx 0.06$. A mean free path which is comparable to ξ_o leads to an enhanced surface resistance,[17] with R_s well above the value given by the Mattis-Bardeen limit. Preliminary calculations indicate that our experiments are in a good semiqualitative agreement with R_s obtained with finite mean free path correc tions. Nevertheless, the experimental results lie somewhat above the calculated curve, implying an excess R_s probably due to grain boundaries and macroscopic inhomogenieties. This is not surprising, and various models which lead to inhomogeniety induced loss mechanisms[18] may also explain our findings.

As discussed earlier, the frequency shift measured is proportional to the penetration depth, and the change of $Q, \Delta(1/Q)$ is proportional to the surface resistance R_s so Eq. (7) can be written as

$$\frac{2\Delta f(T=0)}{\Delta(1/Q)_N} = 1.27 \left(\frac{\hbar\omega}{\Delta}\right)^{1/2} . \tag{14}$$

Using the measured values (for a configuration where H_{ac} is parallel to the layers) of Δf and $\Delta(1/Q)_N$ (measured just above the transition), we obtain

$$2\Delta \simeq 5k_BT_c \quad , \tag{15}$$

a value somewhat exceeding the BCS weak coupling limit.

CONCLUSIONS

Our experiments lead to several conclusions concerning the normal and superconducting state of κ-(BEDT-TTF)$_2$Cu(NCS)$_2$.

The superconducting ground state is dominantly s-wave, as evidenced by the temperature dependence of λ, and the penetration depth is highly anisotropic ($\lambda_\perp/\lambda_\parallel \sim 40$). $\lambda_\perp$ can be described in terms of weakly coupled superconducting layers, for which the coherence length $\xi_{o\perp}$ is comparable to the interlayer separation. $\lambda_\parallel$ can be described well in terms of normal state parameters ω_p and $1/\tau$, and this is highly suggestive of a simple metallic behavior for conduction in the planes.

By comparing the measured surface resistance and surface reactance, we arrive at a gap value which exceeds the weak coupling BCS value, indicating that, perpendicular to the layer, a larger gap may develop.

We have also observed significant flux penetration effects even with only the earth's magnetic field present. These suggest easy flux penetration between the planes, undoubtedly due to the highly anisotorpic superconducting state properties.

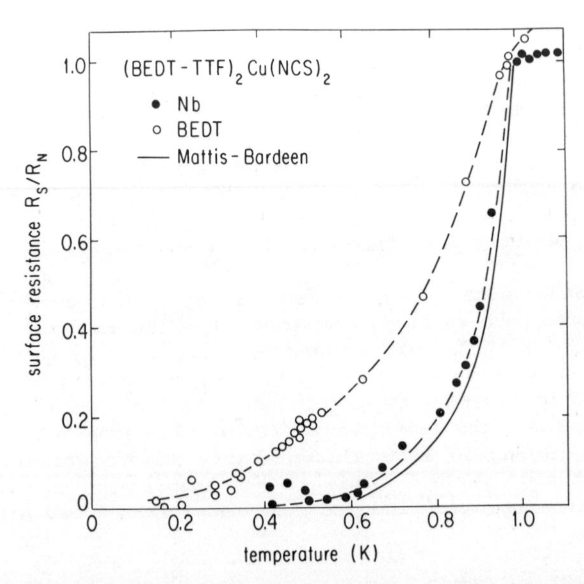

Fig. 3. Temperature dependence of the surface resistance R_S/R_N for Nb and for κ-(BEDT-TTF)$_2$Cu(NCS)$_2$. Full line is the Matthis-Bardeen calculation.[10]

ACKNOWLEDGMENTS

Research at UCLA was supported by the NSF grant DMR 89-13236, at UCSB by the grant DMR 89-20933. We wish to thank D. Scalapino for useful discussions. Some of the early experiments were also conducted with the help of D. Quinlivan.

REFERENCES

1. See for example, "Proceedings of the International Conference on Science and Technology of Synthetic Metals", Santa Fe, June 28 (1988), Synthetic Metals 27, (1988).
2. K. Oshima, Synthetic Metals 27, A419 (1988).
3. T. Takahashi, et al. Synthetic Metals 27, A319 (1988).
4. G. Saito, Workshop on Spin Density Waves, UCLA 1988 (unpublished).
5. Y. Hasegawa and H. Fukuyama, J. Phys. Soc. Japan 56, 877 (1987).
6. D.R. Harshman, et al, Phys. Rev. B64, 1293 (1990).
7. Y.J. Uemura, et al. Nato Advanced Research Workshop on "Dynamics of Magnetic Fluctuations in High Temperature Superconductors" Crete 1989.
8. K. Kanoda, T. Takahaski and G. Saito, Physica C162, 405 (1989).
9. C. Kittel, Quantum Theory of Solids. John Wiley and Sons, New York, 1963.
10. D.C. Mattis and John Bardeen, Phys. Rev. 111, 412 (1958).
11. O. Klein, K. Holczer and G. Grüner (to be published).
12. K. Holczer and G. Gruner (to be published).
13. M. Tinkham, Introduction to Superconductivity (R.E. Kriegler Publ. Co., 1975).
14. Recent optical experiments (K. Kornelsen, et al., Solid State Comm. 72, 475 (1989)) lead to an enhanced effective mass m^* and enhanced relaxation time τ^*. Whether the parameters, or the unrenormalized parameters should be used in evaluating λ, remains to be seen.
15. V. Ambegaokar and A. Baratoff, Phys. Rev. Lett. 10, 486 (1963), A.P. Malozemoff, L. Krusin-Elbaum and J.R. Clem, Physical C162, 353 (1989).
16. K. Oshima, Technical Report of ISSP No. 1910, February 1988.
17. Chen and D.J. Scalapino, Phys. Rev. B40, 4299 (1989).
18. T.L. Hylton, et al, Appl. Phys. Lett. 53, 1343 (1988).

THE INFRARED CONDUCTIVITIES OF SEMICONDUCTING $(TMTSF)_2ReO_4$ AND $(TMTSF)_2BF_4$, COMPARED WITH SEVERAL MODEL CALCULATIONS

Chris Homes and J.E. Eldridge
Physics Dept., University of British Columbia
Vancouver, B.C. Canada V6T 2A6

INTRODUCTION

The $(TMTSF)_2X$ family of salts, where TMTSF is tetramethyltetraselenafulvalene, contain the first organic superconductors, most of which require the application of pressure. At ambient pressure all except the perchlorate undergo a metal-insulator transition as the temperature is lowered. For the perrhenate, $(TMTSF)_2ReO_4$, and the boron tetrafluoroborate $(TMTSF)_2BF_4$ salts, this transition occurs at 177 K and 38 K respectively. The resulting semiconductors have energy gaps of approximately 1700 cm^{-1} and 1100 cm^{-1} respectively. The two compounds are very similar in that, if they are slowly cooled, the energy gap is due to the periodic potential produced by the orientational ordering of the tetrahedral anions. Two orientations are possible and the ordering results in a new unit cell of size $2a \times 2b \times 2c$, for both compounds.

The infrared optical spectrum shows this low-temperature gap, along with many strong resonances below the gap, and antiresonances above, due to the totally-symmetric (a_g) modes of the TMTSF molecule, which are coupled to the charge carriers and thereby become optically active. Two studies of $(TMTSF)_2ReO_4$ have been reported, the first by Jacobsen et al.[1] and subsequently by Bozio et al.[2] No data for $(TMTSF)_2BF_4$ had been reported until our recent preliminary results and assignments for both compounds.[3] Jacobsen et al.[1] fitted their results to a "dimer" model due to Rice[4], whereas Bozio et al.[2] used a modified "tetramer" model, originally developed by Yartsev.[5] Neither model was entirely satisfactory.

We have undertaken a thorough study of this system by measuring, as a function of temperature, the single-crystal reflectivity of both $(TMTSF)_2ReO_4$ and $(TMTSF)_2BF_4$ for $\vec{E}\|\vec{a}$ and $\vec{E}\|\vec{b'}$, and by repeating the measurements with deuterated compounds. Vibrational assignments have been made by comparison with the results of a normal-coordinate calculation. What we present here are the abbreviated highlights of a larger work which will be published subsequently.

Organic Superconductivity, Edited by V.Z. Kresin and
W.A. Little, Plenum Press, New York, 1990

EXPERIMENT

The power reflectivity was measured from single crystals, approximately 3 mm × 1 mm × 0.1 mm, attached to the cold finger of an Air Products Heli-Tran refrigerator. This in turn was mounted in a custom-designed reflectivity unit which fits into an expanded sample chamber of a Bruker IFS-113V rapid-scan Fourier spectrometer. The lowest temperature reached was 25 ± 3K due to room-temperature radiation striking the sample. The entire arrangement and measurement procedures have been described elsewhere,[6] together with details on the sources, beam-splitters, filters and polarizers for the various overlapping spectral ranges.

RESULTS

The reflectivity results for the BF_4 compound, for two polarizations, are shown in figure 1 at 295 K, and in figure 2 at 20 K. The room-temperature spectra have low-frequency extrapolations below 60 cm^{-1}, while the low-temperature measurements extend down to 30 cm^{-1}. The ReO_4 spectra are similar but not shown.

At low temperatures, the far-infrared reflectivity drops, with the formation of the energy gap, and the spectra become busy with activated vibrational multiplets, in both polarizations. The strong multiplet below 100 cm^{-1} for $\vec{E}\|\vec{a}$ has already been reported.[7] It was assigned to a normally-active lattice mode which is split by zone-folding resulting from the new unit cell of 2a × 2b × 2c, and which gains its unusual strength by coupling directly to the charge-density wave (CDW).

The conductivities obtained after Kramers-Kronig analyses of the reflectivity data are shown in figures 3 through 5. Room-temperature and low-temperature spectra are shown in each figure. The transverse polarization is included only for the BF_4 compound.

DISCUSSION

With our high resolution, the low-temperature conductivity spectrum of Fig. 5 for the ReO_4 compound has more features in it than the previously-published data[1, 2] and extends further into the far-infrared. Otherwise, it is similar in shape although significantly different in magnitude from previous work. The semiconducting gap is clearly visible below 1000 cm^{-1} or 1500 cm^{-1}, the exact position being unclear from Fig. 5. A similar but smaller gap is seen in the new BF_4 data for $\vec{E}\|\vec{a}$ in Fig. 3 where the conductivity is effectively zero below 750 cm^{-1}. An interesting result is that the same gaps are present in the $\vec{E}\|\vec{b}$ spectra, (see Fig. 4) even though the conductivity is generally less by a factor of 100. This behaviour is not new, having been seen by us in the octrahedral-anion, AsF_6 and SbF_6, compounds[8]. In reference 8, we found that the conductivity above the gap displayed the expected inverse square-root singularity shape resulting from the one-dimensional density of states, but that the peak in the transverse polarization was much weaker, and was shifted away from the energy gap. We showed that this was mainly the result of different effective masses for each polarization. This is certainly what is again being observed here but the effect is more exaggerated. The BF_4

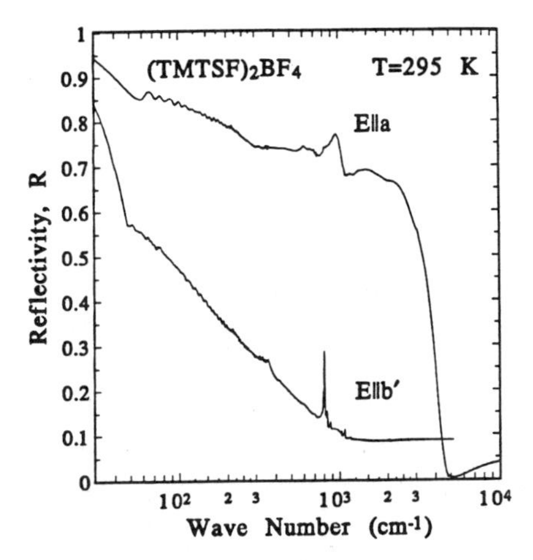

FIG. 1 The reflectivity of (TMTSF-h_{12})$_2$BF$_4$ at 295K for $\vec{E}\|\vec{a}$ and $\vec{E}\|\vec{b}'$. The resolution is 2 cm^{-1}. The data is extrapolated below 60 cm^{-1}.

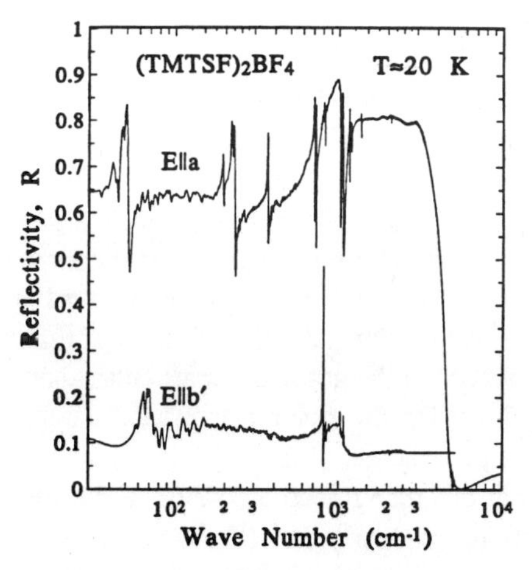

FIG. 2 The reflectivity of (TMTSF-h_{12})$_2$BF$_4$ at 25K for $\vec{E}\|\vec{a}$ and $\vec{E}\|\vec{b}'$. The resolution is 2 cm^{-1} above 100 cm^{-1} and 0.2 cm^{-1} below.

compound shows it best with the expected shape for $\vec{E}\|\vec{a}$ in Fig. 3 and the peak near 1100 cm^{-1}, while for $\vec{E}\|\vec{b}'$ the conductivity starts to rise near 1000 cm^{-1} and peaks later near 1350 cm^{-1} (see Fig. 4). Fitted values of the gaps will be presented in the next section.

MODEL FITTINGS

Dimer Model

The dimer model was originally developed by Rice and Yartsev[4] to study the optical properties of unpaired electron states on molecular dimers, and applied to the one-dimensional semiconductor MEM $(TCNQ)_2$. We have applied it to the low-temperature $\vec{E}\|\vec{a}$ conductivity spectra of both of the compounds, but will present the result obtained only for the BF_4 compound. The formulation is given in detail by Rice et al.[4] and will not be repeated here. The frequencies of the 10 most prominent vibrational features were entered into the program, which then refined the bare frequencies ω_α, electron-molecular vibrational (e-mv) coupling constants g_α, and phonon dampings γ_α, along with the charge-transfer frequency ω_{CT}, the strength of the charge-transfer band ω_p*, defined in Ref. 1, and the electronic interband damping Γ. The results are shown in Fig. 6.

Fig. 6 includes small tic marks at the "bare" phonon frequencies, ω_α, and it is seen that they correspond closely with the seven antiresonance dips and the three resonance peaks. The dashed line is the Lorentzian charge-transfer band with no phonons. On comparing the experimental result of Fig. 3 with Fig. 6 it is clear that the shape of the one-dimensional density of electronic states is poorly represented by the Lorentzian band.

Phase Phonon Model

The dimer model deals with isolated clusters of molecules and not with solids. The optical conductivity of a one-dimensional semiconducting solid was first described by Lee et al.[9] Rice[10] showed that the a_g modes would become activated by charge exchanged between molecules in a stack arising from oscillations of the phase of the charge-density-wave (CDW). Rice et al.[10] applied their "phase-phonon" theory to TEA $(TCNQ)_2$. We have applied the model to the data for both of the present compounds and again show the results for only the BF_4 compound in Fig. 7. On comparing Fig. 7 with the data in Fig. 3 one observes that the one-dimensional density of states of this model fits the data much better than the dimer model above the gap and in particular above 1500 cm^{-1}. The failure of the model is the way in which the electron interband damping Γ is incorporated. In order to obtain the correct peak height of 5000 $(\Omega cm)^{-1}$, a not unreasonable damping of 170 cm^{-1} was required, but this then ruined the shape of the spectrum below the gap and produced a d.c. conductivity of nearly 500 $(\Omega cm)^{-1}$.

One-Dimensional System with Twofold-Commensurate CDW

The phase-phonon model gives the correct shape of the conductivity spectrum since it deals with a one-dimensional solid, but it is unable to correctly treat the electronic

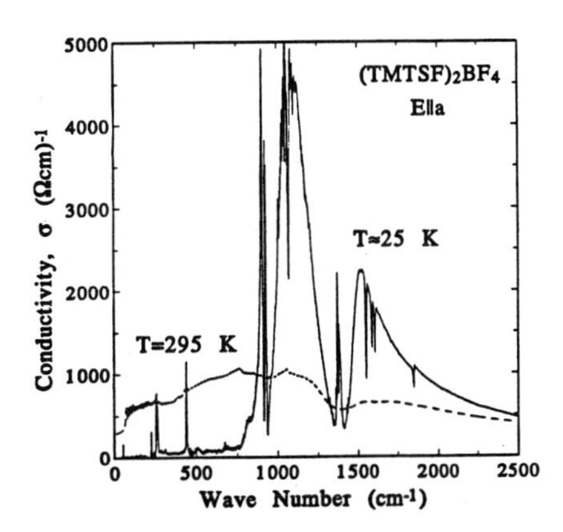

FIG. 3 The conductivity of $(TMTSF\text{-}h_{12})BF_4$ at 25K (full line) and 295K (dashed line) for $\vec{E}\|\vec{a}$.

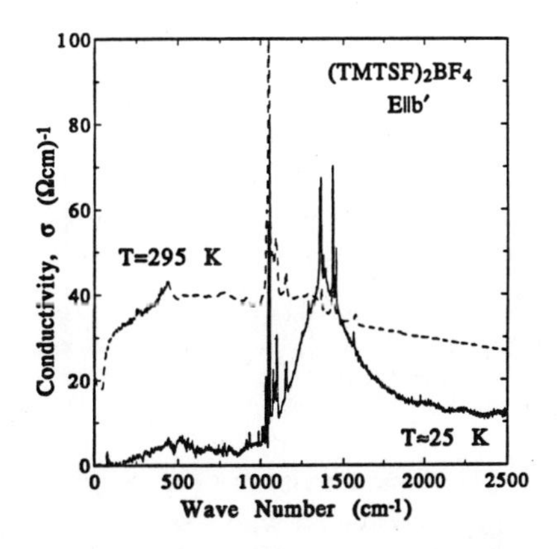

FIG. 4 The conductivity of $(TMTSF\text{-}h_{12})BF_4$ at 25K (full line) and 295K (dashed line) for $\vec{E}\|\vec{b}'$.

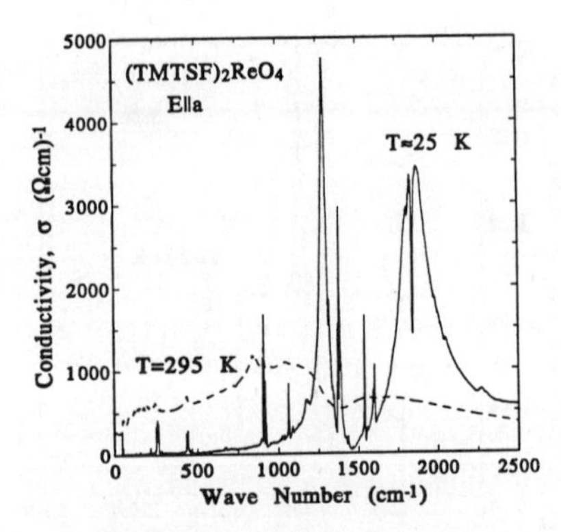

FIG. 5 The conductivity of ($TMTSF\text{-}h_{12})ReO_4$ at 25K (full line) and 295K (dashed line) for $\vec{E}\|\vec{a}$.

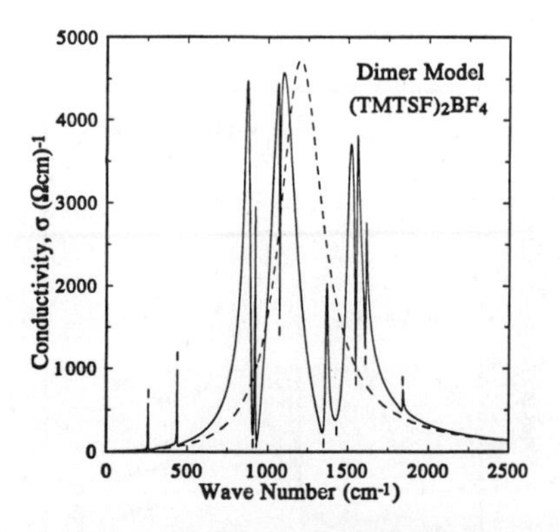

FIG. 6 The calculated 25K $\vec{E}\|\vec{a}$ conductivity of $(TMTSF)BF_4$ using the dimer model of Rice and Yartsev[4], to be compared with Fig. 3. The solid line results from the electron coupling to ten phonons. The frequencies of the phonons are indicated by tic marks. The dashed line is the no-phonon spectrum.

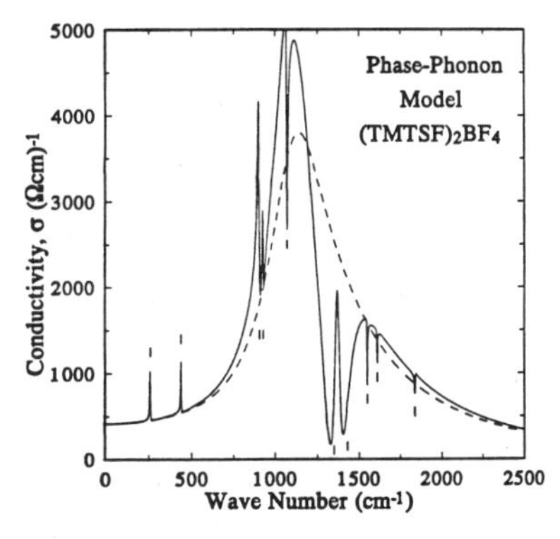

FIG. 7 The calculated 25K $\vec{E}\|\vec{a}$ conductivity of $(TMTSF)_2BF_4$ using the phase-phonon model of RICE[10], to be compared with FIG. 3. The solid line results from electron coupling to ten phonons. The frequencies of the phonons are indicated by tic marks. The dashed line is the no-phonon spectrum.

TABLE 1. Two-fold commensurate CDW model parameters for the fit to the $\vec{E}\|\vec{a}$ optical conductivities of $(TMTSF)_2ReO_4$ and $(TMTSF)_2BF_4$ at 25K.

t 1400 cm^{-1}, a 3.55Å (ReO_4) 3.51 Å (BF_4)

Lattice distortion u_0 = 0.022 $Å^3$, V_m 341 $Å^3$ (ReO_4) 334 $Å^3$ (BF_4)

Gap from anion potential: 1569 cm^{-1} (ReO_4), 969 cm^{-1} (BF_4)

Gap from lattice distortion: 131 cm^{-1} (ReO_4), 151 cm^{-1} (BF_4)

Total gap from CDW centered at bond: 1700 cm^{-1} (ReO_4), 1120 cm^{-1} (BF_4)

Temperature 25K ϵ_∞ 2.6

$\omega_\alpha(cm^{-1})$	$g_\alpha(cm^{-1})$	$\gamma_\alpha(cm^{-1})$	Assignment (a_g modes)
1840	60	2	$\nu_3 + \nu_{11}$
1610	60	2	ν_3
1550	100	2	ν_3
1430	420	10	ν_4
1350	180	5	ν_4
1071	60	2	ν_7
930	190	2	ν_8
908	80	2	ν_8
440	160	2	ν_9
260	170	2	ν_{11}

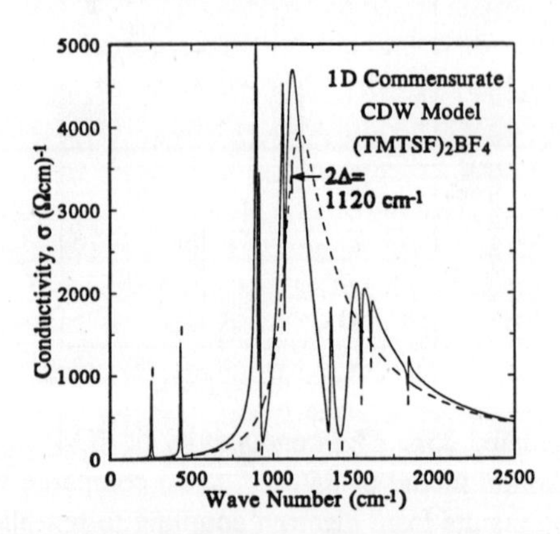

FIG. 8 The calculated 25K $\vec{E}\|\vec{a}$ conductivity of $(TMTSF)_2BF_4$ using the two-fold commensurate CDW model of Bozio et al.[11] to be compared with Fig. 3. The solid line results from electron coupling to ten phonons (see Table 1). The frequencies of the phonons are indicated by tic marks. The dashed line is the no-phonon spectrum. The energy gap is also indicated.

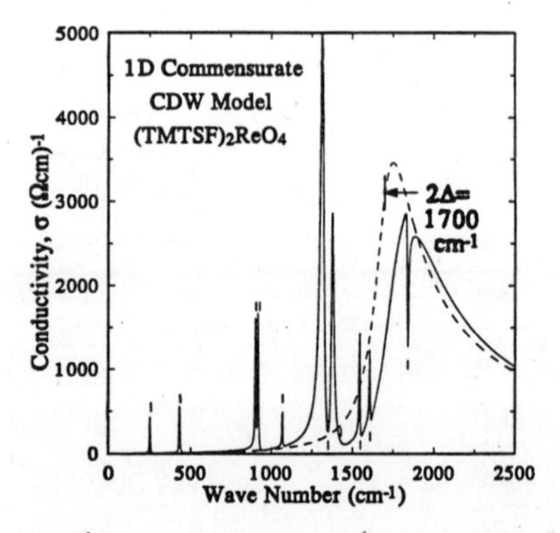

FIG. 9 The calculated 25K $\vec{E}\|\vec{a}$ conductivity of $(TMTSF)_2ReO_4$ using the two-fold commensurate CDW model of Bozio et al.[11] to be compared with Fig. 5. The solid line results from electron coupling to ten phonons (see Table 1). The frequencies of the phonons are indicated by tic marks. The dashed line is the no-phonon spectrum. The energy gap is also indicated.

damping and the phonons near the gap edge. Bozio et al.[11], following the work of Schulz,[12] have developed a comprehensive and versatile model for a one-dimensional system with a two-fold commensurate CDW. This CDW accompanies a periodic lattice distortion (PLD) which may be due to either a lattice dimerization (LD) or an alternating molecular deformation (AMD) or both. The presence of nearby counterion chains is taken into account by introducing an effective potential which can have a component with extremes on the sites (s) and/or a component with extremes on the bonds (b) between the sites. Coupling with an arbitrary number of internal modes and one longitudinal acoustic external mode is possible. Since the authors[11] show that the a_g modes are vibronically actived only by the contribution to the energy gap by the b-CDW, which is made up of contributions from the LD and the b component of the anion potential, these are the non-zero elements we have included in our calculations. We think that the lattice mode that is active at 50 cm^{-1} is not an acoustic mode,[6] and it will not therefore come out of our calculation in which we put the s-CDW component of the energy gap equal to zero.

The formalism is given in Ref. 7 and will not be repeated here, especially since it is extremely involved.

The results obtained for both compounds are shown in Figs. 8 and 9 and in Table 1. The ten input internal phonon frequencies are the same for each compound and are an average of the close frequencies of the most prominent features observed in the two spectra of Figs. 3 and 5. Likewise, the e-mv coupling coefficients g_α, and the phonon damping values γ_α are identical for the two calculations. Apart from the slightly different lattice parameters a and volumes V_m, the calculations are virtually identical, except for the value of the total semiconducting energy gap, 1700 cm^{-1} for $(TMTSF)_2ReO_4$ and 1120 cm^{-1} for $(TMTSF)_2BF_4$. The excellent agreement between the calculated and observed spectra show that this is the only parameter changing significantly between the two compounds, and that the model is capable of reproducing all of the major elements of the spectra.

The dashed line shows the damped electronic conductivity in the absence of phonon coupling, and the energy gap 2Δ is indicated. The sum rules give equal areas under the full and dashed lines. One can see how difficult it would be to extract an energy gap from the spectra with phonons present. No justification for the assignments in Table 1 have been given here due to lack of space. This will be presented in the full report, along with a description of the vibrations responsible for the ν_4 doublet and the higher-order splittings.

REFERENCES

1. C.S. Jacobsen, D.B. Tanner and K. Bechgaard, Phys. Rev. B. 28, 7019 (1983).
2. R. Bozio, M. Meneghetti, D. Pedron and C. Pecile, Synthetic Metals 27, B109 (1988).
3. C.C. Homes and J.E. Eldridge, Synthetic Metals 27, B49 (1988).
4. M.J. Rice, V.M. Yartsev and C.S. Jacobsen, Phys. Rev. B21, 3437 (1980).
5. V.M. Yartsev, Phys. Stat. Sol. (b) 126, 501 (1984).
6. J.E. Eldridge, and C.C. Homes, Infrared Phys. 29, 143 (1989).
7. C.C. Homes and J.E. Eldridge, Phys. Rev. B40, 6138 (1989).
8. J.E. Eldridge and G.S. Bates, Phys. Rev. B34, 6992 (1986).

9. P.A. Lee, T.M. Rice and P.W. Anderson, Solid State Commun. 14, 703 (1974).
10. M.J. Rice, L. Pietronero and P. Bruesch, Solid State Commun. 21, 757 (1977).
11. R. Bozio, M. Meneghetti and C. Pecile, Phys. Rev. B36, 7795 (1987).
12. H.J. Schulz, Phys. Rev. B18, 5756 (1978).

NORMAL AND SUPERCONDUCTING PROPERTIES OF ORGANIC MATERIALS: THERMOPOWER, PRESSURE EFFECTS, STM

An Introduction to Thermopower for Those Who Might Want to Use It to Study Organic Conductors and Superconductors

P.M. Chaikin

Dept. of Physics
Princeton University
Princeton, NJ 08544

INTRODUCTION

Thermopower is one of the easiest experiments that can be performed on conductors. It is also one of the most sensitive probes of the carriers in the system. As cartooned in figure 1, a sample is heated at one end and the voltage difference between the hot and cold side are measured. The ratio between the temperature difference and the voltage (or equivalently the temperature gradient and the electric field) is called the Seebeck[1] coefficient and is an intrinsic property of the material.

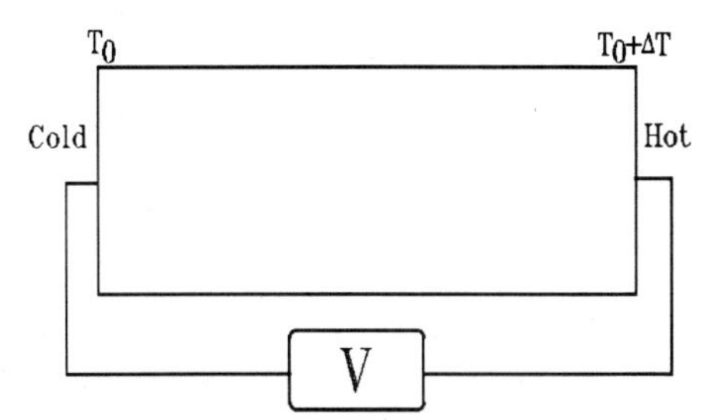

Fig. 1 In the Seebeck Effect a temperature gradient across a material produces a voltage. In the usual geometry the measured voltage is actually the difference in Seebeck coefficients between the sample and the wires going to the voltmeter (since they are also experiencing the temperature drop.)

Many researchers avoid thermopower measurements because they think the interpretation is too difficult. Others plunge ahead and over interpret their data. The philosophy which will guide the presentation to follow is that the thermoelectric power is easy to understand in some simple, maybe idealized, situations. There are characteristic behaviours for metals, semiconductors, superconductors etc. If a sample is found to have one of characteristic behaviours - great, the interpretation is straight

Organic Superconductivity, Edited by V.Z. Kresin and W.A. Little, Plenum Press, New York, 1990

forward and a lot of information can be easily obtained. If the behavior is not ideal, at least the sample may be placed closer to one catagory than another and some information is obtained with some caveats attached. The idea here is not to make a presentation of how to analyze thermopower data. Instead we would like to give some intuition into what thermopower measures.

The second cartoon (figure 2) shows the simplest model in which we can envision the microscopic origins of the thermopower. We have a bunch of charged particles (charge q) in a box. In the presence of the temperature gradient there may be a steady state density gradient which arises from the fact that the particles which are leaving the hot side are moving faster than those which are leaving the cold side. If the particles are electrons (negative charge) then the cold side is negative and the hot side is positive (we assume a uniform background charge of the opposite sign to the carriers so that the system is completely neutral in the absence of the temperature gradient). An electric field develops across the sample. In this simple picture it is clear how the voltage appears and particularly that the sign of the voltage would change if we changed the sign of the mobile carriers. However, it is difficult to proceed further along this line and calculate or intuit the size of the voltage generated.

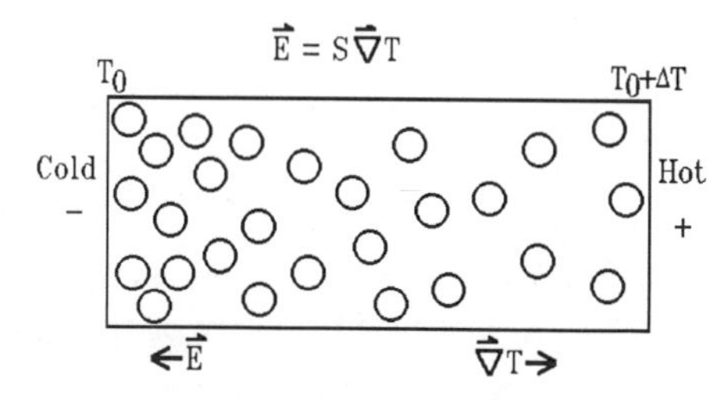

Fig. 2a) The particle density on the hot side may be lower than on the cold side in steady state. In this figure we assume negatively charged particles and find that the electric field is opposite to the temperature gradient, so S gives the carrier sign. Positive carriers would have positive S.

Suppose instead we consider the problem of the same box of particles with no temperature gradient but with a current flowing (cartoon 2b). If the particles also have heat then there is a heat current, $\vec{u}$, flowing whenever an electrical current, $\vec{j}$, is flowing.

$$\vec{j} = ne\vec{v} \qquad \vec{u} = c\vec{v} \qquad \Pi = \frac{cT}{q}. \tag{1}$$

Where:

$$\vec{u} = \Pi\vec{j} \tag{2}$$

is the defining relation for the Peltier Heat[2], Π, the ratio of the heat to the electrical current. Both Π and S are transport coefficients for combined thermal and electrical transport and they are related by an Onsager relation[3] (from the theory of non-equilibrium thermodynamics):

$$S = \frac{\Pi}{T} \tag{3}$$

from equation 1 we have:

$$S = \frac{c}{q} \tag{4}$$

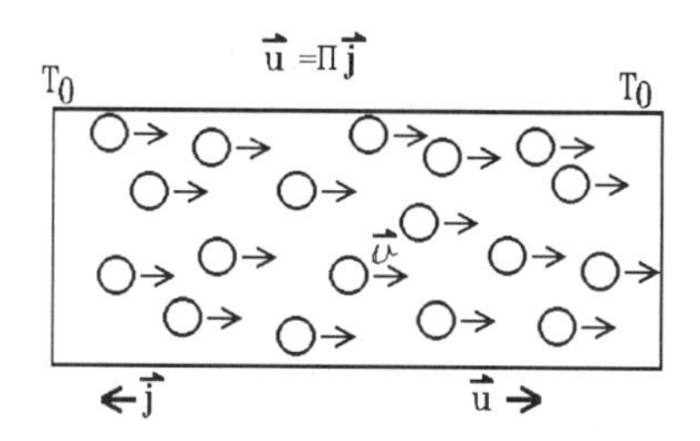

Fig. 2b) In the Peltier Effect a current is flowing through a sample with no temperature gradients. Since there is entropy and heat associated with each particle, a particle current ⇒ a heat current and an electrical current. Again the situation for negative particles is shown.

For a classical gas, each particle has 3/2 k_BT thermal energy and:

$$c \sim \frac{3k_B}{2} \qquad S \sim \frac{3k_B}{2q} \sim \frac{k_B}{q} \tag{5}$$

The main point of this paper is the simple idea that the thermopower is a measure of the heat per carrier over temperature or the entropy per carrier:

$$\mathrm{S} = \frac{\text{“heat” per carrier}}{\text{charge per carrier T}} \quad \text{or} \quad \mathrm{S} = \frac{\text{entropy per carrier}}{\mathrm{q}} \tag{6}$$

The idea is not only intuitively appealling but it directly leads to the dominant thermopower characteristics in almost all cases relating to organic conductors. From these equations we realize that the two pieces of information that we might expect the thermopower will provide are:

- the sign of the charge of the carriers
- the characteristic energy associated with the carriers

We immediately can find the thermopower in several interesting but trivial cases, where the heat or entropy associated with the conduction process is zero.

The thermopower is zero for:

- superconductors[4]
- sliding density wave transport[5]
- transport by filled Landau levels as in the Quantum Hall Effect[6]

THE SIGN OF THE CARRIERS?

The first bit of information we expect from the thermopower is the sign of the carriers. There is another measurement which is supposed to provide the same bit of information, the Hall Effect. Again using the cartoon in figure 3 we see that in the presence of a perpendicular magnetic field an electrical current is diverted and charge accumulates on the transverse edges of a sample. The charge produces a transverse electric field which builds up until it exactly cancels the Lorentz force:

$$\vec{E} = \frac{\vec{v} \times \vec{B}}{c} \qquad \rho_{xy} \equiv \frac{E_y}{j_x} = \frac{B_z}{nqc} \qquad R_H \equiv \frac{\rho_{xy}}{B_z} = \frac{1}{nqc} \tag{7}$$

where we have again taken $\vec{j} = nq\vec{v}$ and introduced the Hall resistance ρ_{xy} and the Hall coefficient R_H. The Hall coefficient should also be able to tell us whether we are dealing with electrons or holes. If we are to believe that either Hall or thermopower can give us the carrier sign, then it is probably reasonable to expect that the two measurements will agree with each other.

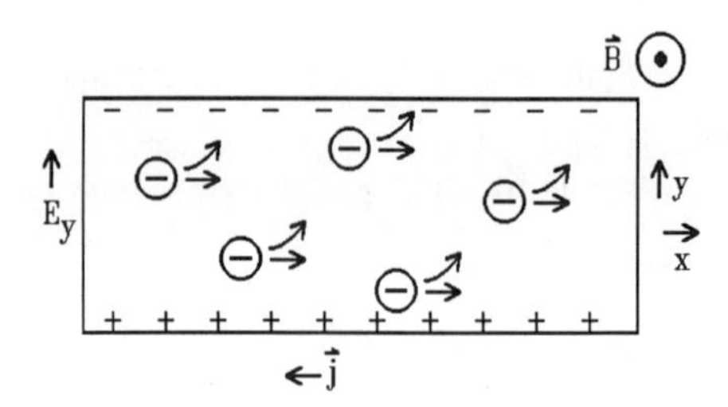

Fig. 3 The Hall Effect results from the establishment of a transverse voltage which builds up in steady state to cancel the effect of the Lorentz force on the moving particles in a mangetic field.

Although an exhaustive study of all conducting materials has not been done, for illustrative purposes the Hall coefficient and the thermopower have been compiled for 24 of the 28 elemental superconductors[7]. A plot of their Hall vs the Seebeck coefficients is shown in figure 4. There are clearly some correlations since the data seem to fall largely into the first and third quadrants. An actual count however, reveals that 16 of the 24 elements in this survey have S and R_H which agree on being electron or hole like whereas in eight cases the answer from Hall and thermopower is opposite (meaning that the Hall must be wrong??). So if you want to determine the sign of the carriers from either measurement you may be right up to 66% of the time.

How can it be that such a basic property as the sign of the carriers can be ambiguous since the two cartoons in figures 2 and 3 seem so clear? The answer is that the simple Drude model[8] that we have been using, and which is the way that virtually everyone wants to imagine transport actually occurs is just slightly too simplified. In any real material the carriers are not all the same. Depending on the energy vs momentum dispersion relation and particularly on the scattering rates, the particles in each cartoon should be viewed as having differing velocities and carrying different charges and heats. The whole question then comes down to how we do the averaging to get the simple pictures from the much more complicated reality.

The answers take some work to get to but are surprisingly simple. We imagine that we have many different kinds of carriers each with the well defined S and R_H cartooned above. The measured thermopower and (low field) Hall coefficients will then be:

$$\mathrm{S} = \frac{\sum_i \sigma_i \mathrm{S}_i}{\sum_i \sigma_i} \tag{8}$$

$$\mathrm{R_H} = \frac{\sum \sigma^2 \mathrm{R_H}}{\sum_i \sigma_i^2} \tag{9}$$

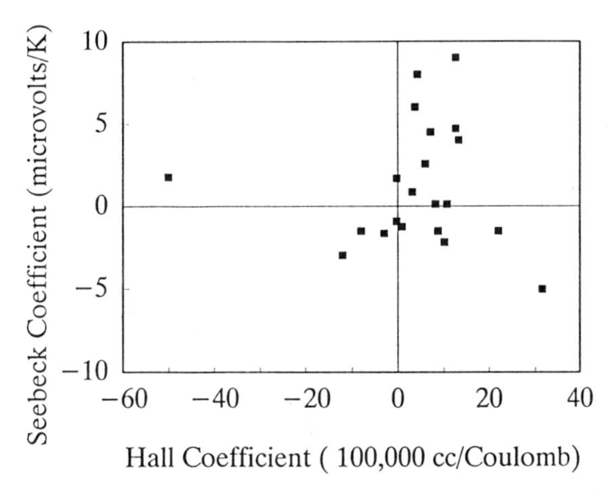

Fig. 4 Hall Coefficient vs. Seebeck Coefficient for most of the elemental superconductors. Note that the signs of the two coefficients are often ($\sim$ 66%) but not always the same.

Thus for anything but the simplest materials, say metals with a spherical Fermi surface, the averages give different results. The monovalent alkali metals (Na, K, Cs ...) have the same sign for S and R_H, but step up to the divalent noble metals (Cu, Ag and Au) and already the thermopower and Hall give different signs.[9]

THERMOPOWER FROM HEAT PER CARRIER

Now we are going to see how far we can go with obtaining semiquantitative expressions for thermopower for a number of different cases by using the notion of heat per carrier. What we will find is that, depending on how inventive we are in constructing the heat or entropy for a particular system, we can readily describe the characteristic behavior of the thermopower[10]. This is true whether the transport is single particle, where conventionally we would write down a Boltzmann transport equation, or for strongly interacting systems where we formally have to consider the velocity heat flux correlation function in a Kubo formalism[11].

Metals

In figure 5 we schematically show the density of states versus energy and the occupation of states in real space for a metal. Specifically we consider a metal as a degenerate electron gas with a temperature much less than the Fermi temperature or the bandwidth. The "heat" per carrier is $\approx c_{el}T$ where c_{el} is the electronic specific divided by the number of electrons, n. The specific heat of a metal is proportional to the temperature and the density of states at the Fermi energy:

$$c_{el} = \frac{\pi^2}{3}\frac{k_B^2 T N(\epsilon_F)}{n} \sim \frac{k_B^2 T}{\epsilon_F}$$

$$S \sim \frac{c_{el}}{q} \sim \frac{k_B}{e}\frac{k_B T}{\epsilon_F} \tag{10}$$

Thus the characteristic thermopower of a metal is decreasing with decreasing temperature and is small. It is much less than $k_B/e \approx 87\mu volts/K$, the thermopower of the classical electron gas. (From our definition of a metal in terms of a degenerate electron gas we have $\epsilon_F \gg k_B T$).

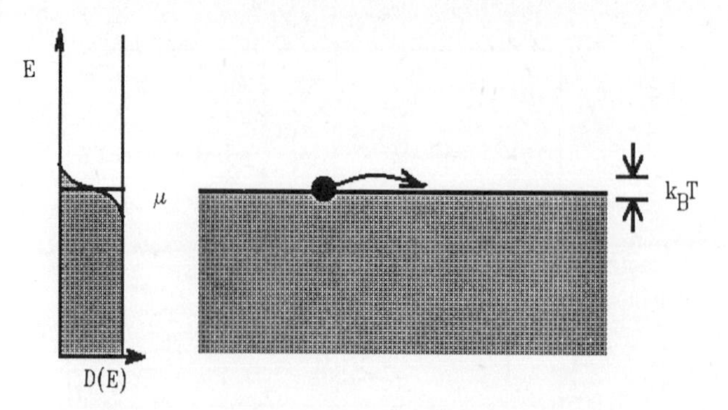

Fig. 5 Density of states vs. energy and allowed states vs. position in real space for a metal. Transport takes place via carriers within k_BT of $\mu \approx \epsilon_F$.

Semiconductors

In figure 6 we illustrate the gap and the transport mechanism for semiconductors. In the simplest case we have only one type of carrier, say electrons in the conduction band. Their energy is essentially E_{gap}. The heat that a particle carries is the difference in its energy from the chemical potential, $\mu \approx \epsilon_F$.

$$\text{"heat"} = < E - \mu > \sim E_g - \mu \sim E_g/2$$

$$S \sim \frac{E_g/2}{eT} \sim \frac{k_B}{e}\frac{E_g}{2k_BT} \tag{11}$$

So the characteristic thermopower for a semiconductor is opposite to that of a metal - it is larger than $87\mu volts/K$ and is increasing with decreasing temperature. Particularly in semiconductors it is important to consider the role of several carriers since in the intrinsic case the number of electrons must equal the number of holes. The weighted average over the conductivities (eq. 7) then yields:

$$S = \frac{k_B}{e}\left\{\frac{b-1}{b+1}\frac{E_g}{k_BT} + \frac{3}{4}\ln\frac{m_e}{m_h}\right\} \qquad b = \frac{\mu_e}{\mu_h} \tag{12}$$

where $\mu_{e,h}$ and $M_{e,h}$ are the mobilities and effective masses of the electrons and holes respectively.

Since thermopower is like entropy one might expect that it must go to zero as $T \to 0$. However, if there are no carriers then their entropy doesn't matter. Thus the real requirement of the third law of thermodynamics is:

$$\sigma S \overset{T \to 0}{\to} 0 \tag{13}$$

A semiconductor for example has divergent S but a faster decrease in σ.

Variable Range Hopping

Disorder can lead to the localization of electron states. If we start with a perfectly periodic system then the electronic states are extended and form into the usual bands separated by band gaps. A small amount of disorder leads to band tailing of localized states into the gap[12]. With sufficient disorder the band tailing can lead to a finite density of states throughout the gap region. With a clean semiconductor ϵ_F is in the gap and the density of states at ϵ_F is zero. With disorder we can have a finite density of states at ϵ_F. However, these states are localized and there may still be a large energy separation between ϵ_F and the extended states. The energy separation between the extended states above and below ϵ_F is the mobility gap, E_μ. Often disordered systems

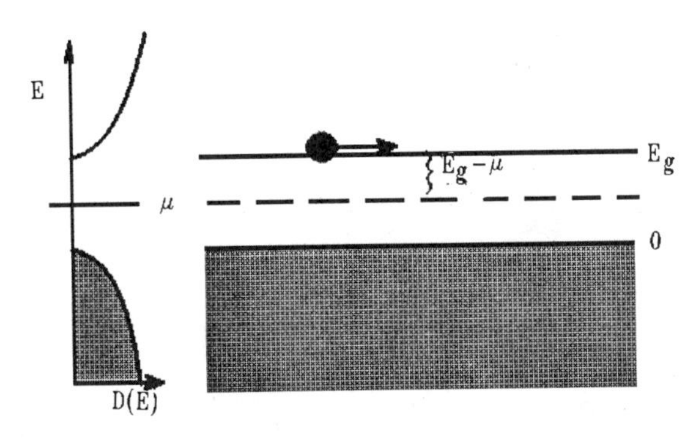

Fig. 6 Density of States and allowed energy states in real space for a semiconductor. Transport takes place above (or below) the energy gap. Each carrier transports a large amount of heat $\sim E_g/2$ even as $T \to 0$.

have a mobility gap but a continuous distribution of energy states, see figure 7. As disorder is increased states further into the band become localized.

At low temperature, $T << E_\mu$, transport takes place within the localized states[13]. To transfer a carrier from one site to another the rate goes down exponentially with distance and exponentially with ΔE (the energy difference between the sites) over the temperature, i.e. the rate goes as $e^{-\alpha r}e^{-\Delta E/k_BT}$. If the density of states is constant then the further we look the more likely we are to find a state with low energy difference. In d dimensions the volume is $\propto r^d$ and the density of states per unit energy per unit volume is $N(\epsilon_F)$. On average the spacing of states in a volume r^d is $1/r^dN(\epsilon_F)$.

$$\sigma \sim e^{-\alpha r}e^{-\Delta E/k_BT} \qquad \Delta E \sim \frac{1}{r^dN(\epsilon_F)} \qquad \sigma \sim e^{-\alpha r - 1/r^dN(\epsilon_F)T} \tag{14}$$

To find the most probable jump we minimize the exponent with respect to r. We then find that the average distance jumped and the average energy difference are temperature dependent. The dimension dependent conductivity for this "variable range hopping" process is:

$$r \sim \frac{1}{T^{\frac{1}{d+1}}} \qquad \Delta E \sim T_0\left(\frac{T}{T_0}\right)^{\frac{d}{d+1}} \qquad \sigma \sim e^{\left(-\frac{T_0}{T}\right)^{\frac{1}{d+1}}} \tag{15}$$

We now want the thermopower from the "heat". We have something like the metal case (section 3.1) in that there is a finite density of states at ϵ_F. But the energy spread of the carriers is ΔE, not k_BT (when weighted by their effectiveness in conduction).

$$\text{"heat"} \sim c_{el}T \sim k_B^2T^2N(\epsilon_F) \overset{T\to\Delta E}{\to} N(\epsilon_F)k_B^2\Delta E^2$$

$$S \sim \frac{N(\epsilon_F)k_B^2\Delta E^2}{eT} \propto T^{\frac{d-1}{d+1}} \tag{16}$$

3 dim $S \propto T^{1/2}$
2 dim $S \propto T^{1/3}$
1 dim S independent of T

Thus although the resistivity is diverging with decreasing temperature the thermopower is typically decreasing for conduction inside the mobility gap. This simply results from the fact that the carriers are collapsing in energy to the Fermi energy. **Thus the thermopower can distinguish between a mobility gap and an energy gap.**[14] The

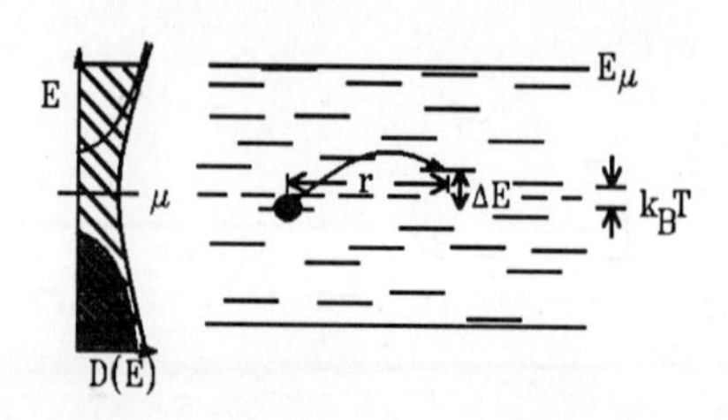

Fig. 7 Density of states for a disordered semiconductor where the gap has been filled in by localized states and the mobility edge (separating localized [diagonalled area] and extended states) has moved into the band. The positions and energies of the localized states are also shown. Transport occurs by jumps of different lengths depending on the temperature.

thermopower only diverges in the presence of an energy gap where the "heat" gets stuck at some minimum value E_{gap} as temperature decreases. One of the nicest illustrations of this behavior is to be found in two dimensional metal films. All films with resistance less than $h/e^2 \approx 26k\Omega/\Box$ have thermopower which goes to zero as $T \to 0$. All films with greater resistance have increasing thermopower with decreasing temperature. However, the films have increasing resistance with decreasing temperature independent of whether their resistance is greater or less than h/e^2. This was one of the first indications that an energy gap opens at the quantum of resistanceh/e^2 [15].

Correlated Hopping, Hubbard Model

We now want to start considering the effects of electron-electron interactions. Previously we have just allowed the electrons to feel the static potential of the lattice. The potential was ~ flat (metals), periodic (semiconductors) or random (for localized states and variable range hopping). Traditionally to treat interactions we should leave the realm of single particle transport and consider the correlation functions in the Kubo formula[11]. However, we will see just how far we can go using our intuition about heat and entropy.

The easiest model for our consideration of the Coulomb interaction is the Hubbard model. It allows electrons to transfer from site to site (with transfer or hopping integral t) and form bands but it also contains a repulsive energy (representing the Coulomb repulsion) which is taken as so short range that it has magnetude U when two electrons are on the same site and zero when they are not.

$$H = t\sum_i (c_i^\dagger c_{i+1} + c_i^\dagger c_{i-1}) + U\sum_i n_{i\uparrow}n_{i\downarrow} \qquad \text{Hubbard Model} \qquad (17)$$

where $c_i^\dagger c_{i+1}$ destroys the electron on site i+1 and recreates it on the neighboring site i and $n_{i\uparrow(\downarrow)}$ tests for the presence of an up (down) spin electron on site i.

When U is small with respect to t then we have a simple band of width 4t. The opposite limit of $U \gg t$ is more interesting. In this case two electrons will not occupy the same site. Lets imagine that the bandwidth 4t is smaller even than k_BT. The picture associated with this situation is shown in figure 8a. The N electrons spend most of their time on one of the N_A individual sites and occassionally jump to neighboring sites. The entropy, Σ, and degeneracy, g, that we associate with this

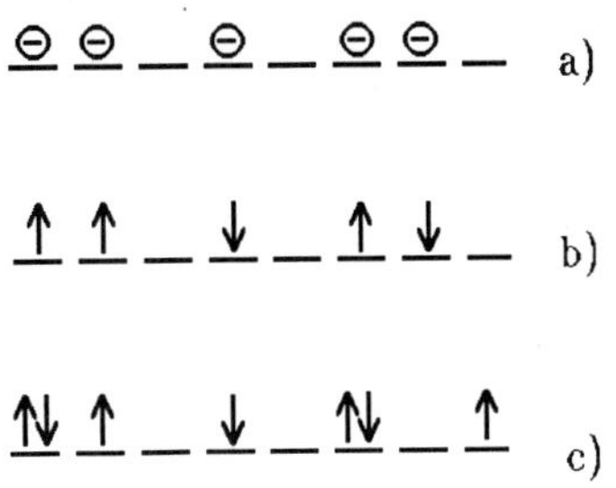

⊖ ⊖ _ ⊖ _ ⊖ ⊖ _ a)

↑ ↑ _ ↓ _ ↑ ↓ _ b)

⇅ ↑ _ ↓ _ ⇅ _ ↑ c)

Fig. 8 a) With negligible bandwidth compared to the temperature the degeneracy is just the number of ways of arranging the carriers on the allowed sites. b) Considering the spin each particle has an additional degenacy of 2 as long as no two occupy the same site (limit of very strong Coulomb correlations). c) Counting the degeneracy is different when double occupancy is allowed for small Coulomb interaction.

situation is just given by the combinatorial problem of N balls in N_A boxes:

$$g = \frac{N_A!}{N!(N_A - N)!} \qquad \Sigma = k_B \ln g \tag{18}$$

The thermopower as entropy per carrier is just the derivative of the entropy Σ with respect to N. We arrive at the Heikes formula[16] for hopping thermopower:

$$S = -\frac{k_B}{|e|} \ln\left(\frac{1-\rho}{\rho}\right) \tag{19}$$

where the electron charge is put in as $-|e|$ to emphasize the fact that the almost empty band has a negative thermopower (electron-like) while the almost filled band has a positive thermopower (hole-like). This traditional formula is only true if we allow only one electron per site. The Pauli exclusion principle legislates up to two electrons per site (assuming ground state wavefunctions), as long as they have opposite spins. The fact that we have allowed only one per site is from our taking the strong U limit of the Hubbard model. the strong onsite correlation limit. In this case we do have only one electron per site, but it has a spin degree of freedom which contibutes an extra $k_B \ln 2$ to the entropy[17]. The result is:

$$g = \frac{N_A!}{N!(N_A - N)!} 2^N \qquad S = -\frac{k_B}{|e|}\left[\ln\left(\frac{1-\rho}{\rho}\right) + \ln 2\right] \tag{20}$$

This is illustrated in figure 8b.

We note that the term: $k_B/e \ln 2 \approx 60 \mu volts/K$ from spin entropy should be added to the configurational term whenever the Coulomb interaction is sufficiently high that the carriers retain their spin degrees of freedom, not just in the atomic limit of the Hubbard model as above. Thus if we have a highly Coulomb correlated metal we would simply add this term to the usual metal bandstructure term found in eq. 10[19]. To complete the picture we can also derive the thermopower when double occupation of a site is allowed as in figure 8c:

$$S = -\frac{k_B}{|e|} \ln\left(\frac{2-\rho}{\rho}\right) \tag{21}$$

and emphasize that equation 18 is only valid for spinless carriers or in the limit of infinite magnetic field when the spin degrees of freedom are quenched.

Some time ago it was noted that all of the organic conductors based on TCNQ with 1/4 filled bands, $\rho = 1/2$, had just the $-60\mu volts/K$ associated with the spin and no orbital contribution since $\ln(1-\rho)/\rho = 0$. Now the real test of whether the correlated hopping model is the correct description of a system is to be found in the magnetic field dependence of the thermopower. The magnetic field dependence of the entropy of free spins and the initial change in the thermopower for the free spin contribution is:

$$\Sigma = k_B[\ln(e^{-\mu H\beta} + e^{\mu H\beta}) - \mu H\beta \tanh(\mu H\beta)] \qquad \beta = \frac{1}{k_B T}$$

$$\Delta S \sim 17\mu\text{volts}/K \times \left(\frac{B(\text{Tesla})}{\text{T(Kelvin)}}\right)^2 \tag{22}$$

Experiments at high magnetic field on the TCNQ's demonstrated this behavior[18].

Recently it has been suggested that the high temperature superconductors based on cuprate planes could also be described in terms of the large U Hubbard model[20]. The crucial test, the magnetic field dependence was performed and it was found that the thermopower was virtually independent of field[21]. The high T_c superconductors (in the normal state) do not have the free spin contribution to their thermopower. The reason may be that the spin is quenched by a large antiferromagnetic exchange field, or more interestingly by haivng charge carriers which lack spin. Treatments of the extended Hubbard model with further neighbor repulsion can be found from similar counting arguments[17].

Quantum Hall Effect

Electrons in a magnetic field undergo circular motion with a frequency $\omega_c = eB/mc$ which is independent of the energy or size of the orbit. This is the classical cyclotron frequency. Since in quantum mechanics periodic motion is quantized, the energy in the plane perpendicular to the field takes on the discrete values:

$$E_N = (N + \frac{1}{2})\hbar\omega_c \tag{23}$$

The effects of this "Landau quantization" are most dramatic in two dimensions (where the dispersion along the field does not broaden the density of states from a set of δ functions at E_N). With a fixed number of electrons the Fermi level may lie in Landau level or in the gap between Landau levels, see figure 9. In most cases ϵ_F can only lie between Landau levels if there are localized states caused by the presence of some disorder. When ϵ_F is in a gap then the electrical current is carried by the filled Landau levels and there is no dissipation, the resistance is zero in the direction of current. However, the electric field in the direction perpendicular to the current is non zero and the conductance $\sigma_{xy} = j_x/E_y$ is quantized in units of the quantum of conductance $e^2/h \sim 1/(26K\Omega)$[6,22].

Digression: The area of a quantized orbit is exactly that which will give a flux quanta, $\phi_o \equiv hc/e$ in the applied field B. Therefore the degeneracy, g, of a Landau level is the number of times the area ϕ_0/B fits into the unit area $g = B/\phi_0 = eB/hc$. The Hall resistance $\rho_{xy} \equiv R_H B$ is proportional to the reciprocal of the density of carriers, eq. 6. When N levels are completely filled the density of carriers is $n = Ng = NB/\phi_0$ and the Hall resistance is:

$$\rho_{xy} = \frac{B}{nec} = \frac{B}{\frac{NB}{\phi_0}ec} = \left[\frac{\phi_0}{\mathbf{e}}\right]\frac{1}{Nc} = \frac{h}{Ne^2} \tag{24}$$

ρ_{xy} only has this value at the specific field B where N levels anre completely full and all others are empty. The reason there are plateaus in ρ_{xy} is that ϵ_F can remain between Landau levels for some range of field if there are disorder induced localized states. However, the real quantization that comes into the problem is that of flux and charge in the ratio ϕ_0/e.

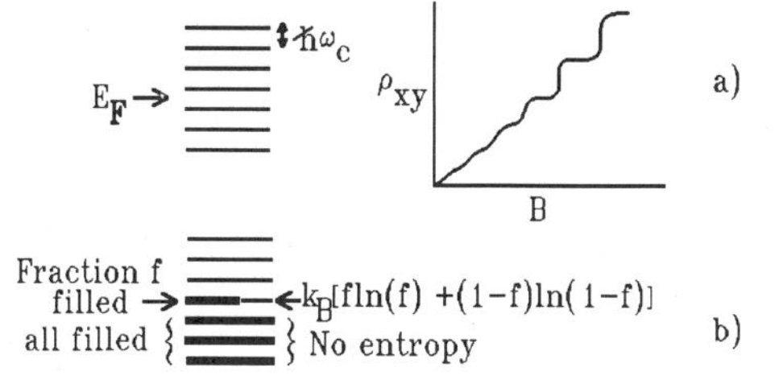

Fig. 9 a) Landau level spacing ($\hbar\omega_c$) is Proportional to the applied field. As field is increased the levels move past the Fermi level. Each time ϵ_F is between levels there is a plateau in ρ_{xy}. b) When ϵ_F is in a Landau level it is only partially filled and can contribute to the entropy.

Now for the thermopower by entropy. Suppose we have N Landau levels completely filled and the $(N+1)^{th}$ filled a fraction f. The filled levels contribute no entropy (each can only be filled in one way). The entropy of the partially filled level is:

$$\Sigma = gk_B[f\ln f + (1-f)\ln(1-f)] \tag{25}$$

The total number of carriers however, is from all of the filled levels plus the partially filled level: $n = g(N+f)$ and the thermopower is:

$$S = \frac{\text{total entropy}}{e \times \text{total number of carriers}} = \frac{-k_B[f\ln f + (1-f)\ln(1-f)]}{e(N+f)} \tag{26}$$

$$S = \begin{cases} 0 & \text{when f=0} \\ \frac{k_B}{e}\frac{\ln 2}{(N+\frac{1}{2})} & \text{when f} = \frac{1}{2} \end{cases} \tag{27}$$

On the Hall plateaus the thermopower is zero (as advertised in the introduction), then it oscillates to maxima at the half integer values. Since the fractional QHE state is supposed to be a macroscopic condensate we would expect it to have vanishing entropy and thermopower[23].

It is interesting to contrast the calculation of the thermopower in the case of hopping, section 3.4, to the Landau level case. In the former the entropy associated with partially filling a band was calculated and its *derivative* taken to get "entropy per carrier". This is consistent with the idea that filled bands do not contribute to the conductivity, an almost filled band should have a hole-like thermopwer, and a half filled band should have zero thermopower from electron-hole symmetry. In the Landau level case we took the total entropy from the unfilled Landau level (or band) and simply divided by the total number of carriers to get the "entropy per carrier". In this case each filled Landau bands *does* contribute to the current with the same sign.

Polarons

One of the clearest distinctions between the "gaps" measured by thermopower and conductivity can be seen in the case of transport of small polarons. Suppose that we have a relatively low density of charge carriers per available site and that the sites are deformable as one might have for a molecule. The undistorted molecules may all have the same electronic energy eigenstate, but when an electron is placed on the molecule it can distort, lower the electronic energy and trap the electron[25].

In order for the electron to hop to the next site the molecule must become partially undistorted and the next molecule partially distorted, see figure 10. The probability for these fluctuations, which limit the transfer rate and conductivity, is governed by a Boltzmann factor $e^{-E_p/2k_BT}$ so the conductivity is activated. However, the final configuration of the tranferred electron has no additional energy gain. Therefore, the heat carried by the electrons is just their configurational entropy which in the low carrier limit is just $k_B \ln \rho$ as in eq. 18. If the carriers which are present come from excitation across an energy gap, E_g (as for a semiconductor) then this heat $\sim E_g/2$ does contribute to the thermopower, but again the rate controlling polaronic energy does not[13].

$$\text{polaronic states at } \epsilon_F \qquad \sigma \sim e^{-E_p/2k_BT} \qquad S \sim \frac{k_G}{e} \ln \rho$$

$$\text{activation to polaronic states} \qquad \sigma \sim e^{-(E_g+E_p)/2k_BT} \qquad S \sim \frac{k_B}{e} \frac{E_g}{2k_BT} \tag{28}$$

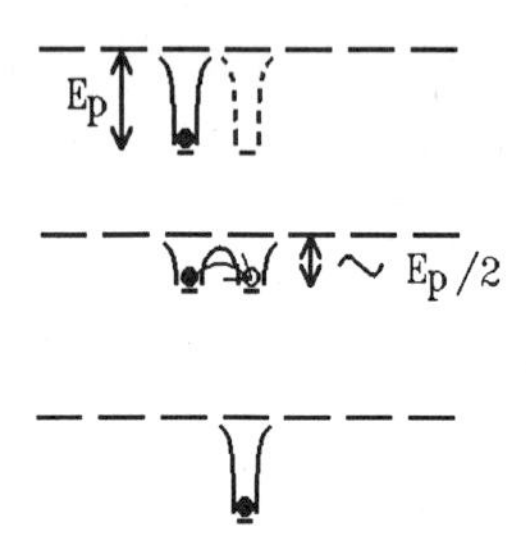

Fig. 10 A molecule can distort to trap an electron when it is on the site. The polaron binding energy is E_p. In order to hop to the next site there must be an excitation of order $E_p/2$ but this energy is not carried with the particle.

Phonon Drag

In all of the cases we have treated above the heat was carried by the charge carriers themselves. We know that the charge carriers can interact with other excitations of the system some of which have no charge associated with them. The primary example is the scattering of electrons by phonons which gives rise to the usual resistance we see in most materials. The situation is schematized in figure 11. The charged particles are moving with a velocity $\vec{v}$, the uncharged particles are moving with velocity $\vec{V}$. Suppose that the only relaxation, or momentum scattering, took place between these two sets of particles. Whatever momentum came out of the charged system would have no place to go except back to the charged system. In steady state the two systems would be moving at the same velocity. In this case the conductivity would be infinite. An electric field would accelerate the electrons and the electrons would drag the phonons with them.

The key to finite conductivity is held by the phonon relaxation processes. Suppose that the phonons relax to the electrons (via electron-phonon scattering) with a rate τ_{e-p}^{-1} and that they relax to everything else (impurities, Umklapp, boundaries, all of which have 0 velocity) with rate τ_{other}^{-1}. The phonons will have an average velocity between 0 and the electron velocity controlled by the ratio of the e-p scattering to the total scattering:

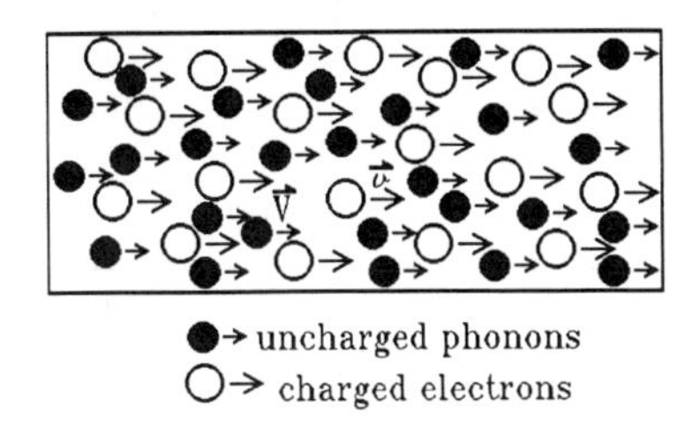

Fig. 11 If the charge carriers collide with neutral particles or excitations then there is an additional heat current as the neutrals are "dragged" with the charges.

$$\vec{V} = \vec{v}\frac{\tau_{e-p}^{-1}}{\tau_{e-p}^{-1} + \tau_{other}^{-1}} \tag{29}$$

The phonons carry heat and there is a phonon current whenever there is an electrical current. Thus there is a new contribution to the heat current which accompanies an electrical current:

$$\vec{u} = C_{phonons}T\vec{V} = C_{phonons}T\frac{\vec{j}}{ne}\left(\frac{\tau_{e-p}^{-1}}{\tau_{e-p}^{-1} + \tau_{other}^{-1}}\right) \tag{30}$$

where $C_{phonons}$ is the lattice specific heat which goes as $234n_{atoms}k_B(T/\theta_{Debye})^3$ at low temperature, and is constant $\sim 3n_{atoms}k_B$ at high temperature. The lattice specific heat per atom is usually much higher than the electronic specific heat except at very low temperature (and there are often more atoms than conduction electrons). Therefore the "phonon drag" contribution to the thermopower:

$$S_L = \frac{C_{phonons}}{ne}\left(\frac{\tau_{e-p}^{-1}}{\tau_{e-p}^{-1} + \tau_{other}^{-1}}\right) \tag{31}$$

often dominates the direct electronic part which is called the diffusion thermopower. The characteristic temperature dependences are:

$$S \propto \begin{cases} T^{-1} & T < \theta_D \quad \text{dominated by Umklapp effecting } \tau_{other}^{-1} \\ T^3 & T \ll \theta_D \quad \text{dominated by } C_{phonons} \propto T^3 \end{cases} \tag{32}$$

As Peierl's pointed out in the original treatment of phonon drag[24] the resistivity is also effected. Assuming that the electrons only relax to the phonons, there is a "bottleneck" in the momentum relaxation, the phonon gas can't get rid of the momentum as fast as it is being dumped by the electrons without going to some intermediate velocity. The resistivity is reduced by the shift of the phonon velocity from rest:

$$\rho = \rho_0 + \rho_{e-p}\left(1 - \frac{\tau_{e-p}^{-1}}{\tau_{e-p}^{-1} + \tau_{other}^{-1}}\right) \tag{33}$$

A similar heat current will arise from coupling of the electrons to any other Bose-like excitations of the system. Most notably, we can expect magnon drag in magnetic systems.

CONCLUSIONS

The results of the simple arguments we have given for the thermopower in most of the cases that are likely to be found in organic superconductors and like materials are given in the table below. Hopefully, these will prove useful in qualitatively understanding the thermopower of interesting systems. For the more complex and rigorous treatments which in particular contain the nasty energy dependent scattering terms, rather than simply the thermodyamic terms, try references 9, 10 and 13.

Table 1

Summary of Characteristic Thermopower Behaviors	
General	$\frac{c_{el}}{q}$ or $\frac{\text{"heat"}}{qT}$
Classical	$\frac{k_B}{q}$
Metals	$\frac{k_B}{q}\frac{k_B T}{\epsilon_F}$
Semiconductors	$\frac{k_B}{q}\frac{E_g}{2k_B T}$
Variable Range Hoping	$\frac{k_B}{q}\left(\frac{T}{T_0}\right)^{\frac{d-1}{d+1}}$
Hubbard Large U Hopping	$\frac{k_B}{q}(\ln 2 + \ln \frac{1-\rho}{\rho})$
Polarons	$\frac{k_B}{q}\ln \rho$
Phonon Drag	$\frac{k_B}{q}\frac{1}{\rho}\left(\frac{T}{\theta_D}\right)^3$ Low T $\frac{k_B}{q}\frac{1}{\rho}\left(\frac{\theta_D}{T}\right)$ High T
Superconductor	0
Sliding Density Wave	0
Quantum Hall	0 filled level $\frac{k_B}{e}\frac{\ln 2}{(N+1/2)}$ $\frac{1}{2}$ filled

ACKNOWLEDGEMENTS

We would like to thank M. Ivker for permission to use figure 4 from his junior paper, R. C. Yu, and S. Girvin for interesting discussions and the National Science foundtation, Division of Material Research for support under grants DMR88-13724 and DMR88-22532.

References

[1] T. J. Seebeck, ABH Akad. Wiss. Berlin 289 (1821), Pogg. Ann. **6**, 133 (1826).

[2] J. C. Peltier, Ann. Chim. Phys. **56**, 371 (1934).

[3] L. Onsager, Phys. Rev. **38**, 2265 (1931).

[4] K. Steiner and P. Grassman, Phys. Zeit, **36**, 527 (1935), for more recent work see papers in THERMOELECTRICITY IN METALLIC CONDUCTTORS, ed. by F. J. Blatt and P. A. Schroeder, (Plenum Press, New York, 1978).

[5] P. M. Chaikin, in THE PHYSICS AND CHEMISTRY OF LOW DIMENSIONAL SOLIDS, ed. By L. Alcacer (NATO ASI D. Reidel, Dorrecht, Holland, 1980), p. 53.

[6] P. Streda, J. Phys. **C 16**, L369 (1983), M. Jonson and S. M. Girvin, Phys. Rev. **B29**, 1939 (1984), H. Obloh, K. von Klitzing, K. Ploog and G. Weimann, Surf. Sci. **170** 292 (9186).

[7] M. Ivker, Junior Paper, Princeton Physics, 1989, unpublished.

[8] P. Drude, Annalen der Physik **1**, 566 and **3** 369 (1900).

[9] F. J. Blatt, P. A. Schroeder, C. L. Foiles and D. Greig, THERMOELECTRIC POWER OF METALS, (Plenum Press, New York, 1976).

[10] J. M. Ziman, ELECTRONS AND PHONONS, (Oxford University Press, Oxford, 1972).

[11] R. Kubo, J. Phys. Soc. Japan, **12**, 1203 (1957).

[12] P. W. Anderson, Phys. Rev. **109**, 1492 (1958).

[13] Mott and Davis, ELECTRONIC PROCESSES IN NON-CRYSTALLINE MATERIALS, 2nd edition, (Clarendon Press, Oxford, 1979).

[14] M. J. Burns, and P. M. Chaikin, "Interaction Effects and Thermopower in Low Temperature Hopping", J. Phys. C **18**, L743 (1985).

[15] M. J. Burns and P. M. Chaikin, "Thermoelectric Power of Two Dimensional Pd and Pd-Au Films", Phys. Rev. **B27**, 5924 (1983).

[16] R. R. Heikes, J. Chem Phys. **26** 582 (1957), R. R. Heikes and R. W. Ure, THERMOELECTRICITY, p. 81 (Interscience, New York, 1961), R. Heikes, Buhl International Conference on Materials, edited by E. R. Shatz (Gordon and Breach, New York, 1974)

[17] P. M. Chaikin and G. Beni, Phys. Rev. **B13**, 647 (1975).

[18] P. M. Chaikin, J. F. Kwak and A. J. Epstein, Phys. Rev. Lett. **42**, 1178 (1979).

[19] J. F. Kwak and G. Beni, Phys. Rev. **B13**, 652 (1975).

[20] H. Ishii, H. Sato, N. Kanazawa, H. Takagi, S. Uchida, K. Kitazawa, K. Kishio, K. Fueki, and S. Tanaka, in PROCEEDINGS OF THE XVIII YAMADA CONFERENCE ON SUPERCONDUCTIVITY IN HIGHLY CORRELATED FERMION SYSTEMS ed. by M. Tachiki, Y. Muto and S. Maekawa (North Holloand, Amsterdam, 1987), R. S. Kwok, S. E. Brown, J. D. Thompson, Z. Fisk, and G. Gruner, *ibid.*, J. R. Cooper, B. Alevi, L.-W. Zhou, W. Bravermann and G. Gruner, Phys. Rev. **35**, 8794 (1987). R. C. Yu, X. Yan, M. J. Naughton, C. Perry, S. Strieb, J. Stuart, P. M. Chaikin and P. Davies, Rev. Solid State Sci., **1**, 181 (1987).

[21] R. C. Yu, M. J. Naughton, X. Yan, P. M. Chaikin, F. Holtzberg, R. L. Greene, J. Stuart and P. Davies, , Phys. Rev. **B37**, 7963 (1988).

[22] K. von Klitzing, G. Dorde and M. Pepper, Phys. Rev. Lett. **45**, 494, (1980).

[23] D. C. Tsui, H. L. Stormer and A. C. Gossard, Phys. Rev. **B25**, 1408 (1982).

[24] R. E. Peierls, Ann. Phys. **12** 154 (1932).

[25] T. Holstein, Annals of Physics, **8** 329 (1961) and **8** 343 (1961).

PRESSURE IN THE STUDY OF ORGANIC SUPERCONDUCTORS

J.E. Schirber

Sandia National Laboratories
Albuquerque, NM 87185

H.H. Wang and J.M. Williams

Argonne National Laboratory
Argonne, IL 60439

INTRODUCTION

There is probably no area in materials science where pressure has had such a profound effect upon the field as in the study of organic superconductivity. The initial discovery of superconductivity at <1.6K in $(TMTSF)_2PF_6$ by Jerome, Bechgaard and coworkers[1] at high pressure initiated a world-wide activity that has resulted in the course of a decade in an increase in T_c from 1.6K to near[2] 12K and identification of more than thirty organic superconductors. This near order of magnitude improvement in T_c has involved a number of breakthroughs in synthesis, in several cases aided by insights from pressure studies.

A convenient classification of organic superconductors is that introduced by Brossard et al[3] who grouped the then-known superconductors into five families. The first family is the Bechgaard salts consisting of the TMTSF (tetramethyl-tetraselenafulvalene) cation and closed shell anions.[4] The above mentioned first organic superconductor, $(TMTSF)_2PF_6$, is a much studied example. The second family is the $(BEDT\text{-}TTF)_nX_m$ series[5] (where (BEDT-TTF)=(ET)=bis-(ethylenedithio) tetrathiofulvalene) which at present has the published record high T_c of ~10K with $X=Cu(SCN)_2^-$ and n = 2, m = 1. The third family is $(DMET)_2X$ (where unsymmetric DMET = dimethyl (ethylenedithio)-diselenadithiafulvalene) with X = I_3^-, I_2Br^-, IBr_2^-, $Au(CN)_2^-$, AuI_2^-, $AuBr_2^-$, and $AuCl_2^-$. The unsymmetric donor is a hybrid[6] of TMTSF and BEDT-TTF with a low T_c ~0.5K. The fourth family is characterized[7] by $(MDT\text{-}TTF)_2\ AuI_2^-$ (where MDT-TTF = methylenedithio tetrathiafulvalene which also has an unymmetric donor molecule as does the third family, but has a T_c = 4.5K. The fifth family is based on the $D(M(dmit)_2)_n$ compounds[3] which are composed of organo-sulfur donor molecule D or tetramethylammonium cations[8] and acceptor metal M(dmit) complexes (dmit = 1,3-dithia-2-thrione-4, 5 dithiolato).

A new family of superconductors has recently been identified,[2] which are characterized by incorporation of oxygen in the reduced cation.[9]

Organic superconductors are typically very compressible and anisotropic materials compared to their inorganic counterparts. Since at low temperatures, all pressure media are solid, this has resulted in some

Organic Superconductivity, Edited by V.Z. Kresin and
W.A. Little, Plenum Press, New York, 1990

controversy and disagreement among various investigators using different pressure generation techniques. Herein we will briefly review the spectrum of materials which have been studied under purely hydrostatic conditions and using single crystal materials because we feel this is the most valid way to make intercomparisons between the various families of these materials.

EXPERIMENTAL

As stated above, all materials discussed here were single crystals whose syntheses have been discussed in the references cited.

The solid He pressure technique[10] has been shown to result in hydrostatic conditions even with highly anisotropic materials and thus is the technique of choice for the study of organic superconductors. The main shortcoming of the method is that it is limited for all practical purposes to pressures less than 10 kbar. Therefore its utility is in establishing the initial behavior of the superconducting properties with pressure. Briefly, this technique involves careful isobaric freezing of He around the sample. In a manner similar to Bridgman growth of single crystal, the experiment is lowered slowly into the temperature gradient above a liquid He bath. The temperature is monitored with a thermocouple during this process. Below the freezing point of He^4, the cooling is assumed to be at a constant volume so the final pressure is calculated from the known phase diagram of He^4. Absolute values of the pressure are achieved to better than ± 2%.

Superconducting transition temperatures were detected in the studies reviewed here by an rf impedance technique discussed earlier.[11] The advantage of this technique for pressure studies on crystals which are often in the submilligram range is that no electrical leads need be attached to the sample. The signature for T_c given by this method has been shown to be identical to the onset of diamagnetism measured by a SQUID magnetometer. Carbon glass resistance thermometry anchored at the sample level outside the pressure vessel, regularly calibrated against known superconductors and the vapor pressure of He provided the temperature determination of T_c in the 2-12K temperature range germane to this study.

RESULTS

a. The first family, the Bechgaard salts, are characterized in their behavior under pressure by small negative[4], pressure derivatives dT_c/dP of near -0.1K/kbar. There is general agreement among investigators on dT_c/dP, but substantial disagreement as to the pressure at which the Peierls transition of the seminal superconductor $(TMTSF)_2PF_6$ occurs; probably because of different pressure techniques and calibration. There is still a very active field of research in the study of the magnetic quantum Hall-like transitions which emphasizes the 2-D character of these materials.

b. The second family $(ET)_nX_m$ has been the most prolific provider of superconductors with record high values of T_c of over 10K. The initially reported[12] superconductivity in this family in $(ET)_2ReO_4$ was found above 6 kbar. $(ET)_3Cl_2(H_2O)_2$ has recently been shown to be superconducting above 16 kbar.[13] The $(ET)_2X$ group with X = I_3^-, IBr_2^- and AuI_2^- have T_c values varying[14] from ~1.5K to 8K but display[15] very similar values of dT_c/dP of -1K/kbar. This behavior is unexpected since it is usually $d\ln T_c/dP$ that is related to the change in electronic properties with pressure. However, it is interesting to note that this same unexpected behavior is observed[16] in the high-T_c copper oxide superconductors where in spite of variations in T_c from 20 to 125K, dT_c/dP is nearly constant value of about 0.1-0.2K/kbar.

This constant negative value for dT_c/dP is violated in the $(ET)_nX_m$ system in a spectacular way by $(ET)_4Hg_{2.98}Br_8$. Here we find[17] that dT_c/dP is initially positive at a value of about 1K/kbar. The transition temperature rises to near 7K at 4 kbar and then begins to decrease at nearly the familiar rate of -1K/kbar, presumably due to a reversible structural phase change.

The 10K member of this family κ-$(ET)_2Cu(NCS)_2$ has the largest pressure dependence for T_c ever reported.[18,23] dT_c/dP = -3K/kbar initially although the derivative seems to decrease to a value more compatible with its family members above 4 kbar (by which time it has been depressed below 1K).

c. The third family $(DMET)_2X$ have T_c values below 1K making pressure measurements considerably more difficult. Therefore no information on dT_c/dP is as yet available.

d. The fourth family of $(MDT\text{-}TTF)_2AuI_2$ has been studied [19] by use of the solid He technique and yielded a value of dT_c/dP of the by now familiar -1K/kbar.

e. The fifth family of $D(M(dmit)_2)bn$ has unusual pressure behavior and thus far superconductivity is only observed under pressure. For M = Pd in $TTF(Pd(dmit)_2)$ superconductivity has been detected[3] near 20 kbar and T_c decreases with pressure. For M=Ni, the compound is superconducting with, as noted above, a very unusual positive pressure derivative causing T_c to rise from 1.1K at 2 kbar to 1.6K at 7 kbar.[20]

f. The recently discovered organic superconductor in the $(BEDO\text{-}TTF)_nX_m$ family is β_n-$(BEDO\text{-}TTF)_3Cu_2(NCS)_3$. These materials are intriguing because of the success in incorporating oxygen giving us a class of "organic oxide" metallic materials in which to seek superconductivity. The possible analogy with the high T_c copper oxides is provocative but thus far only β_n-$(BEDO\text{-}TTF)_3Cu_2(NCS)_3$ has been shown to superconduct at ambient pressure very near 1K. Our attempts to induce superconductivity in $(BEDO\text{-}TTF)_2ClO_4$, $(BEDO\text{-}TTF)_2I_3$ and $(BEDO\text{-}TTF)_2AuI_2$ with hydrostatic pressures to 7-8 kbar have been thus far unsuccessful.

CONCLUSIONS

We have briefly summarized the results of hydrostatic pressure studies on single crystal organic superconductors for the several families discovered to date. Pressure has played a key role since the initial discovery of superconductivity in organic metals and promises to guide the syntheses of yet more materials.

While various correlations between unit cell volume and anion size have been suggested[21], no microscopic model for these systems has yet emerged to aid the materials synthesis in the quest for higher-T_c candidates. The obvious approach of changing anions or cations in the indicated size directions too often resulted in changing entirely the crystal structure, invariably to one which was not superconducting. The extraordinarily large pressure derivatives in the $(ET)_nX_m$ family would seem to offer the possibility of less radical changes such as alloying which might retain the favorable structural arrangement and still increase T_c appreciably. The counter argument to this approach is that alloying (unless resulting in an ordered structure) may introduce disorder which considerable evidence shows is detrimental to superconductivity in the $(ET)_2X$ family.[22]

Intriguing parallels between the pressure behavior and dimensionality of the organic and copper oxide based superconductors are noted, but await increased theoretical understanding of both classes of materials before a firm connection can be made.

ACKNOWLEDGEMENTS

The work discussed in this paper dealt with extensive collabortive efforts involving a large number of investigators at Sandia and Argonne National Laboratories. Major contributors included E.L. Venturini, L.J.Azevedo, J.F. Kwak, D.L. Overmyer, M.A. Beno, K.D. Carlson, A.M. Kini and U.W. Geiser.

Work at Sandia National Laboratories and Argonne National Laboratory is sponsored by the U.S. Department of Energy (DOE), Office of Basic Energy Sciences, Division of Materials Sciences, under contracts DE-04-76DP00789 and W-31-104-ENG-38, respectively.

REFERENCES

1. D. Jerome, A. Mazaud M. Ribault and K. Bechgaard, J. Phys. (Paris) Lett. 41:L195 (1980).
2. M.A. Beno, H.H. Wang, A.M. Kini, K.D. Carlson, U.Geiser, W.K. Kwok, J.E. Thompson and J.M. Williams, Inorg. Chem. (Communication) 29:1599 (1990).
3. L. Brossard, M. Ribault, L. Valade and P. Cassoux, J. Phys. 50:1521 (1989) and references therein.
4. D. Jerome, Physica 109 and 110:B1447 (1982) and references therein.
5. J.M. Williams, H.H. Wang, T.J. Emge, U. Geiser, M.A. Beno, P.C.W. Leung, K.D. Carlson, R.J. Thorn and A.J. Schultz, Prog. Inorg. Chem. 35:51 (1987).
6. K. Kikuchi, K. Murata, Y. Honda, T. Namiki, K. Saito, K. Kobayashi, T. Ishiguro, and I. Ikemoto, J. Phys. Soc. Jpn. 56:3436 (1987).
7. G.C. Papavassiliou, G.A. Mousdis, J.S. Zambounis, A. Terzis, A.Hountas, B. Hilti, C.W. Mayer, J. Pfeiffer, I.C.S.M. 88, Proc. Synth. Met. 27:B379 (1988).
8. K. Kajira, Y. Nishio, S. Moriyama, R. Kato, H. Kobayashi, W. Sasaki, A. Kobayashi, H. Kim and Y. Sasaki, Sol. State Commun. 65:361 (1988).
9. T. Suzuki, H. Yamochi, G. Srdanov, K. Hinkelmann and F. Wudl, J. Amer. Chem. Soc. 111:3108 (1989).
10. J.E. Schirber, Cryogenics, 10:418 (1970).
11. L.J. Azevedo, J.E. Schirber, J.M. Williams, M.A. Beno and D.R. Stephens, Phys. Rev. B30:1570 (1984).
12. S.S.P. Parkin, E.M. Engler and R.P. Schumaker, Phys. Rev. Lett. 50:270 (1983).
13. T. Mori, P. Wang, K. Imaeda, T. Enoki, H. Inokuchi, F. Sakai and G. Saito, Synth. Met. 27:A451 (1988).
14. J.M. Williams, M.A. Beno, H.H. Wang, U.W. Geiser, T.J. Emge, P.C.W. Leung, G.W. Crabtree, K.D. Carlson, L.J. Azevedo, E.L. Venturini, J.E. Schirber, J.F.Kwak and M-H. Whangbo, Physica 136B:371 (1986).
15. J.E. Schirber, L.J. Azevedo, J.F. Kwak, E.L. Venturini, P.C.W. Leung, M.A. Beno, H.H. Wang and J.M. Williams, Phys. Rev. B33:1987 (1986).

16. J.E. Schirber, E.L. Venturini, B. Morosin and D.S. Ginley, Physica C162-164:745 (1989).
17. J.E. Schirber, D.L. Overmyer, E.L. Venturini, H.H. Wang, K.D. Carlson, W.K. Kwok, S. Kleinjer and J.M. Williams, Physica C161:412 (1989).
18. J.E. Schirber, E.L. Venturini, A.M. Kini, H.H. Wang, J.R. Whitworth and J.M. Williams, Physica C152:175 (1988).
19. A.M. Kini, M.A. Beno, D. Son, H.H. Wang, K.D. Carlson, L.C. Porter, U. Welp, B.A. Vogt, J.M. Williams, D. Jung, M. Evain, M-H. Whangbo, D.L. Overmyer and J.E. Schirber, Sol. State Commun. 69:503 (1989).
20. J.E. Schirber, D.L. Overmyer, J.M. Williams, H.H. Wang, L. Valade and P. Cassoux, Physics Lett. 120A:87 (1987).
21. T.J. Emge, P.C.W. Leung, M.A. Beno, H.H. Wang, M.A. Firestone, K.S. Webb, K.D. Carlson, J.M. Williams, E.L. Venturini, L.J. Azevedo and J.E. Schirber, Mol. Cryst. Liq. Cryst. 132:363 (1986).
22. T.J. Emge, H.H. Wang, M.A. Beno, P.C.W. Leung, M.A. Firestone, H.C. Jenkins, J.D. Cook, K.D. Carlson, J.M. Williams, E.L. Venturini, L.J. Azevedo and J.E. Schirber, Inorg. Chem. 24:1736 (1985).
23. W. Kang, D. Jerome, C. Lenoir, P. Batail, J. Phys. Condens. Matter 2:1665 (1990).

UNIAXIAL STRESS EFFECTS ON ORGANIC SUPERCONDUCTOR κ-$(BEDT\text{-}TTF)_2Cu(NCS)_2$

T. Ishiguro, H. Ito, Y. Nogami, Y. Ueba*, and H. Kusuhara*

Physics Department, Kyoto University, Kyoto 606, Japan
* R&D Group, Sumitomo Electric Ind., Osaka 554, Japan

INTRODUCTION

The electronic structures of the organic conductors and superconductors are characterized by the low-dimensionality and the sensitivity to externally applied stresses. The restricted dimensionality is inherent to the molecular assembly, where the bonding between the molecules is highly anisotropic. To materialize the organic conductors, the planar structures are adequate to stabilize the π electrons within constituent molecules.

The binding energies are classified into two groups; intramolecular ones and intermolecular ones, where the former is much stronger than the latter. Naively, the response of the organic molecular crystals to externally applied stress is characterized by the relative displacements of the rigid molecules solidified by the intramoleculer energies. Then the displacements are anisotropic reflecting the styles of the molecular assemblies. For example, the $(TMTSF)_2X$ salt which consists of columns formed by stacked planar molecules exhibits three types of intermolecular interactions, within columns, among adjacent columns and across counter anions. When the intermolecular bonding is soft, the spacing is changed substantially with moderate stress. As a result the electronic structure for current carrying electrons, which is principally determined by the overlap integrals among the molecules, changes sensitively to the applied pressure. In fact the superconductivity was discovered in $(TMTSF)_2PF_6$ under limited range of pressure and the temperature-pressure phase diagram is one of the characteristic features of the organic superconductors. Generally speaking, the superconducting transition temperature in any organic compounds changes remarkably with pressure below 1 GPa [1].

The pressure experiments so far have been carried out by using hydrostatic one. The pressure effect should be different for the types of bonding or assembly. Then, to elucidate the role of individual intermolecular interaction to the electronic state, it is essential to modify the intermolecular spacings selectively, e.g., uniaxially. It is substantial to discriminate the nature of the interactions in order to get definite knowledges on mechanisms of the material function and to design materials enhancing a certain property, e.g., the superconducting

Organic Superconductivity, Edited by V.Z. Kresin and W.A. Little, Plenum Press, New York, 1990

transition temperature, T_c. For κ-(BEDT-TTF)$_2$Cu(NCS)$_2$ which is investigated in this paper, the basic electronic structure is of the quasi two-dimensional nature [2]; in other words, it is a kind of layered compound, where the roles of the intralayer and the interlayer interactions are different.

On applying the uniaxial stress, we cannot overlook the limitations on the shapes and brittleness of available crystals, which are grown by segregation from solutions. One cannot shape a grown crystal into a sample with desired dimensions without causing substantial damage. Furthermore, the size of the ordinary crystals is so small to apply external stresses directly by ordinary means.

SUPPRESSION OF THERMAL CONTRACTION

In order to apply a tensile stress we first throw attention to the large thermal contraction of the crystal [3]. If such crystal is sticked at both ends to a holder with smaller thermal contraction, the tensile strain is induced by cooling. To apply the strain further or to see the strain dependence at constant temperature, one can add the strain through the holder. In this procedure one tames the externally applied stress so that it is transformed into moderate level appropriate for delicate crystals. Generally speaking, however, the thermal contraction of the organic molecular assembly is so large that an extraordinary strain, which brings about fracture for ordinary crystals, is applied by cooling in some cases. This does not necessarily suggest that the organic crystals are elastically soft, but we cannot rule out the possibility of introducing dislocation-like faults within the crystals: we should evaluate the degree of damage of the strained crystal and its effect to the electronic properties.

It is interesting to adopt the tensile stress as uniaxial stress by the following reason. The hydrostatic pressure reduces the T_c monotonically, i.e. the value of T_c decreases with decreasing the spacing between donor molecules for various BEDT-TTF salts [4]. Also, through the comparison of T_c and crystal structure for β-(BEDT-TTF)$_2$X (X=I_3, IBr_2, AuI_2), it is concluded that T_c is related to the intermolecular spacing along the stacking direction [5]. For κ-type crystals with the checkered arrangements of the BEDT-TTF molecules, this model cannot be applied directly, but, in any ways, if one can expand the intermolecular spacings dominating the electronic structures, the T_c is expected to increase. For κ-(BEDT-TTF)$_2$Cu(NCS)$_2$, one can get thin planar crystals whose flat surfaces are parallel to the conducting layer. This is adequate for the tensile stress application within the plane.

According to the measurements of the lattice parameters down to 104 K, κ-(BEDT-TTF)$_2$Cu(NCS)$_2$ salt possesses large and anisotropic thermal contraction, e.g. $\Delta b/b$=0.45 %, $\Delta c/c$=2.2 % while $\Delta a/a$=-0.82 %, where Δa, Δb and Δc are the differences of the lattice constants of a, b and c, respectively between room temperature and 104 K, where the minus sign denotes the expansion by cooling. By smooth extrapolation of the change in lattice parameters down to 10 K, we can get following values: $\Delta b/b$=0.95 %, $\Delta c/c$=2.7 %, and $\Delta a/a$=-0.90 %. The stress corresponding to the cooling-induced strain is estimated by using the compressibility data at room temperature by Chasseau et al. [6] as follows; +0.38, +0.75 and -0.90 GPa along the b, c and a axes, respectively. Here the plus sign indicates the compression while the minus sign the elongation. Although the thermal contraction is anisotropic as causing the extraordinary sign reversals for different directions, the compressibility has the same signs for any direction, although the magnitude is anisotropic. When a thin crystal, whose planar face corresponds to the b-c plane, is stuck on a holder with smaller thermal contraction, such as Cu or fused quartz, tensile stresses are applied in the plane by cooling.

CHANGE IN T_c

By cooling the fixed on a Cu holder at both ends in the b direction, the thin planar crystal was elongated along the b axis [7]. Since there is a distritution of T_c from sample to sample and the associated shift is not so large compared to the width of the distribution, we have to evaluate the shift in T_c with uncertainty. From the measurements with crystals in the same lots, the shift is estimated to be +0.6(±0.5) K. On the other hand when the thin planar crystal is fixed all over the surface on the holder, we regard that the contractions along the b and the c directions are suppressed. The shift in T_c in this case is rather small and represented as 0.0(±1.0) K.

To interpret the experimental results, we compare the observed with the calculated by the Yamaji's theory on the mechanism of superconductivity in TTF-analog molecule [8], in which the correlation between the HOMO and the totally symmetric intramolecular vibration leads important attractive interactions between π-electrons [9]. By taking account of the coupling with the acoustic phonons and the effect of the Coulomb repulsions also, the formula for T_c is given as follows :

$$\ln \frac{2\gamma h\nu_1}{\pi k_B T_c} = \cfrac{1}{\lambda_1 - \cfrac{1}{x_2 - \cfrac{1}{\lambda_2 - \cfrac{1}{x_3 - \cfrac{1}{\cdots \quad \cdots \cfrac{1}{x_m - \cfrac{1}{\lambda_m}}}}}}}$$

$$x_i = \frac{1}{2} \ln \frac{\min(h\nu_i, D_1) \cdot \min(h\nu_i, D_2)}{\min(h\nu_{i-1}, D_1) \cdot \min(h\nu_{i-1}, D_2)} \quad (i \geq 2)$$

with min(x,y) = the smaller between x and y,

$$\lambda = 2g_i^2 h\nu_i N_F/N \quad ,$$

where γ=1.78107, ν_i and g_i denotes the frequency and the coupling costant for i-th mode, N is the number of TTF-analog molecules. It is assumed that the electronic state density N_F per spin is constant in the whole band extending $-D_1$ and D_2, which can be estimated through the band structure calculation by the extended Hückel method [7]. The coupling constant with the acoustic phonon is denoted by λ_1 and the Debye frequency by ν_1, while λ_c is given by the on-site Coulomb interaction I_0 by

$$\lambda = -N_F I_0/N \quad .$$

For κ-$(BEDT\text{-}TTF)_2Cu(NCS)_2$ we have not useful data to calculate T_c. Then by the similarity with the TTF molecule in its basic structure, we calculate T_c based on the electron molecular-vibration coupling data for the TTF shown in Table 1 and plausible parameters such as band width, λ_1(=0.15) and $h\nu_1/k_B$(=52 K) in order to know the roles of various factors to the superconductivity of κ-$(BEDT\text{-}TTF)_2Cu(NCS)_2$.

Table 1 Frequency ν_i and coupling constant λ_i of totally symmetric (a_g) intermolecular vibrations for TTF molecule. (From ref. 8)

i	2	3	4	5	6	7	8
ν_i (cm^{-1})	253	472	740	1077	1518	1559	3099
λ_i	0.002	0.287	0.061	0.010	0.200	0.028	0.001

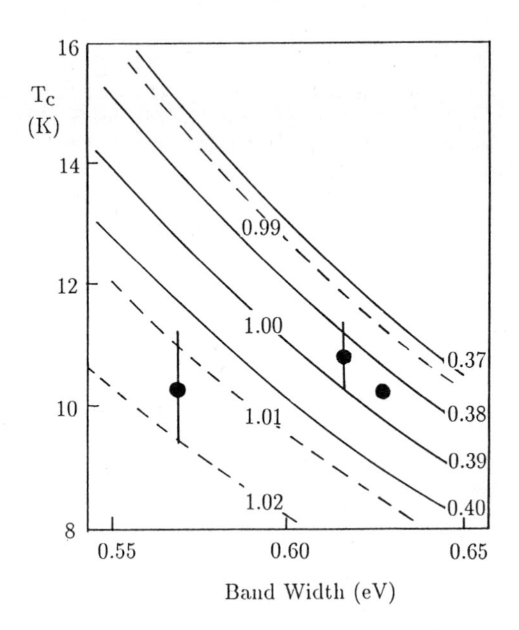

Fig. 1 Band width dependence of T_c as functions of the on-site Coulomb energy in eV (shown by solid lines) and the fraction of the frequency shift in the intramolecular vibration (shown by the broken lines). The T_c's with frequency shifts of 1 % decrease and 1 % and 2 % increases are calculated for I_0=0.39 eV.

In Fig. 1, the relation between the band width and T_c is shown by solid lines for different I_0, so that T_c falls in the region of 10 K at the band gap width of 0.627 eV, which is the calculated for strain-free κ-(BEDT-TTF)$_2$Cu(NCS)$_2$. Note that with increase of I_0, T_c is reduced. In the figure we plotted the obsedrved shift in T_c by setting that T_c in the strain-free sample is 10.2 K. According to the band calculation [7], the corresponding band widths are given by 0.617 eV and 0.569 eV for the b axis elongation and the b-c plane elongation, respectively. Since the experimental data are far from lying in a line, they cannot be explained in terms of the change in the band width, in other words, relative position of the constituent molecules within a conducting layer.

To interpret the observation, the smallness in the shift in T_c for the b-c elongation, we consider first the effect of the change in the Coulomb energy, accompanied by the modification of the molecular assembly. For this case, its increase by 10 % is required to account for the deviation from the line representing the band width dependence for a given I_0 values. On the other hand, the effect of the frequency hardening by the applied strain can give more plausible explanation. The effect is taken into account through the shift in ν_i itself λ_1 and the associating change in $\lambda_i \propto \nu_i^{-2}$ [8]. Since the a_g mode is accompanied by the variation of the molecular volume, it is expected to be sensitive to the intermolecular spacing along its planar direction. At present, however, we have not the detailed information on the shift in each mode. Then, to evaluate its influence, we calculated T_c assuming the uniform shifts in ν_i and the results are shown in Fig. 1 by broken lines. It is remarkable that 1 % increase in $\nu_i (i=2, \ldots 8)$ decreases T_c by 1.4-1.9 K and the large deviation for the b-c planar elongation can be explained on assumption of 2 % increase in the intramolecular frequencies. Thus the interpretation of the observed T_c under tensile stress by the Yamaji theory indicates that the T_c in κ-(BEDT-TTF)$_2$Cu(NCS)$_2$ is determined sensitively by the modes of the intramolecular vibration as well as the band structure and the on-site Coulomb energy. This is consistent with an interpretation of the isotope effects [10]: the deuteration of BEDT-TTF molecule accompanying the frequency softening results in the rise of T_c by 0.6 K [4].

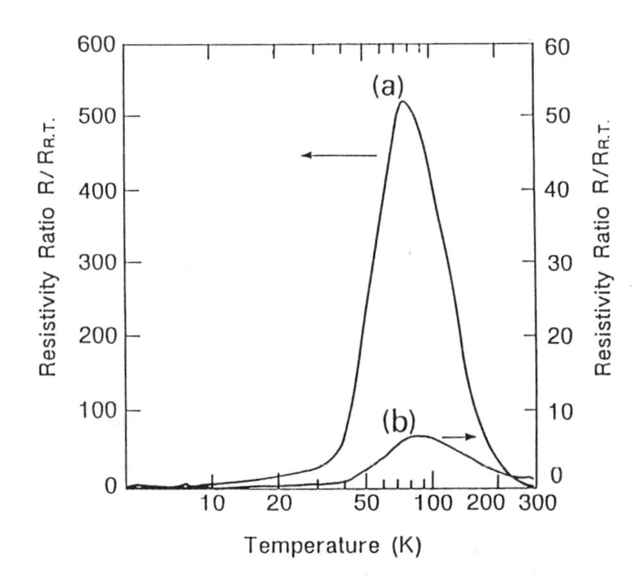

Fig. 2 Temperature dependence of the resistivity in κ-(BEDT-TTF)$_2$Cu(NCS)$_2$ (a) with tensile stress along b axis and (b) without stress. (Reproduced from ref. [12]).

The significance of the interlayer interaction in determining T_c has been pointed out by Whamgbo et al. [11]. They have proposed that the lattice softness in relation to the hydrogen-like bonding should be important in determining T_c, but the physical implication to T_c has not been given explicitly. The aforementioned effect of the frequency hardening (or softening) to T_c can be an answer.

ENHANCEMENT OF RESISTANCE PEAK

The tensile stress induced by cooling causes a dramatic enhancement of the resistance maximum appearing near 80 K in this compound, as shown in Fig. 2 [12]. The origin of the resistance peak has attracted strong interests [13]. In our knowledge, there are three models on the mechanism of the resistance maximum such as the effects of the structural transition [14], the metal-insulator transition with regard to variation of the band structure [15,16] and the polaron mechanism [17,12].

The possibility of the structural transition with respect to any superstructure has not yet been ruled out, but there is no direct support for that. Recently, Toyota and Sasaki asserted that the band splitting due to the intersite Coulomb interaction is suppressed by the increase of the subband width through the increase in the transfer integrals, resulting in the non-metal to metal transition [15]. Since the band structure consists of the hole band and the electron one [18] and hance semimetallic, a drastic change may not appear in the Hall effect [19]. Another way to interplet the resistance peak has been given in view of the scattering mechanism, assuming that the carriers are coupled to the lattice strongly resulting in the formation of the small polaron [17], which exhibits a change in the motion from the hopping in high temperature regime to the coherent movement [20]. Recent experiments externally varying tensile stress at a temperature show a surprizing behavior [21]: the resistivity increases with the tensile stress at room temperature, while it is not much affected by the externally applied strains in the temperature region where the resistance peak appears as illustrated in Fig. 3. The latter is contrary to expectation and open to question, since the enhancement of the resistance maximum is caused by the effective elongation by cooling.

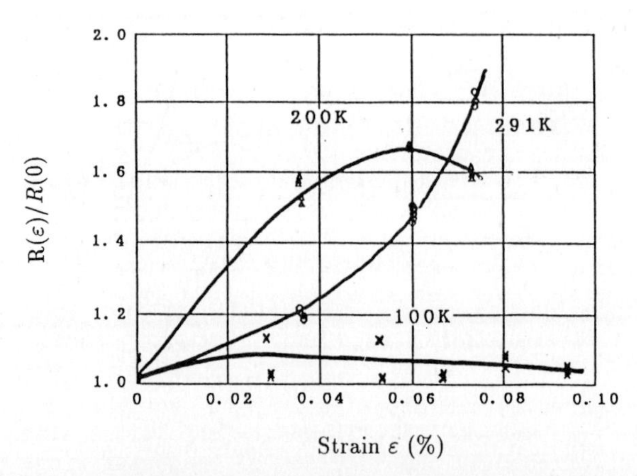

Fig. 3 Strain ε dependence of the normalized resistance R at different temperatures for κ-$(BEDT\text{-}TTF)_2Cu(NCS)_2$.

The reason for this difference, in the resistivity changes due to the suppression of the thermal contraction and to the external elongation, is not clear yet. We conjecture that this is related to the fact that for cooling the a axis expands [3], while for elongation the a axis may be compressed more or less by the Poisson ratio to minimize the elastic energy. It is noted here that the band structure is insensitive to the separation between the conducting layer as far as the extended Hückel calculation stands for [18]. The physical properties which may be affected by the a axis separation are the vibrational modes and, possibly, the intersite Coulomb energy.

EFFECTS OF DEFECT FORMATION

We should remind of the fact that the resultant expansion by suppressing the contraction through cooling is very large. For ordinary inorganic crystals, the strain larger than 10^{-3} results in break or fracture. For κ-(BEDT-TTF)$_2$Cu(NCS)$_2$ the induced strain reaches as high as 3×10^{-2} but most of the crystals were not broken. This however does not mean that the crystals are elastically deformed in wide range. On the contrary we cannot rule out the possibility of local release of the induced strain by forming defects. On cooling, the samples exhibit resistance jumps suggesting the formation of local break, reminiscent of the resistance jumps in (TMTSF)$_2$X crystals [22]: in some cases the drastic increase of the resistivity indicates the formation of partial fracture of the sample. However the large strain does not necessarily cause plastic deformation as far as the electrical resistivity is concerned. This is proved by the fact that the ratio of the resistivities at room temperature before and after the cooling becomes a little higher than one and not so big, in successful cases. When the break occurred substantially the ratio becomes much higher than one and the resistivity at room temperature was not reproducible.

It is to be noted that the intrinsic features of the salt such as the superconducting transition near 10 K and the resistance maximum are retained even in the sample exhibiting the remarkable breaks. This result implies that the defect caused by the external strain appears locally and the microscopic structure of the rest is conserved even in the sample suffering apparent break locally. This contrasts with the suppression of the superconductivity by the anion disorder [23] which occurs uniformly and microscopically.

CONCLUDING REMARKS

The pressure modifies the electronic ground states of the organic superconductors such as (BEDT-TTF)$_2$X salt. With respect to the anisotropy of the intermolecular interactions which is directly connected to the crystal orientation, the study by using the uniaxial external strain is indispensable. Then we applied tensile strain to crystals uniaxially: this experiment is attractive since the elongation of the intermolecular spacing may enhance the superconductivity. However, the experiment shows that the increase in T_c is not as expected from the band structure variation by the strain. The result, interpreted in terms of the Yamaji theory, shows the importance of the intramolecular vibrational frequency in determining T_c. This suggests that the enhancement of T_c in (BEDT-TTF)$_2$X compared to (TMTSF)$_2$X is in part ascribed to the reduction of the vibrational frequency of the TTF-analog molecules, since the relevant frequencies are softened by the addition of the outer rings for the former. We should mention, however, the limitaiton of the discussion with the Yamaji model. First we did not use the experimental data of BEDT-TTF molecule itself but invoked those for TTF, on both the electron molecular vibrational coupling data and the Debye frequency. Secondly the density of states of the electronic band is assumed to be consant to make the theory tractable. It is well known that the detailed structure of the density of states at the Fermi level participates in determining T_c. Thirdly it does not take account of the band splitting explicitly, which accompanies the semimetallic state [16]. Even if these shortages are taken into account, we may conclude this paper by pointing out that the sensitivity of the current organic superconductors to lattice structures which are modified by either pressure or chemical substituiton is strongly related to the vibrational modes coupling to the π electrons.

ACKNOWLEDGEMENTS

The authours are obliged to K. Yamaji (Electrotechnical Lab.) for useful discussion and N. Toyota for sending preprint. This work was supported in part by Grant-in-Aid for Special Project Research on Mechanisms of Superconductivity from the Ministry of Education, Science and Culture of Japan.

REFERENCES

1. See for example, T. Ishiguro and K. Yamaji: "Organic Superconductors", Springer Series in Solid State Sciences 88 (Springer-Verlag, Heidelberg) (1990).
2. H. Urayama, H. Yamochi, G. Saito, K. Nozawa, T. Sugano, M. Kinoshita, S. Saito, K. Oshima, A. Kawamoto and J. Tanaka, Chem. Lett. 1988, 55.
3. G. Saito and H. Urayama : Solid State Physics 23 198 (1988) in Japanese.
4. G. Saito, H. Urayama, H. Yamochi and K. Oshima, Synth. Metals. 27 A331 (1988).
5. M. Tokumoto, K. Murata, H. Bando, H. Anzai, K. Kajimura and T. Ishiguro, Physica 143B 338 (1986).
6. D. Chasseau, J. Gaultier, H. Houbib, M. Rahal, L. Ducasse, M. Kurmoo and P. Day, to be published.
7. H. Kusuhara, Y. Sakata, Y. Ueba, K. Tada, M. Kaji and T. Ishiguro, in "The Physics and Chemistry of Organic Superconductors", ed. by G. Saito and S. Kagoshima (Springer-Verlag) to be published.
8. K. Yamaji, Solid State Commun. 61 413 (1987).
9. N.O. Lipari, C.B. Duke and L. Pietronero; J. Chem. Phys. 65 1165 (1976).
10. M. Tokumoto, K. Murata, N. Kinoshita, K. Yamaji, H. Anzai, Y. Tanaka, Y. Hayakawa, K. Nagasaka, Y. Sugawara; Mol. Cryst. Liq. Cryst. 181 295 (1990).
11. M.-H. Whangbo, J.J. Williams, A.J. Schults, T.J. Emge and B.A.Beno, J. Am. Chem. Soc. 109 90 (1978).
12. H. Kusuhara, Y. Sakata, Y. Ueba, K. Tada, K. Kaji and T. Ishiguro, Solid State Commun. 74 251 (1990).
13. See for example "The Physics and Chemistry of Organic Superconductors" ed. by G. Saito and S. Kagoshima (Springer-Verlag) in press.
14. I.D. Parker, R.H. Friend, M. Kurmoo, P. Day, C. Lenoir, and P. Batail; J. Phys. Cond. Matter 1 4479 (1989).
15. H. Muller, C.P. Heidmann, A. Lerf, R. Sieburger and K. Andress, to be published in "The Physics and Chemistry of Organic Superconductors" ed. by G. Saito and S. Kagoshima.
16. N. Toyota and T. Sasaki, Solid State Commun. 74 36 (1990).
17. K. Yamaji, Synth. Met. 27 A115 (1988).

18. K. Oshima, T. Mori, H. Inokuchi, H. Urayama, H. Yamochi and G. Saito; Phys. Rev. B38 938 (1988), Synth. Metals 27 A165 (1988).
19. K. Murata et. al., private communication.
20. T. Holstein, Annals of Physics 8 325 (1959).
21. H. Ito, Y. Nogami, T. Ishiguro, H. Kusuhara and Y. Ueba; unpublished.
22. T. Ishiguro, K. Murata, K. kajimura, N. Kinoshita, H. Tokumoto, M. Tokumoto, T. Ukachi, H. Anzai and G. Saito; J. Phsique 44 C3-831 (1983).
23. M. Tokumoto, H. Anzai, K. Murata, K. Kajimura and T. Ishiguro; Japan J. Appl. Phys. Supplement 26-3 1977 (1987).

PRESSURE DEPENDENCE OF THE QUANTUM HALL EFFECT AND GIANT OSCILLATIONS IN $(TMTSF)_2PF_6$

S. T. Hannahs,[a] J. S. Brooks,[b] A. S. Perel,[b]
A. G. Swanson[b] , W. W. Kang,[c]
and L. Y. Chiang[d]

[a] Francis Bitter National Magnet Laboratory, M.I.T., Cambridge, MA
[b] Boston University, Boston, MA
[c] Princeton University, Princeton, NJ
[d] Exxon Research Corporation, Annandale NJ

ABSTRACT

The compound $(TMTSF)_2PF_6$ exhibits a complicated pressure dependent Hall effect. We have measured this pressure dependence at low temperatures and high magnetic fields. Near the critical pressure where the ambient pressure Spin Density Wave (SDW) state is supressed leading to superconductivity, the Hall effect shows a complicated structure. At higher pressures and low temperatures the Integer Quantum Hall Effect (IQHE) is observed along with a series of Field Induced SDW (FISDW) states. At higher temperatures we ovserve giant oscillations in the magneto-transport. We will discuss the pressure dependence of both the onset and frequencies of the FISDW and the giant oscillations.

INTRODUCTION

The Bechgaard salts, $(TMTSF)_2X$, where X is the anion ClO_4, ReO_4, or $PF_6$and TMTSF is tetrameythl-tetraselenafulvalene have interesting behavior at high magnetic fields. Though $(TMTSF)_2PF_6$ was the first discovered organic superconductor, it still serves as a model system to help understand the underlying magnetic, transport, and thermodynamic properties of the other organic superconductors. It appears to be the best candidate to be explained by the "standard model" that has evolved to explain the phases of this family of compounds.[1]

The anions in this system are of cubic symmetry and hence centro-symmetric. Thus for $(TMTSF)_2PF_6$ there is no "ordering" transition as in compounds such as $(TMTSF)_2ClO_4$ where the ions have tetrahedral symmetry. $(TMTSF)_2PF_6$ has an interesting phase diagram as a function of pressure and temperature. Fig. 1 shows that at ambient pressure there is a Fermi surface instability leading to a SDW insulating state. A further transition to a second SDW state has been observed in NMR data.[2]

Organic Superconductivity, Edited by V.Z. Kresin and
W.A. Little, Plenum Press, New York, 1990

PHASE DIAGRAM

Application of hydrostatic pressure completely supresses this SDW state and a metallic or superconducting state occurs above 8 kbar. Above this pressure and below about 1.1 K a superconducting region exists. The T_c of this superconducting transition gradually decreases with pressure. Our measurements give $\mathrm{d}Tc/\mathrm{d}P$ = -0.034 K/kbar which is significantly smaller than that reported previously.[3] However the scatter in the data and the pressure dependence of the coefficient would certainly account for the discrepancy.

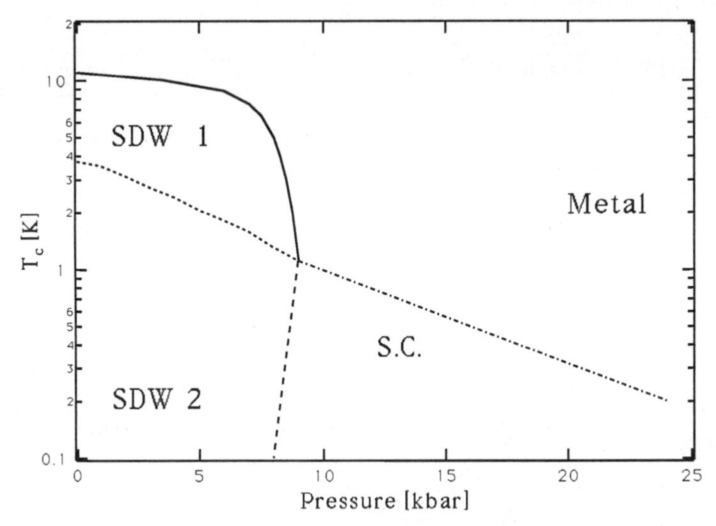

Figure 1. The Phase diagram of $(TMTSF)_2PF_6$ at zero magnetic field

A proposed explanation by Gor'kov and Lebed[4] shows that the tendency to SDW formation in a quasi-one dimensional metal increases with increasing magnetic field. This material is a quasi two dimensional open orbit metal when the SDW state is supressed by pressure. The application of a magnetic field will further reduce the dimensionality leading to the reappearance of the SDW states supressed by pressure. The magnetic field leads to a series of field-dependent Landau bands and gaps in the vicinity of the Fermi energy. The system lowers its energy by adjusting the wave vector of the SDW, q, so that E_F lies in the gap between Landau levels at a given field. As the field is increased, the Fermi energy will jump to a new gap with one fewer Landau levels filled. This process continues until there are no remaining filled Landau bands and the system reaches a semi-conducting state at high field ($n = 0$). Figs. 2 and 3 show data taken as a function of magnetic field and temperature for $(TMTSF)_2PF_6$ at ambient pressure. The transition from the metallic state to the first SDW regime shows up clearly in both measurements as an increase in resistance or a decrease in magnetization. There is no indication of the transition from the first to second SDW states in either measurement. Fig. 3 shows the resistance increasing several orders of magnitude and clearly entering an insulating state.

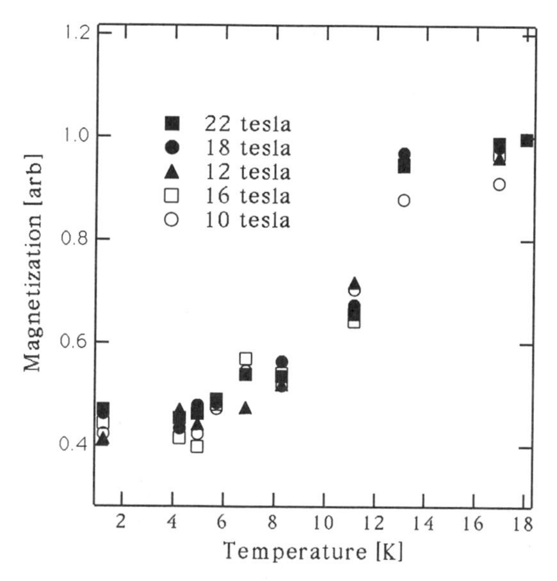

Figure 2. Magnetization of $(TMTSF)_2PF_6$ at ambient pressure as a function of temperature.

Hydrostatic pressure is applied to the samples in a miniature Be-Cu clamp with tungsten carbide pistons. A single crystal of about 2 mm*times*1 mm*times*0.1 mm is mounted with a standard 6 terminal configuration with silver paint to 25μ gold wires on a epoxy feed-through in a Be-Cu end cap. The 300 K resistance of the sample is usually a few milliohms and the contact resistance of the silver paint is between 1 and 10 ohms. The crystal is aligned such that the c axis or least conducting direction is parallel to the magnetic field. Subsequent determination of the sample orientation after pressurization show that the alignment is maintained to less than 10°. The sample is then inserted into a Teflon cup filled with pressure fluid and inserted into the clamp. The pressure is measured at room temperature by a force gauge when pressure is applied and at low temperature by measuring the T_c of a small piece of Pb placed near the sample. The pressure medium is a fluorocarbon fluid which gives good hydrostatic results as measured by the sharpness of the superconducting transition of the Pb manometer. Typically there is a pressure difference of about 2 kbar between 300 K and 7 K. The force gauge has been calibrated against a manganin wire manometer at 300 K. The uncertainty in the pressure measurement is $\approx \frac{1}{2}$ kbar. The entire clamp has a diameter of 11 mm so that it can be accomodated in the bore of a

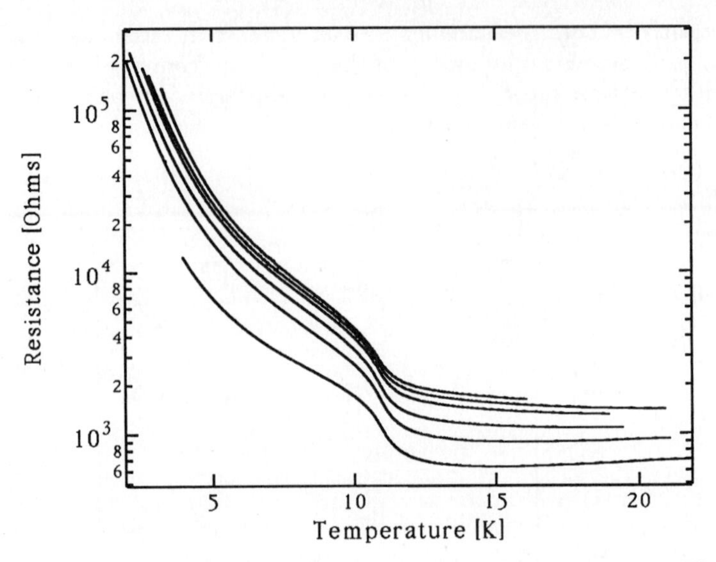

Figure 3. Resistance of $(TMTSF)_2PF_6$ as a function of temperature. The magnetic field from top to bottom is 23 T, 21 T, 18 T, 12 T, 6 T, and 0 T. Note that the the SDW transition is independent of magnetic field and occurs at about 11 K.

top loading dilution refrigerator and into a 20 T magnet at the Bitter National Magnet Laboratory. This clamp has been used at pressures up to 15 kbar, temperatures down to 0.1 K, and fields up to 30 tesla.

INTEGER QUANTUM HALL EFFECT

Until now the IQHE has been observed unambiguously only in the single layer quantum wells formed in silicon MOSFETS and semiconductor heterostructures or a few weakly coupled layers.[5] The only bulk materials that have shown QHE-like behavior are the Bechgaard salts $(TMTSF)_2X$[6] Theoretical work indicates that the FISDWs should be associated with QHE behavior,[7] and experimental work,[8] in $(TMTSF)_2ClO_4$ led to the hope that the QHE could be clearly observed in $(TMTSF)_2PF_6$. The magneto-transport at 8 kbar pressure shown in Fig. 4 shows a definite threshold transition at about 5 T and then a series of sharp peaks ending with a transition to an semi-conducting state. To obtain the Hall resistance the field is reversed and the signal subtracted. A typical Hall resistance at low temperatures is shown in Fig. 5. There is zero Hall resistance below the threshold field seen in the magnetoresistance measurements. Above this threshold field, the Hall resistance shows a sharp upturn and a series of clear flat steps. Above the final transition at about 18 T, there is a large fluctuation in the Hall signal due to the subtraction of two large numbers. At this point ρ_{xx} has increased by 100 fold. From a large number of measurements of various samples, at several temperatures and pressures, we believe that the Hall resistance is close to zero. It has certainly decreased subtantially below that of the $n = 1$ state.

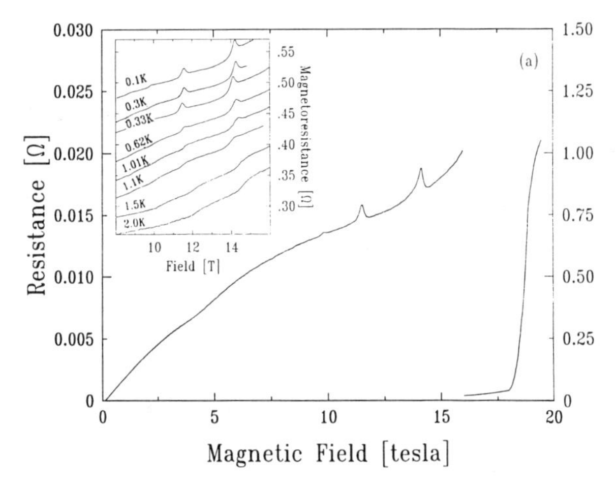

Figure 4. Original trace of mixed magnetoresistance and Hall signal at 0.1 K. Scale above 16 T is reduced by a factor of 50. The inset shows the temperature evolution of the corrected magnetoresistance for (top to bottom) 0.1, 0.3, 0.33, 0.62, 1.01, 1.1, 1.5, and 2.0 K.

The step heights in the Hall signal are in the ratio of $1 : \frac{1}{2} : \frac{1}{3} : \frac{1}{4} \cdots$ and spaced equally in $1/H$. After the final step the system enters the $n = 0$ state where the resistance has increased dramatically and the Hall voltage drops to zero. This extreme quantum limit has been investigated in fields up to 30 T and no further transitions have been observed. The steps in the Hall resistance correspond exactly to the peaks in the magnetoresistance as is expected in the IQHE. However in the region of the steps where the Hall resistance is flat, the magnetoresistance is non-zero. This fact and the existence of an offset threshold field indicate that the standard QHE alone will not fully explain all of the data.

The temperature dependence of these steps is show in the inset in Fig. 5. As the temperature is increased the steps are not as well defined and finally only the last transition to the extreme quantum limit remains. The position of the steps is quite constant up to the point where they disappear. The step height of the $n = 1$ step is 6.4 ± 0.8 k$\Omega/\square$/layer or about $h/4e^2$. This is a factor of two less than the expected value of $h/2e^2$ that is expected for the IQHE with a spin degeneracy of 2. Since the sample is modeled as a set of layered resistors in parallel, there may be large inaccuracies in estimating the number of layers. Also microcracks or defects may disrupt the layering of the sample.

The $n = 0$ or extreme quantum limit regime has a large thermally activated resistance. Thus at high fields, heating causes the phase boundary to be crossed back to the metallic state. ρ_{xy} is near zero in this state. For these reasons we have labeled this region of the phase diagram with the $n = 0$ index.

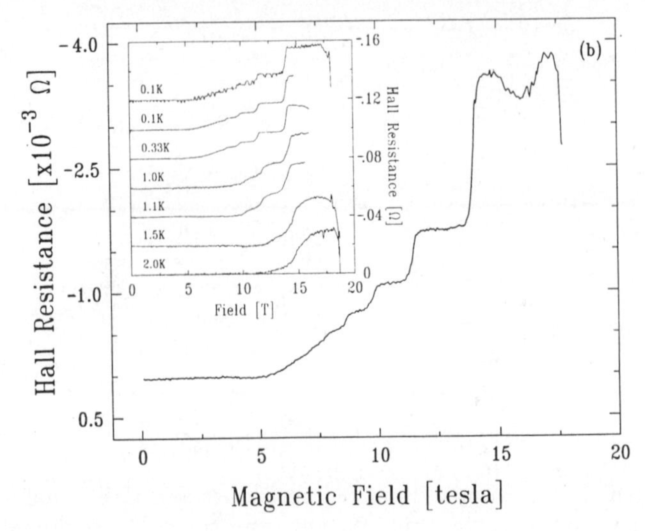

Figure 5. Hall signal at 0.1 K. Uncertainties above 15 T originate from errors in subtraction of digitized signals with low resolution and the rapidly increasing magnetoresistance terms. The inset shows similar traces for 0.1, 0.1, 0.33, 1.01, 1.1, 1.5, and 2.0K. Note the negative sign of the Hall signal.

Fig. 7 shows the Hall conductance of the steps as a function of index number n. The straight line character indicates the integer ration of the step heights. Note that the line passes extremely close to zero showing that the $n = 1$ step is labeled correctly aside from a possible overall scale factor. From the inset in Fig. 5 we see that the height of the IQHE steps is constant as temperature is increased up to the point where the second order transition to the metalic state occurs and the Hall step disappears. The deviation from the straight line at high index number, n, is due to these steps being near the threshold line and thus the height being reduced.

Fig. 8 shows the index of the steps as a function of the inverse field at which the step occurs. The slight curvature of the points in the graph is due to the sample being at non-zero temperature. Again as temperature is increased and the threshold field approaches the IQHE transition, the QHE transition is shifted to slightly higher fields again causing a slight deviation from exact linear behavior.

The phase diagram of these FISDW states in $(TMTSF)_2PF_6$ is similar to that seen in $(TMTSF)_2ClO_4$ but with important differences. There are no interveneing phases between the consecutive integer phases, no sign reversals in the Hall resistance, and no fractional phase after the $n = 1$ phase. In $(TMTSF)_2ClO_4$ the high field state is associated with the $n = 1/3$ quantum number.[9]

HEAT CAPACITY AT HIGH HYDROSTATIC PRESSURE

We have also measured the heat capacity of $(TMTSF)_2PF_6$ at 8 kbar pressure and 0.2 K. This was done with a standard ac calorimetric technique. The several single crystals of $(TMTSF)_2PF_6$ were aligned and placed between a thin film thermometer and heater and the entire assembly placed in a Be-Cu pressure bomb. The heater

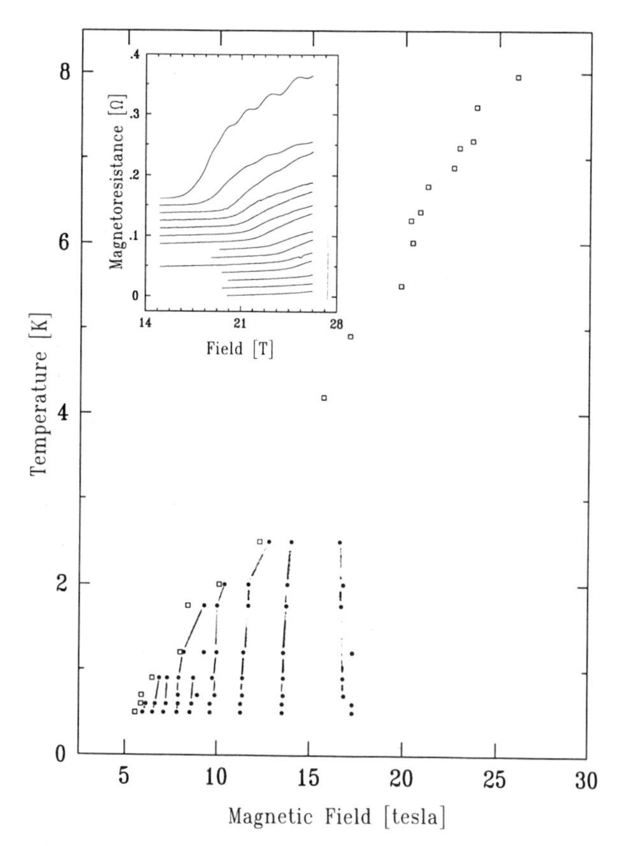

Figure 6. B-T phase diagram produced from the positions of the peaks in the magnetoresistance. Threshold field and phase boundary are represented by squares. Inset: magnetoresistance measurements used to determine the phase boundary at high temperatures.

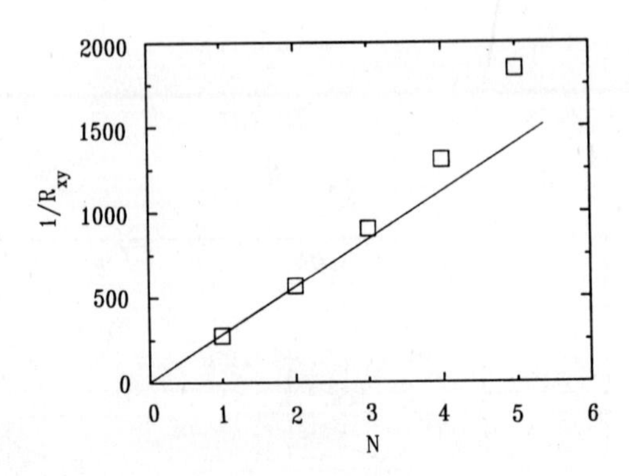

Figure 7. Plot of $1/R_{xy}$ *vs.* FISDW phase index at low temperatures. The solid line is a linear fit to the $i = 1$ and 2 values.

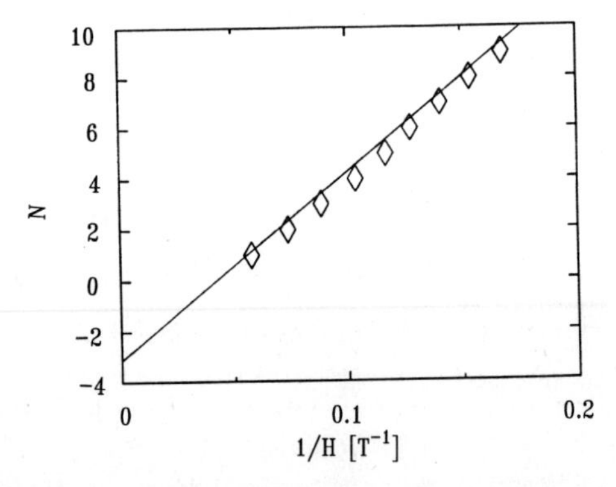

Figure 8. Plot of FISDW transition index *vs.* inverse field at low temperatures. The line is a fit to all points. Note the negative intercept of -3.5.

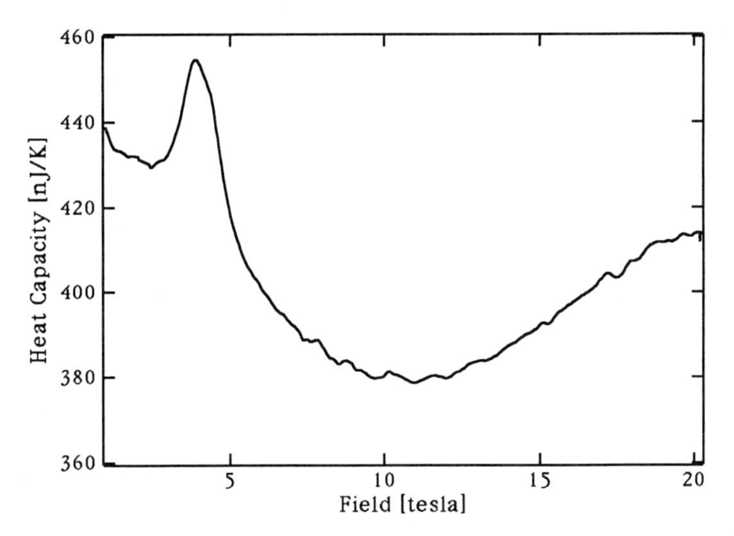

Figure 9. The heat capacity of $(TMTSF)_2PF_6$ at a hydrostatic pressure of 8.5 kbar and a temperature of 0.2 K as a function of magnetic field. The prominent peak at 4 T corresponds to the threshold field observed in transport measurements.

was driven with a sinusoidal ac current and the thermal response was measured by the signal from the thermometer at twice the drive frequency with a phase sensitive detector. The thermal response is inversely proportional to the heat capacity. And the thermal conductivity appears as a response at a phase shift of 90° from the drive frequency. By selecting an appropriate drive frequency, we find that the thermal response is due to the heat capacity and thermal conductivity of the sample with minimal effects due to the background signal from the surrounding pressure fluid. The results are shown in Fig. 9. Note the peak position corresponding to the threshold field for the onset of the Hall resistance and the second order transition to the FISDW regime.

Heat capacity measurements on $(TMTSF)_2ClO_4$ show a similar structure and peaks corresponding to the first order SDW transitions. Slightly better resolution may enable the observation of these transitions. In the extreme quantum limit regime we see a general rise in the heat capacity similar to those reported recently[10] for $(TMTSF)_2ClO_4$ but do not see any signature for the entrance to the $n = 0$ state. $(TMTSF)_2ClO_4$ has been seen to have a reentrance to a metallic state at extremely high fields.[9,11]

GIANT OSCILLATIONS

Fig. 10 shows the magneto-resistance of $(TMTSF)_2PF_6$ under pressure at several different temperatures. The striking feature is the appearance of "giant" oscillations at a temperature around 2 K. These magneto-oscillations have a high quasi de Haas-van Alphen frequency. These oscillations have also been observed in both transport and thermodynamic measurements of the ClO_4 and ReO_4 salts but were previously attributed to the effects of anion ordering in those materials.[12] In the case for $(TMTSF)_2PF_6$ the anion is centro-symmetric precluding any orientation ordering phenomena.

These rapid oscillations have been seen to exist in the exteme quantum limit state and at pressures below the onset of superconductivity. The observation of these oscillations in several different salts and in such widely varying regimes of conductivity indicate that they are a result of a fundamental property of this family of salts. The

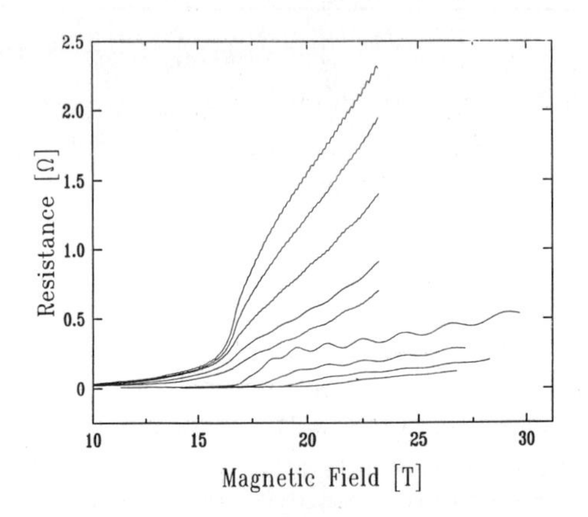

Figure 10. Magneto-resistance of $(TMTSF)_2PF_6$ at 8 kbar pressure and various temperatures. From top to bottom T = 0.5 K, 0.6 K, 0.9 K, 1.0 K, 1.3 K, 1.8 K, 2.3 K, 3.5 K, 4.3 K.

origin of these oscillations remains one of the most intriguing questions left. The maximum amplitude is at 2 K, and this temperature is either independent or only weakly dependent on pressure.

QUASI FREQUENCIES

In this material there are two features which scale with the inverse magnetic field. These are the FISDW transitions and the giant oscillations. Both of these phenomena can therefore be parameterized by a quasi de Haas-van Alphen frequency. These frequencies have been measured at low temperatures for several different pressures. These results are summarized in Fig. 11. From the figure it appears that in

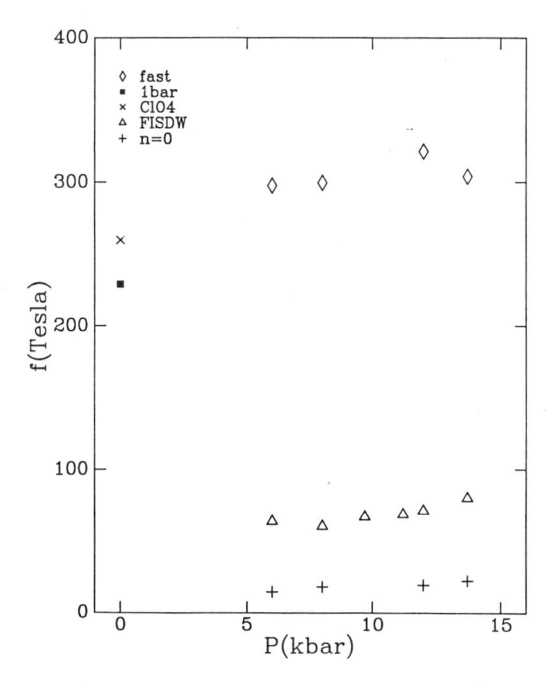

Figure 11. The quasi frequencies of both the FISDW and giant oscillations for $(TMTSF)_2PF_6$ as a function of pressure. The ambient pressure data for $(TMTSF)_2$-ClO_4 and the field for the transition to the $n = 0$ state in $(TMTSF)_2PF_6$ is shown for comparison.

$(TMTSF)_2PF_6$, there is a slight positive pressure dependence of both the frequencies of the FISDW and the giant oscillations. Uncertainty in the determination of pressure and the need for extremely high fields to determine the frequency accurately complicate the measurements. Theoretical speculation[13] is that a possible reentrant phase will appear at the field corresponding to the frequency given by the FISDW. For $(TMTSF)_2ClO_4$ this frequency is 31 T and the reentrant phase was observed at about 28 T. For $(TMTSF)_2PF_6$ this critical field is between 70-80 T. Since this frequency appears to increase with temperature, application of even higer pressures will not allow the reentrant phase to be observed. There is no curvature evident in the second order threshold field line in the phase diagram of Fig. 6. Thus there is either no reentrant phase or it is well above the dc magnetic fields available.

At pressures near p_c, the critical pressure for the appearance of superconductivity, both the Hall and magnetoresistance measurements show various anomalies. These include, negative Hall steps, double peaks in the magnetoresistance, large background resistances, and large low magnetic field Hall signals. These effects can be reduced or eliminated if the applied pressure is increased for the same sample. Some of these effects are quite reproducible, and a much better control of the sample pressure is needed to investigate this region thoroughly.

CONCLUSIONS

We have made extensive studies of the high magnetic field phases of the $(TMTSF)_2PF_6$ compound. There measurments include magneto-transport, heat capacity and magnetization. This material shows almost complete agreement with the expectations for a system undergoing a series of FISDW transitions. Unlike $(TMTSF)_2ClO_4$ there are no anomalies or additional transitions to complicate the basic theoretical description of the $(TMTSF)_2$ family of materials. $(TMTSF)_2PF_6$ has many features similar to $(TMTSF)_2ClO_4$ but the presence of the $n = 0$ state instead of the $n = 1/3$ state and the clear existance of QHE features make it a much better example of the theoretical standard model.The IQHE has now been observed in a "real material" and is no longer only observed in two-dimensional semiconductor devices.

ACKNOWLEDGEMENTS

We would like to express thanks for assitance to the staff of the Francis Bitter Magnet National Laboratory (supported by the National Science Foundation) where these measurements were made. We would like to thank P. M. Chaikin for many valuable discussions. This work was supported by the National Science Foundation at BU under DMR88-18510.

REFERENCES

1. P. M. Chaikin, Phys. Rev. B **31**, 4770 (1985); G. Montambaux, M. Heritier, P. Lederer, J. Phys. Lett. **45**, L-533 (1984); M. Heritier, G. Montambaux, and P. Lederer, J. Phys. Lett. **45**, L-943 (1984); K. Yamaji, J. Phys. Soc. Japan **54**, 1034 (1985); M. Ya. Azbel, Per Bak and P. M. Chaikin, Phys. Lett. **A117**, 92 (1986); K. Maki, Phys. Rev. B **33**, 4826 (1986).
2. T. Takahashi, H. Kawamura, T. Ohyama, Y. Maniwa, K. Murata, and G. Saito, J. Phys. Soc. of Japan **58**, 703 (1989).
3. H. Kawamura, T. Ohyama, Y. Maniwa, T. Takahashi, K. Murata, and G. Saito, *Proc. 18th Int. conf. on Low Temperature Physics*, Kyoto, 1987, Jap. J. Appl. Phys. **suppl. 26-3**, 583 (1987).
4. L. P. Gor'kov and A. G. Lebed, J. Physique Lett. **45**, L-433 (1984).
5. K. von Klitzing, G. Dorde and M. Pepper, Phys. Rev. Lett. **45**, 494, (1980); D. C. Tsui, H. L. Stormer and A. C. Gossard, Phys. Rev. B **25**, 1408 (1982); H. L. Stormer, *et al.*, Phys. Rev. Lett. **56**, 85 (1986).
6. D. Jerome, and H. J. Schultz, Adv. in Phys. **31**, 299 (1982); See review articles by M. Ribault, P. M. Chaikin, G. Montambaux, M. Herritier in NATO ASI Series, *Low Dimensional Conductors and Superconductors*, ed. by D. Jerome and L. G. Caron, (Plenum Press, New York, 1987); See articles by K. Yamaji, P. Lederer, F. Pesty, X. Yan, G. Montambaux, P. M. Chaikin, P. Garoche, J. Brooks, M. J. Naughton in *Proceedings of the International Conference on Science and Technology of Synthetic Metals*, Ed. A. J. Heeger Mol. Cryst. Liq. Cryst. **27B**, (1989).
7. D. Poilblanc *et al.*, Phys. Rev. Lett. **58**, 270 (1987).

8. M. Ribault *et al.*, J. Phys. Lett. **44**, L-953 (1983); P. M. Chaikin *et al.*, Phys. Rev. Lett. **51**, 2333 (1983); R. V. Chamberlin *et al.*, Phys. Rev. Lett. **60**, 1189 (1988).
9. M. J. Naughton *et al.*, Phys. Rev. Lett. **61**, 621 (1988).
10. N. A. Fortune, J. S. Brooks, M. J. Graf, G. Montambaux, L. Y. Chiang, Jos A. A. J. Perenboom and D. Althof, Phys. Rev. Lett. **64**, 1054 (1990).
11. V. M. Yakovenkom, Zh. Eksp. Teor. Fiz. **93**, 627 (1987) [Sov. Phys. JETP **66**, 355 (1987)].
12. H. Schwenk *et al.*, Phys. Rev. Lett. **56**, 667 (1986); T. Osada, N. Miura, and G. Saito, Solid State Commun. **60**, 441 (1986); Physica **143B**, 403 (1986); J. P. Ulmet *et al.*, Physica **143b**, 400 (1986); X. Yan *et al.*, Phys. Rev. B **36**, 1799 (1987).
13. V. M. Yakovenko, Z.E.T.F. **93**, 627 (1987) [Soviet Physics JETP **66**, 355 (1987)].

SCANNING TUNNELING MICROSCOPY AND SPECTROSCOPY OF QUASI ONE-DIMENSIONAL ORGANIC CONDUCTORS

R. Fainchtein

Applied Physics Laboratory
Johns Hopkins University
Laurel, MD 20723-6099, USA

S. Pan and A. L. De Lozanne

Department of Physics
University of Texas at Austin
Austin, TX 78712-1081, USA

ABSTRACT

High resolution images and spectroscopy of different quasi one-dimensional organic conductors and superconductors have been obtained with the scanning tunneling microscope (STM). Single crystals of TTF-TCNQ and related molecular compounds have been studied in air and in ultra high vacuum (UHV) covering a temperature range from 300K to 10K. The images reveal the position and orientation of the molecules that constitute the solid and can be directly related to the structure of the material. The corrugations appearing in the images are in excellent agreement with the crystallographic information. The normalized tunneling conductance of three anion substitutes show similarities in the bands for energies below the Fermi level and differences for energies above the Fermi level. At temperatures below ~80K images on TTF-TCNQ reveal corrugations with wavelength of double the interchain lattice parameter. This modulation is commensurate with the underlying lattice and can be attributed to a direct observation of the charge density wave (CDW) on this type of material. Low temperature tunneling spectroscopy data show a energy gap between 40-50mV, which confirms the metal-semiconductor transition.

INTRODUCTION

Until now, the majority of structural studies of TTF-TCNQ and related organic materials have been confined to diffraction experiments. Not many systematic studies have been made on the real space surface structure and aggregation of the molecules that

Organic Superconductivity, Edited by V.Z. Kresin and W.A. Little, Plenum Press, New York, 1990

make these materials. Part of the reason is the lack of tools required to probe surfaces in real space. The STM can provide structural and electronic information at the sub-nanometer scale both in air and in UHV environments, and at different temperatures.

The technique is based on the quantum mechanical phenomena of tunneling electrons through a nonconducting gap at the application of a small potential bias. The tunneling current can be expressed as the product of the electron density of states between the two electrodes forming the tunneling junction weighted by their appropriate Fermi distribution[1]. In the case when one electrode ha the configuration of a sharp tip, ideally with one atom at the end, the images represent contours of equal electronic density of states[1]. For spectroscopic measurements over small voltages the density of states of the tip is constant, thus the measurements presented here show structure due to the sample.

We have used the STM as a tool in an attempt to resolve the molecular structure of TTF-TCNQ and related compounds as well as the presence of the charge density wave in real space. The STM not only can provide images of the crystals with atomic resolution but can provide spectroscopic information of the density of states near the Fermi level. As previously reported[2], TTF-TCNQ was the first organic solid imaged with the STM. However, TTF-TCNQ is only one example of charge transfer organic salts in which the molecules arrange themselves in columnar stacks. It is relatively easy to make chemical substitutions in the constituent molecules of these solids. The salts made with these substitutions provide different geometric arrangements of the stacks in the solids they form. In this way, the degree of crosstalk between the stacks that form the solids can be controlled, going from a system with almost non-interacting chains such as TTF-TCNQ to a system with a higher degree of chain coupling such as TMTSF-TCNQ. In this paper we will present STM images of TTF-TCNQ and related chemical compounds. We will also present images and tunneling spectroscopy data taken on TTF-TCNQ at low temperatures in which we were able to directly observe the CDW and the metal-semiconductor transition associated with it.

EXPERIMENT

Samples

The samples consisted of single crystals of the TTF-TCNQ family of organic charge transfer compounds. The crystals have typical dimensions in the range of 1-3mm long by 0.2-0.5mm wide and 0.01-0.1mm thick. They were grown by two methods: thermal diffusion in a solution of acetonitrile[3], and by controlled electrochemistry of solutions of acetonitrile and methylene-chloride in an H shaped diffusion tube, as described in related publications[4,5]. The crystal structure of this class of compounds under this study has been previously analyzed by X-ray diffraction[6-8]. The results are summarized in Table 1. For the ambient measurements the crystals were mounted on cleaved mica substrates coated with gold. Cu substrates were used in the UHV STM. The crystals were then fixed to the substrates with conductive Ag paint.

Room Temperature STM in Air

The room temperature ambient measurements were made with a commercial STM[9]. The STM was operated in the constant current mode (using feedback which was fast compared to the scanning speed) as well as in the constant height mode (using slow feedback compared to the scanning speed). Both modes of operation revealed the same structural features in the images. The tunneling current was set at values less than or equal to 2nA; 1nA was used for most images. The scans were made under positive and negative bias voltages in the range of -1 to 1V. On all of these conductive samples, changes in the bias voltage within the specified range produced no apparent change in the structures revealed by the images. However, increasing the bias voltage (and thus increasing the junction resistance) in either polarity caused a decrease in resolution in some cases. Line scanning rates were 156Hz when operating in the constant current mode, and 4Hz when operating in the constant height mode. Images consist of 200 lines with 200 resolved elements per line. Unless specified, no manipulation other than background plane subtraction was applied to the images. Pt with 20% Ir tips were used. The tips were either mechanically[9] or electrochemically sharpened[10]. Better results were obtained with electrochemically sharpened tips. All the images presented are direct photographs of the computer screen.

Low Temperature STM in UHV

The low temperature images and I-V characteristics were obtained with a UHV Low Temperature STM. This STM has a base pressure of 10^{-11} torr and variable temperature range from 10K to 400K. The sample temperature can be controlled and monitored with good accuracy. We have calibrated the thermometer against a precalibrated LakeShore thermometer and also against the transition temperature of a superconductor sample. The tip and sample can be transferred without breaking vacuum through a loadlock, and they can be ion milled and annealed in-situ. Other analysis and preparation tools (Auger, LEED, thermal evaporation) are housed in the main chamber but were not used in the experiments described here.

The STM tip was electrochemically etched from a 0.25mm Pt/Ir(90%/10%) wire. It was then cleaned with distilled water and placed in the load-lock chamber, where it was ion milled and then introduced into the UHV STM. The TTF-TCNQ samples are similar to those described above. Since the organic crystals have irregular shape we mounted them with silver paint on a copper substrate that fits well into our STM. The sample was not ion-milled to avoid possible surface damage.

RESULTS

STM images in air at room temperature

The large flat phase of the whisker shape crystal was imaged in most cases. The STM image of TTF-TCNQ Fig.2a reveals the existence of large rows that extend along the entire sample. The rows are evenly spaced throughout the image. By looking at a line scan

Table 1. Crystal structure parameters of the three different organic compounds under study. The lattice parameters a,b and c are given in Å, the angles α, β and γ are given in degrees[6-8].

Compound	Structure	a	b	c	α	β	γ
TTF-TCNQ	monoclinic	12.298	3.819	18.468	-	104.46	-
TMTTF-TCNQ	monoclinic	18.82	3.85	15.08	-	103.7	-
TMTSF-TCNQ	triclinic	3.883	7.645	18.846	77.34	89.6	94.63

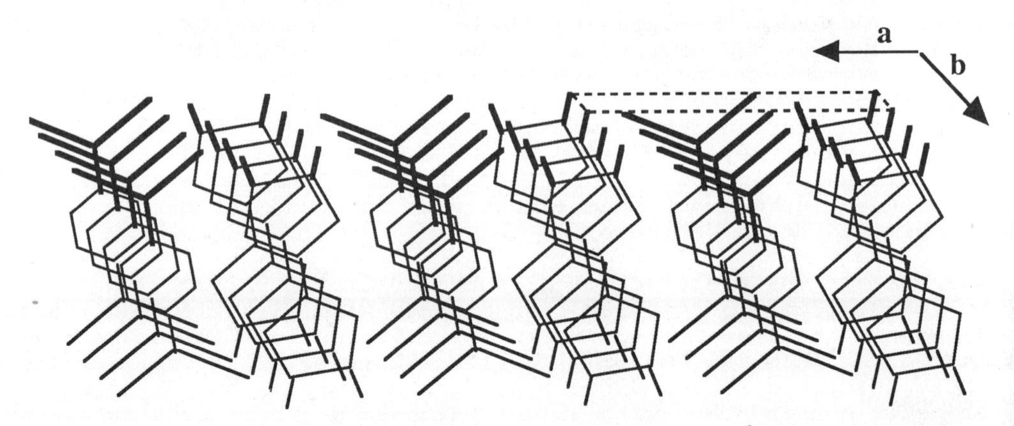

Figure 1. Schematic representation of the ab plane of TTF-TCNQ.

a

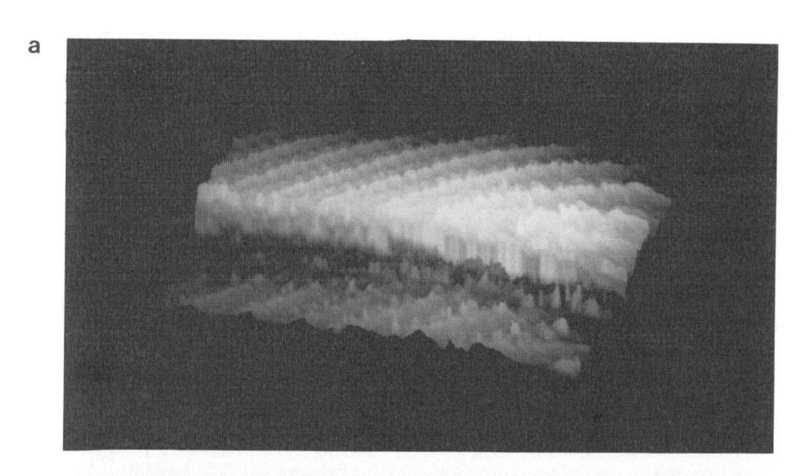

b

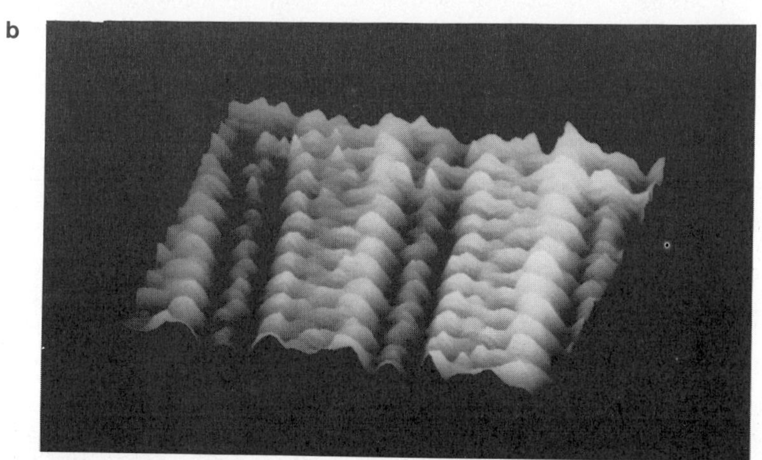

Figure 2. Topographical image of the ab plane of TTF-TCNQ taken with a STM at room temperature in air. The tip was held at -10mV and the tunneling current was 2nA. The scanning area of (a) is 99.1Å x 99.1Å. The step is 9.3Å high. The scanning area of (b) is 35.7Å x 35.7Å. The W shape features correspond to the TCNQ molecules.

along the direction perpendicular to the rows a spacing of 12.2Å between similar features of repeated rows can be obtained. A similar line scan along the rows Fig.2a reveals a periodic feature of 3.8Å separation distance along the elements forming the rows. Comparing these measured periodic distances with the lattice parameters of Table 1 leads us to conclude that Figs. 1 and 2 show the ab plane of the crystal. Rotations of the crystal with respect to the scanning axis produced corresponding rotations in the images allowing us to identify the direction of the rows with respect to the macroscopic orientation of the crystal. The b axes, is coincident with the long direction of the whisker-shaped crystal. A line scan along the step shown in Fig.2a provided a height of 9.3Å. This corresponds to the dimension of a monomolecular step which is that of the c/2 crystal parameter.

A more detailed image of the ab plane can be seen in Fig.2b. The image shows a combination of double and triple features adjacent to one another that repeat themselves to form columns. The triple W shaped features can be identified as the CN radicals at the extreme of the TCNQ molecule. The double features adjacent to the TCNQ molecules as the H atoms at the end of the TTF molecule. Comparison of the image with the diagram of the ab plane shown in Fig.1 makes identification of the STM images easier

Further rotations of the crystal allowed us to study other crystallographic planes. Fig.4 corresponds to a rotation by ~90° along the b axes. The image shows columns 4.9Å apart, believed to be due to the CN extremes of the TCNQ molecule.

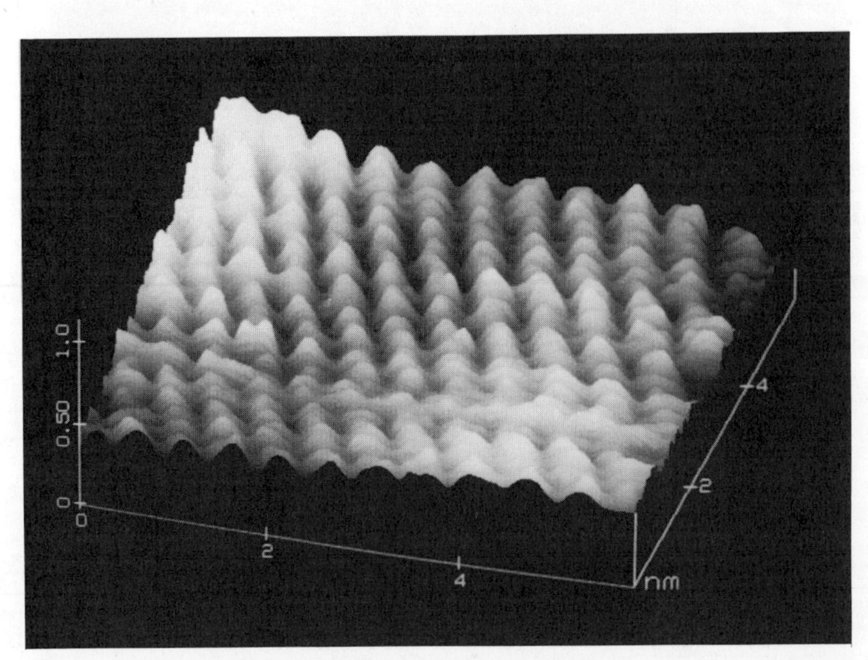

Figure 3. Topographical images of TTF-TCNQ taken with a STM at room temperature in air. The crystal was rotated by ~90° with respect to the ab plane along the b axis.The tip was held at -10mV and the tunneling current was 2nA.

When CH_3 groups replace the hydrogen atoms at the extreme of the TTF molecule, TMTTF-TCNQ is formed. Fig.4 shows an STM image of this compound. A corrugation of 15Å between the chains and of 4Å along the chain leads us to conclude that the image corresponds to the bc plane of the crystal. The monomolecular step on the image is 9.3Å high and corresponds to the a/2 crystallographic parameter. A careful look at the corrugations of both terraces seem to indicate that the lower terrace is composed of TCNQ molecules while the upper terrace is composed of TMTTF molecules.

In TMTSF-TCNQ the S in the TMTTF molecule is replaced by Se. An STM image of this compound is shown in Fig.5. Interchain corrugations of 7.7Å and intrachain corrugations of 3.8Å present in the image helped to identify it as that corresponding to the ab plane. A step of 19Å was also observed, in this case the step corresponds to the full c lattice parameter.

Tunneling spectroscopy curves for the three compounds were collected at room temperature. The normalized conductance is shown on Fig.6. All samples show typical metallic behavior. Differences in the curves are apparent at energies above the Fermi level.

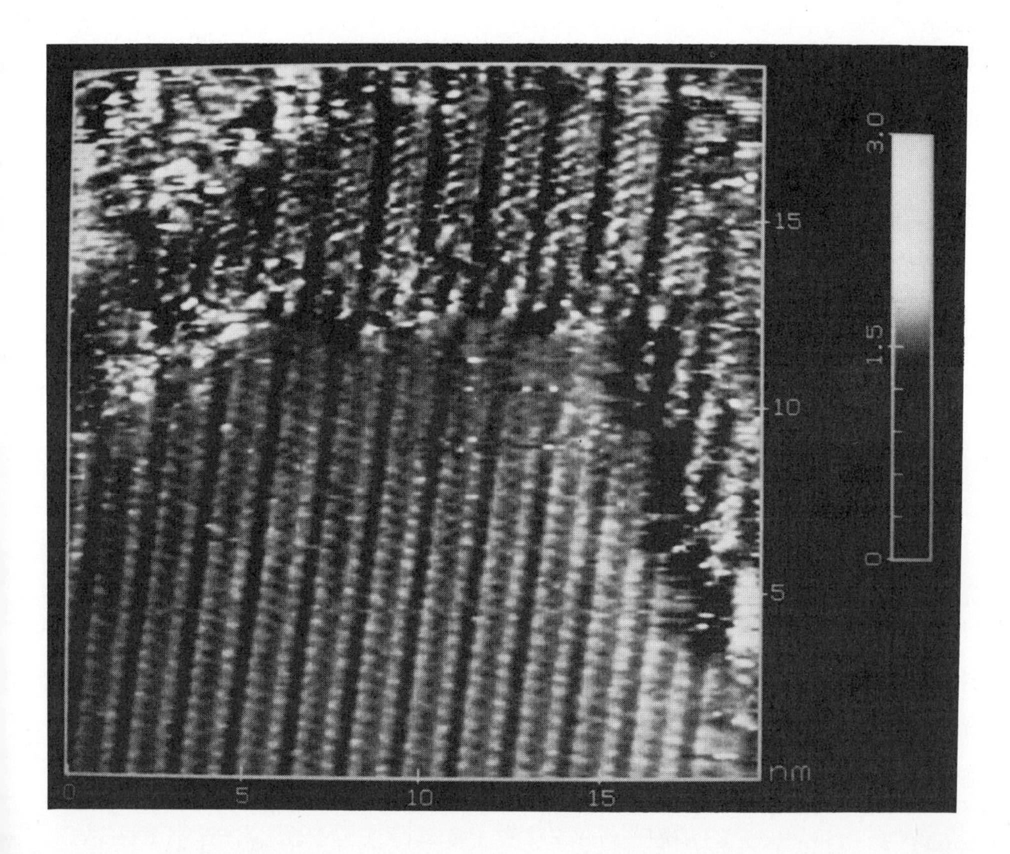

Figure 4. STM image of the bc plane of TMTTF-TCNQ in air at room temperature. The tip was held at 50mV and the tunneling current was 1nA.

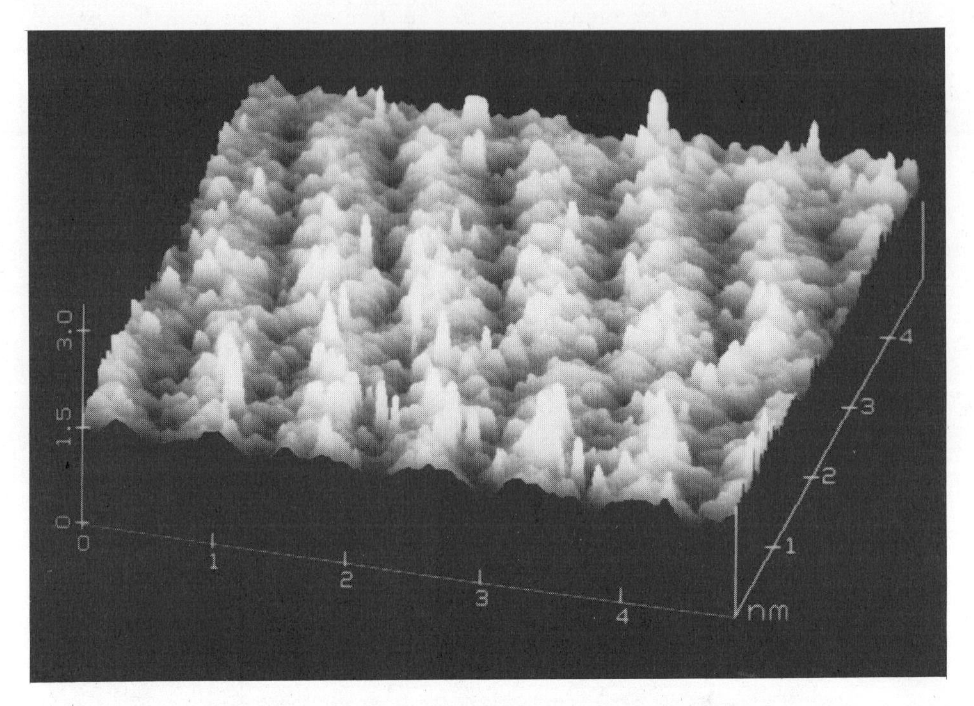

Figure 5. STM image of the ab plane of TMTSF-TCNQ in air at room temperature. The tip was held at -89mV and the tunneling current was 1nA.

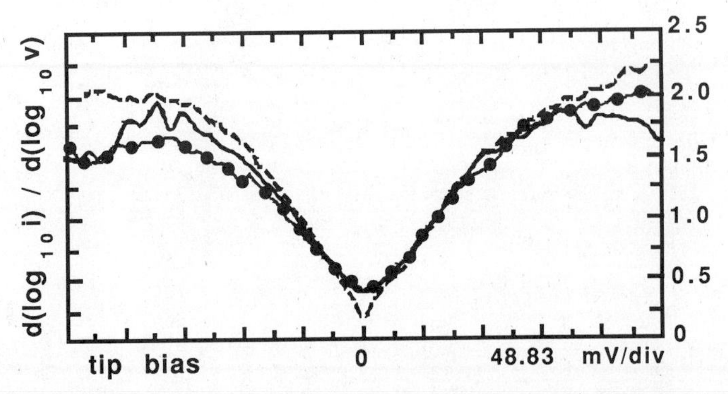

Figure 6. Normalized tunneling conductance for TTF-TCNQ (solid line), TMTTF-TCNQ (dashed line) and TMTSF-TCNQ (solid line with dark circles) in air at room temperature.

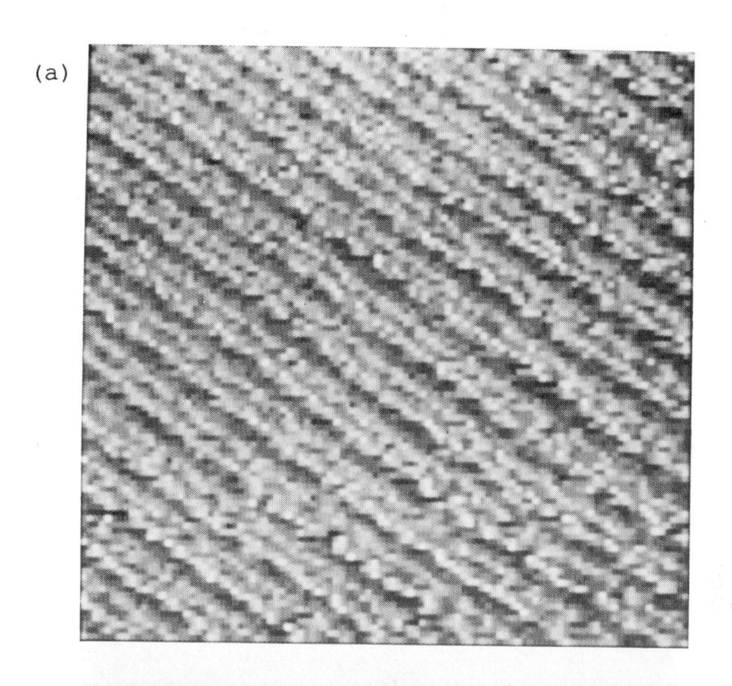

(a)

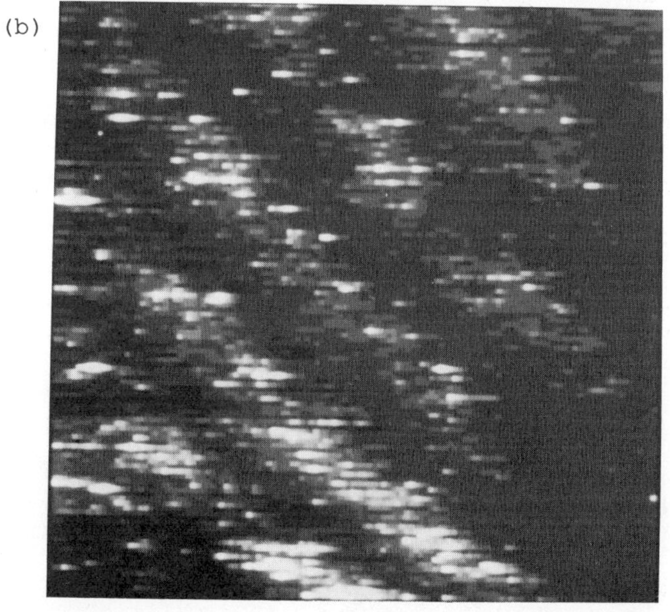

(b)

Figure 7. Topographical image of the ab plane of TTF-TCNQ taken with a STM (a) at room temperature in UHV and (b) at 87K. The tip was held at -50mV and the tunneling current was 0.5nA. The scanning area of both images is 125Å x 125Å.

STM images in UHV at low temperature

We first imaged the sample in UHV at room-temperature. The images obtained agree very well with the results obtained in air. We have seen large flat areas with one dimensional chain structures, as shown in Fig.1. The one-dimensional chains with a period of 12.2Å are clearly distinguished. These images also reveal steps and terraces.

After we cooled down the sample to about 80K imaging became more and more difficult. The conductance became lower (this may be due to a loss of carriers in the TCNQ chains as they undergo a phase transition). Above the first transition temperature (TH=54K) we were still able to obtain images, but the quality of the images was not as good as the ones obtained at room temperature. Nevertheless, we could still obtain atomic resolution, and we still observed steps, terraces and the same one dimensional chain structures with unchanged periodicity. Moreover, in some regions, a one dimensional double period of 24.5Å started to appear, as shown in Fig.2. In most cases the double period superstructures are not developed completely, but we could clearly see that the double period superstructure is commensurate with the underlying lattice. We attribute the double period superlattice to a CDW. When the double period superlattice was well developed, it became dominant, and we could not resolve individual atoms. This might due to the TCNQ chains completly localizing their carriers and becoming nonconductive. This image was taken at 78K, in constant current mode, with 50mV tip negative bias and 0.5nA tunneling current.

We also measured the I-V characterastics of the same sample. At temperatures above about 11K, the I-V characteristic is basically metallic. Bellow 11K the I-V characteristic starts changing, and an energy gap starts to open up. Fig.8a shows I-V curves at 44K and 10K, while Fig.8b shows the corresponding derivatives, both clearly displaying a gap at the lower temperature. This gap was observed at many locations on the surface and reproducibly disapeared and reappeared as the sample was warmed up and cooled down. The size of the gap is about 40mV, which agrees with results from DC conductivity measurements. We therefore believe that this is the CDW energy gap. Nevertheless, more experiments are needed to determine why this gap appears only below 11K.

CONCLUSIONS

We have obtained high resolution images of different the surfaces of quasi one-dimensional conductors of the family of TTF-TCNQ. The images reveal the alignment and orientation of molecules that constitute the solids. Tunneling spectroscopy data on TTF-TCNQ and TTF substituted compounds is able to show differences in the electron density of states at energies above the Fermi level. Low temperature images reveal for the first time in real space the presence of the CDW. The low temperature tunneling spectroscopy confirms a metal-semiconductor transition.

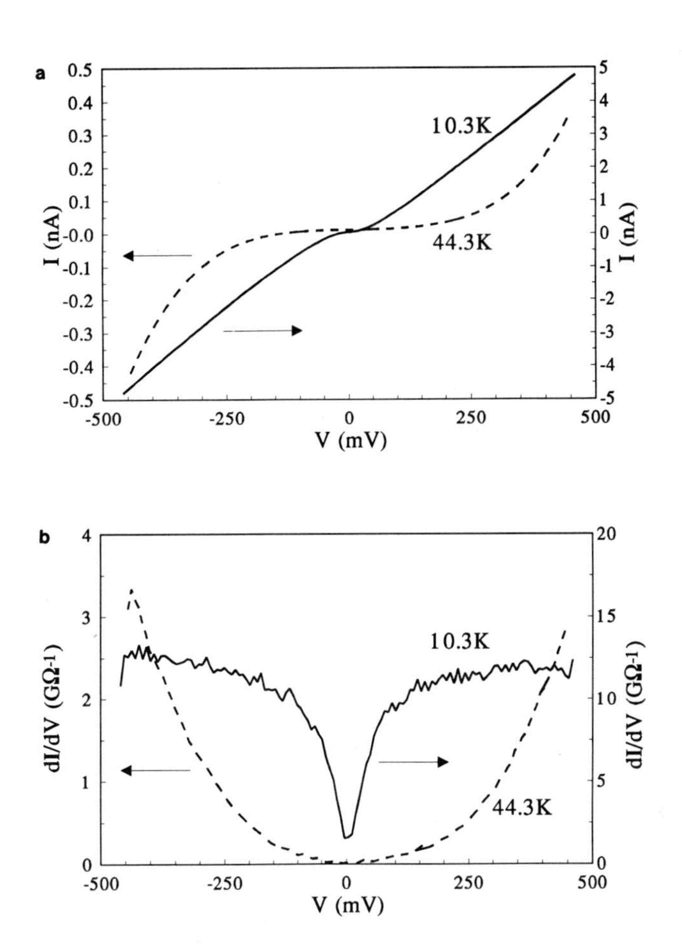

Figure 8. Tunneling characteristics (a) I vs V and (b) dI/dV vs V of TTF-TCNQ at 44K (dashed curve) and at 10K (solid curve).

ACKNOWLEDGEMENTS

We would like to thank Dr. T. O. Poheler for providing us the samples. The authors acknowledge support for this work to be provided (RF) in part by the U.S. Department of the Navy under contract N00039-89-C-5301, and (SP and ALdeL) by the US National Science Foundation under grant (DNR-8553305).

REFERENCES

1. Y. Kuk and P. J. Silverman, Rev. Sci. Instrum. 60:165 (1989).

2. T. Sleator and R. Tyco, Phys. Rev. Lett. 60:1418 (1988).

3. M. L. Kaplan, J. Crystal Growth 33:161 (1976).

4. K. Beachgaard, D. O. Cowan and A. N. Bloch, Chem. Commun. 937 (1974).

5. K. Beachgaard, D. O. Cowan, A. N. Bloch and L. Kenriksen, J. Org. Chem. 40:746 (1975).

6 T. J. Kistenmacher, T. E. Phillips and D. O. Cowan, Acta Crystallogr. Sect. B 30:763 (1974).

7. K. Beachgaard, T. J. Kistenmacher, A. N. Bloch and D. O. Cowan, Acta Crystallogr. Sect. B 33:417 (1977).

8. T. E. Phillips, T. J. Kistenmacher, A. N. Bloch and D. O. Cowan, Acta Crystallogr. Sect. B 33:422 (1977).

9. Nanoscope II from Digital Instruments Inc., Santa Barbara California.

10. E. W. Muller and T. T. Tsong, in "Field Ion Microscopy" Elsevier, New York (1969).

ABSENCE OF FIELD INDUCED SPIN DENSITY WAVE TRANSITIONS IN $(TMTSF)_2NO_3$

W. Kang[1], L.Y. Chiang[2], R. Upasani[1,2], P.M. Chaikin[1,2]

[1]Dept. of Physics
Princeton University
Princeton, NJ 08544

[2]Exxon Research and Engineering Co.
Annandale, NJ 08801

INTRODUCTION

The pressure-temperature phase diagram of the Bechgaard salts $(TMTSF)_2X$, where TMTSF is tetramethyltetraselenafulvalen and X = PF_6^-, ClO_4^-, ReO_4^- etc., is marked by the existence of both spin density wave (SDW) and superconducting phases [1]. Both phases are also observed in the magnetic field-temperature phase diagram. Superconductivity is destroyed in a moderate magnetic field, and in high magnetic field a transition to the SDW state or field-induced SDW (FISDW) state is found [2]. Whereas the currently accepted theoretical model of the FISDW [3,4] known as the "Standard Model" has tended to neglect the role of the superconductivity, in a recent theory proposed by Yakovenko the superconductivity in the Bechgaard salts plays a non-trivial role [5]. The purpose of this work is to search for the FISDW in $(TMTSF)_2NO_3$, a salt that is not found to superconduct, and to clarify some of the outstanding puzzles of the FISDW.

ROLE OF PRESSURE

The prototypical Bechgaard salt, $(TMTSF)_2PF_6$, undergoes a SDW transition at low temperature. Application of pressure suppresses the SDW transition until superconductivity is found above a critical threshold pressure. In the Standard Model, the role of the pressure is to change the band structure until the Fermi surface deviates from perfect nesting and the spin density wave is destroyed [4]. The existence of the superconductivity is simply ignored. In the alternate theory proposed by Yakovenko, the pressure modifies the interaction coupling constant, g, of the SDW until it changes sign and superconductivity is found [5].

MAGNETIC FIELD, RAPID OSCILLATIONS AND PREDICTIONS ON $(TMTSF)_2NO_3$

In both models the role of magnetic field is to undo the effect of pressure. In the Standard Model, application of the magnetic field makes the system more one-dimensional, and the nesting of the Fermi surface is enhanced, leading to a spin density wave instability. Thus, under the framework of the Standard Model, the key to the FISDW is the nesting of the Fermi surface and the FISDW is independent of the superconductivity. A competition of two periodicities of the system, the SDW wave vector, $2\vec{k}_f$, and the

Table 1. A comparison of some members of the Bechgaard salts.

Anion	T_{SDW}	AO	T_{AO}	SC	$P_{M-ISuppression}$	FISDW	RO
ClO_4	5 K[a]	Yes	24 K	Yes	1 bar[b]	Yes	Yes
PF_6	12 K	No	—	Yes	6-8 kbar	Yes	Yes
ReO_4	20 K[c]	Yes	180 K	Yes	8-10 kbar	Yes	Yes
NO_3	12 K	Yes	45 K	No	8 kbar	No	No

a-only for quenched samples

b-for well relaxed samples

c-under recycled pressure [14]

reciprocal of the magnetic length, π/λ, leads to a series of first order transitions. In very high magnetic field a transition to a semi-conducting(n = 0) state is expected.An excellent qualitative agreement is found with the experiment in the PF_6 salt [6] and in the ClO_4 salt for H < 8 tesla [2]. However, in its greatest shortcoming, it fails to predict the reentrance to the metallic state at 27 tesla in $(TMTSF)_2ClO_4$ [7].

In Yakovenko's model the effect of the magnetic field is to renormalize the attractive interaction, which gives rise to the superconductivity, to produce an interchain electron-hole pairing interaction that results in the formation of the FISDW. At extremely high magnetic field the electrons are localized on the independent chains and the system becomes truly one-dimensional. Fluctuations reduce the FISDW transition temperature and the reentrance to the metallic state is predicted. Thus, no FISDW should be present if no superconductivity exists. A simple test for the Yakovenko's model is to look for FISDW in a non-superconducting member of the Bechgaard salts. The Standard Model, on the other hand, predicts that field-induced transitions should be found when the SDW insulating state is completely suppressed, regardless of the superconductivity.

Now let's turn to the puzzle of the "rapid oscillations." Magneto-oscillations are observed in both transport and thermodynamic quantities [8]. The rapid oscillations are observed throughout the phase diagram in $(TMTSF)_2ClO_4$[8] and in the n=0 state in $(TMTSF)_2PF_6$[6]. Anion ordering has been invoked as the possible cause of rapid oscillations in the ClO_4 salt by several authors [9,10]. Although $(TMTSF)_2NO_3$ also undergoes an anion ordering transition, the situation in $(TMTSF)_2NO_3$ is somewhat complicated by the fact that its anion ordering wave vector is different than that of $(TMTSF)_2ClO_4$. On the other hand, the open-orbit edge state model [11] predicts that the rapid oscillations be present in the $(TMTSF)_2NO_3$ as well. Thus, the NO_3 salt is an ideal ground for testing various tenets of the different theoretical models.

EXPERIMENTAL RESULTS

We have performed magnetotransport measurements on a non- superconducting member of the Bechgaard salts, $(TMTSF)_2NO_3$. Being the best conductor of the family, $(TMTSF)_2NO_3$ shares many similar features with the other FISDW salts. Some of its properties are outlined in the Table 1. It undergoes a SDW insulator transition at 12 K, similar to the PF_6 salt. An anion ordering transition along the wave vector $\vec{q} = (1/2, 0, 0)$ is observed at 45 K [12]. Although the metallic state is stabilized above 8 kbar, no superconductivity has been observed up to 24 kbars of pressure [13]. The experiment was performed using a miniature pressure clamp. Pressure was gradually increased on successive days until the SDW transition was completely suppressed. Typical cooling rate through the anion ordering transition was about 1 K/min.

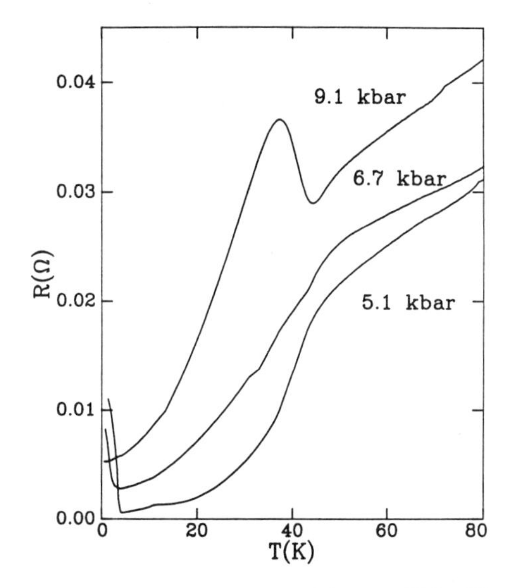

Figure 1. Anion ordering anomaly at 45 K for different pressures. An upturn in resistance is observed at 9.1 kbar. 12 K ambient pressure SDW transition is suppressed under increasing pressure. At 9.1 kbar the SDW transition is completely suppressed but no superconductivity is observed.

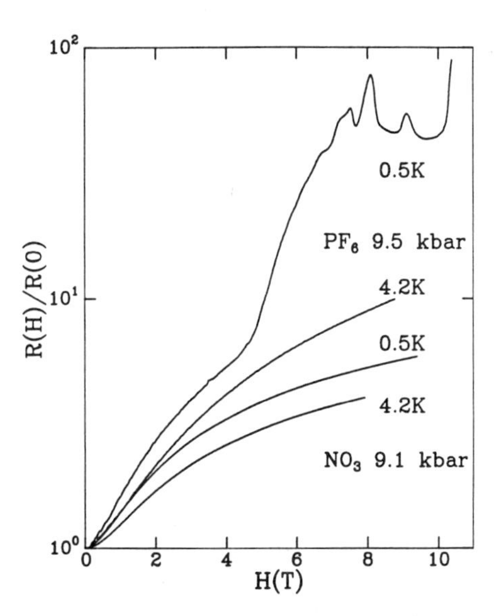

Figure 2. Magnetoresistances for $(TMTSF)_2NO_3$ and $(TMTSF)_2PF_6$ at 9.1 and 9.5 kbar, respectively. For $(TMTSF)_2PF_6$ the transition to FISDW is observed near 4.5 tesla at 0.5 K. The magnetoresistance in $(TMTSF)_2NO_3$ is similar to the normal state in $(TMTSF)_2PF_6$, but no FISDW transition is observed unlike the PF_6 salt.

In Figure 1 resistance through the anion ordering transition of a $(TMTSF)_2NO_3$ sample is shown for various pressure. The anion ordering transition can be seen from the sharp decrease in the resistance at 45K. As the pressure is increase, the depression in the resistance is inhibited until an upturn in resistance is observed.It may be a signature of an anion-ordering induced partial metal-insulator transition. For a different sample showing similar upturn in resistance, the sample was warmed up and then cooled slowly through the anion ordering transition with the cooling rate of about 30 mK per minute. Virtually no difference of resistance between two cooling rates was observed.

In Figure 2 the magnetoresistance of the sample at a pressure of 9.1 kbar at 0.5 K is shown along with the magnetoresistance of a $(TMTSF)_2PF_6$ sample at 9.5 kbar as a comparison.Broad, saturating behavior of the magnetoresistance in $(TMTSF)_2NO_3$ is observed, similar to the normal state magnetoresistance in $(TMTSF)_2ClO_4$ and $(TMTSF)_2PF_6$. Although the magnetoresistance at 10 tesla is about 10 times larger than that of zero-field resistance, no sign of any FISDW transition is observed. The difference between the magnetoresistances of $(TMTSF)_2PF_6$ and $(TMTSF)_2NO_3$ is striking.Whereas the FISDW transition is marked by a sharp increase in magnetoresistance and accompanying "oscillations", which are actual transitions, the magnetoresistance is $(TMTSF)_2NO_3$ is featureless and seems to be display a saturating behavior. In addition, no evidence of rapid oscillations was observed in the range of fields where the experiment was performed.

DISCUSSION

What is the relationship between the superconductivity and the FISDW? We observe no superconductivity in $(TMTSF)_2NO_3$, and in the range of temperature ($T < 1$ K), magnetic field ($H < 10$ T), and pressure ($P \sim 10$ kbar), we observe no sign of FISDW. While it may be argued that the FISDW may appear at higher field, it should be noted that for the ClO_4 salt at ambient pressure and PF_6 and ReO_4 salts at any pressure where the FISDW has been observed, the cascade of FISDW transitions always started below 10 tesla. Thus, Yakovenko's argument that they are interconnected may deserve a careful look in lieu of the preliminary results reported in this paper.

The puzzle of rapid oscillations, however, remains as baffling as ever.All the explanations for rapid oscillations suggest that they should be observable in the NO_3 salt. However, within the FISDW's the rapid oscillations do not always show up in the same field region for different Bechgaard salts. It remains to be seen whether the rapid oscillations will be seen in higher fields as in the PF_6 salt where the oscillations were observed only in the n=0 state [6].

CONCLUSION

In conclusion, we have performed magnetotransport measurement on a non-superconductin member of the Bechgaard salt $(TMTSF)_2NO_3$. No signature of the rapid oscillations nor the FISDW is observed in the metallic state of $(TMTSF)_2NO_3$ for fields up to 10 tesla. The coexistence of the superconductivity and the FISDW may not be accidental. Further work in higher field is currently under progress.

ACKNOWLEGEMENT

We wish to thank S.T. Hannahs for helpful discussions on the pressure apparatus.This work is supported by the NSF under Grant No. DMR 88-22532.

References

[1] D. Jerome and H.J. Schultz, Adv. Phys. 31, 299 (1982), R.L. Greene and P.M. Chaikin, Physica 126B, 431 (1984).

[2] See, for example,*Low Dimensional Conductors and Superconductors*, edited by D. Jerome and L. G. Caron, NATO Advanced Study Institute, Series B, Vol. 155 (Plenum, New York, 1987).

[3] L. P. Gor'kov, A. G. Lebed, J. Phys. Lett. 45, L-433 (1984), P. M. Chaikin, Phys. Rev. B 31, 4770 (1985), G .Montambaux, M. Heritier, P.Lederer, J. Phys. Lett. 45, L-533 (1984), M. Heritier, G. Montambaux, and P. Lederer, J. Physique Lett. 45, (1984) L-943, K. Yamaji, J.Phys. Soc. Japan 54, 1034 (1985), M. Ya Azbel, Per Bak and P. M. Chaikin, Phys. Lett. A117, 92 (1986), K. Maki, Phys. Rev. B 33, 4826, (1986).

[4] G. Montambaux, Phys. Rev. B 38, 4788 (1988).

[5] V. M. Yakovenko, Zh. Eskp. Teor. Fiz. 93, 627 (Soviet Physics JETP 66, 355(1987).

[6] J. R. Cooper, W. Kang, P. Auban, G. Montambaux , D. Jerome, and K. Bechgaard. Phys. Rev. Lett. 63, 1984 (1989), S. T. Hannahs, J. S. Brooks, W. Kang, L. Y. Chiang, and P. M. Chaikin. Phys. Rev. Lett. 63, 1988 (1989).

[7] M. J. Naughton, R. V. Chamberlain. X. Yan, P.M. Chaikin, S .Y. Hsu, L. Y. Chiang, M. Y. Azbel, Phys. Rev. Lett. 61, 621 (1988).

[8] H. Schwenk et al., Phys. Rev. Lett. 56, 667 (1986), T. Osada et al., Solid State Commun. 60, 441 (1986); Physica 143B, 403 (1986), J. P. Ulmet et al., Physica 143b, 400 (1986), X. Yan et al., Phys. Rev. B36, 1799 (1987).

[9] X. Yan et al., Proceedings of the 1988 International Conference on Science and Technology of Synthetic Metals, Vol. 27(Elsevier Sequoia, Lausanne, 1989).

[10] A.G. Lebed and P. Bak, Phys. Rev. B 40, 11433 (1989).

[11] M. Ya Azbel and P.M. Chaikin, Phys. Rev. Lett. 59, 926 (1987); M. Ya. Azbel, Phys. Rev. Lett. 64, 1282(1990).

[12] J. P. Pouget, R. Moret, R. Comes, and K. Bechgaard, J. Phys. Lett. 42,L-5203 (1981), J. P. Pouget, G. Shirane, K. Bechgaard, and J. M. Fabre, Phys. Rev. B27, 5203 (1983).

[13] A. Mazaud, 3rd Cycle Thesis, Orsay (1981), unpublished.

[14] S. Tomic, unpublished.

FERMIOLOGY

FERMI SURFACE STUDY OF ORGANIC SUPERCONDUCTORS $(BEDT\text{-}TTF)_2X$: A WONDERLAND OF TWO-DIMENSIONAL MOLECULAR METALS

M. Tokumoto,[a,b] A. G. Swanson,[b] J. S. Brooks,[b] C. C. Agosta,[c] S. T. Hannahs,[c] N. Kinoshita,[a] H. Anzai,[a] M. Tamura,[d] H. Tajima,[d] H. Kuroda[d] and J. R. Anderson[e]

[a]Electrotechnical Laboratory, Tsukuba, Ibaraki 305, Japan; [b]Department of Physics, Boston University, Boston, MA 02215, U.S.A.; [c]Francis Bitter National Magnet Laboratory, Massachusetts Institute of Technology, Cambridge, MA 02139, U.S.A.; [d]Department of Chemistry, Faculty of Science, The University of Tokyo, Hongo, Tokyo 113, Japan; [e]Department of Physics, University of Maryland, College Park, MD 20742, U.S.A.

INTRODUCTION

Organic charge transfer salts $(BEDT\text{-}TTF)_2X$, where BEDT-TTF (or ET) is bis(ethylenedithio)tetrathiafulvalene and X stands for monovalent anions such as I_3^-, IBr_2^-, AuI_2^-, $Cu(SCN)_2^-$ and so on, constitute the most promising family among organic superconductors at present. They provided us with plenty of organic metals, which are mostly characterized by layered crystal structure, two-dimensional (2D) electronic structure and ambient-pressure superconductivity with relatively high critical temperature, including the record high T_c among organics. In order to reach an understanding of the superconductivity mechanism in organic metals, it is extremely important to study not only the superconducting characteristics but also the electronic band structure, in particular, the Fermi surface.

Until very recently, however, there were virtually no experimental data on the Fermi surface of these organic superconductors. The absence of such data is due primarily to the absence of single crystals of reasonably high quality which most of experimental methods of studying the Fermi surface require. Progress in the technique of growing such crystals has recently led to the discovery of magnetic quantum oscillations such as Shubnikov-de Haas (SdH) effect on $(BEDT\text{-}TTF)_2X$ with $X=Cu(NCS)_2$,[1] β-I_3,[2-5] β-IBr_2,[2,3,6,7] β-AuI_2,[8] β''-$AuBr_2$,[9,10] and $KHg(SCN)_4$,[11-13] and de Haas-van Alphen (dHvA) effect on $(BEDT\text{-}TTF)_2X$ with $X=\beta$-AuI_2,[8] $Cu(NCS)_2$,[14] β''-$AuBr_2$,[10] and θ-I_3.[15] ***Fermiology*** of organic superconductor started in 1988 and has now burst into bloom as demonstrated by the number of reports shown in chronological order in Table I.

Organic Superconductivity, Edited by V.Z. Kresin and W.A. Little, Plenum Press, New York, 1990

Table I. Fermi Surface Study of $(BEDT\text{-}TTF)_2X$.

κ-$(ET)_2Cu(NCS)_2$	SdH		Oshima *et al.*[a]	(1988)
β-$(ET)_2IBr_2$	SdH		Kartsovnik *et al.*[b]	(1988)
β_L-$(ET)_2I_3$	SdH		Murata *et al.*[c]	(1988)
β-$(ET)_2IBr_2$	SdH		ibid.	(1988)
β-$(ET)_2IBr_2$	SdH		Toyota *et al.*[d]	(1988)
β_L-$(ET)_2I_3$	SdH		ibid.	(1988)
κ-$(ET)_2Cu(NCS)_2$	SdH		ibid.	(1988)
β-$(ET)_2AuI_2$	SdH,	dHvA	Parker *et al.*[e]	(1988)
β"-$(ET)_2AuBr_2$	SdH		Pratt *et al.*[f]	(1988)
β-$(ET)_2IBr_2$	SdH		Kartsovnik *et al.*[g]	(1988)
β_H-$(ET)_2I_3$	SdH		Kang *et al.*[h]	(1989)
β_L-$(ET)_2I_3$	SdH		Kartsovnik *et al.*[i]	(1989)
κ-$(ET)_2Cu(NCS)_2$		dHvA	Swanson *et al.*[j]	(1990)
$(ET)_2KHg(SCN)_4$	SdH		Osada *et al.*[k]	(1990)
$(ET)_2KHg(SCN)_4$	SdH		Tokumoto *et al.*[l]	(1990)
$(ET)_2KHg(SCN)_4$	SdH		Sasaki *et al.*[m]	(1990)
θ-$(ET)_2I_3$		dHvA	Tokumoto *et al.*[n]	(1990)
β"-$(ET)_2AuBr_2$	SdH,	dHvA	Swanson *et al.*[o]	(1990)

[a]Ref. 1, [b]Ref. 6, [c]Ref. 2, [d]Ref. 3, [e]Ref. 8, [f]Ref. 9, [g]Ref. 7, [h]Ref. 4, [i]Ref. 5, [j]Ref. 14, [k]Ref. 11, [l]Ref. 12, [m]Ref. 13, [n]Ref.15, [o]Ref. 10.

Observation of magnetic quantum oscillations is a standard and well-established technique in the study of the Fermi surface of ordinary metals. It provides us with various informations about the electronic band structure, in particular about the Fermi surface. The periodicity in the inverse magnetic field $\Delta(1/H)$ (T^{-1}) gives us the frequency of oscillation F(T) which is related to the extremal cross-sectional area S_{FS} of the Fermi surface normal to the magnetic field direction by the equation $F=1/\Delta(1/H)=hcS_{FS}/2\pi e$. In addition to such informations about topology of the Fermi surface, we can also obtain important physical quantities such as the cyclotron mass m_c and the Dingle temperature $T_D(=h/2\pi k_B\tau,$: τ is the relaxation time).

Now, a question arises. Do we have something attractive or exotic in the "fermiology" of these organic metals? The answer is "Yes". It seems that we are at the entrance of a vast treasure-house of two-dimensional electronic system. A typical example demonstrating such fascinating phenomena where the two dimensionality of the energy dispersion can be responsible, would be the observation of "**giant**" SdH oscillations. Kang *et al.*[4] observed an extraordinary large amplitude of the magnetoresistance oscillations in the "high-T_c" phase (β_H) of β-(BEDT-TTF)$_2$I$_3$ with T_c=8 K. It was attributed to the two-dimensional nature of the cylindrical Fermi surface exhibiting only a small warping along the direction of lowest conductivity.

In the following, we would like to introduce two new examples which also exhibit extraordinary features of magnetic quantum oscillations observed in two-dimensional organic metals (BEDT-TTF)$_2$X. One is (BEDT-TTF)$_2$KHg(SCN)$_4$, where a clear "**spin-splitting**" of SdH oscillations was observed directly.[12,13] The other is θ-(BEDT-TTF)$_2$I$_3$, where a "**saw-tooth**" dHvA oscillation was observed after a "**magnetic breakdown**".[15] The "saw-tooth" variation of magnetization demonstrates that organic metals can provide an ideal example of highly two-dimensional and extraordinary clean electron system with a cylindrical Fermi surface. It is surprizing that crystals of organic metals are so clean that we can observe these phenomena in magnetic field conditions far from the quantum limit.

(BEDT-TTF)$_2$KHg(SCN)$_4$

The title compound, (BEDT-TTF)$_2$KHg(SCN)$_4$, is a newcomer to the (BEDT-TTF)$_2$X family. It was synthesized by Oshima *et al.*[16] with an intention to realize a modification of the superconductor κ-(BEDT-TTF)$_2$Cu(SCN)$_2$ with T_c=10.4 K. It was reported to show a metallic behavior down to 0.5 K without any superconducting or metal-insulator transitions.[16] The high-field magnetoresistance of (BEDT-TTF)$_2$KHg(SCN)$_4$ was measured by Osada *et al.*[11] under pulsed magnetic fields up to 40 T. They observed magnetoresistance with a negative slope above 10 T, a sharp kink structure at 22.5 T, and a large enhancement of the SdH oscillations above the kink structure. Originally, we planned to reproduce the magnetoresistance measurement under the DC magnetic fields up to 23 T, but we found a new aspect in the SdH oscillation waveform, i.e. splitting of each oscillation due to spins, as described below.[12]

We present here a direct observation of the spin-splitting in the SdH oscillations of (BEDT-TTF)$_2$KHg(SCN)$_4$ under the magnetic fields below 23 T, where more than 30 Landau tubes are embedded below the Fermi level. The observation of direct spin-splitting of SdH oscillations in such a condition, i.e.

far from the quantum limit, is very unusual and has never been reported so far in organic metals.

Experimental

Single crystals of $(BEDT\text{-}TTF)_2KHg(SCN)_4$ were grown at Electrotechnical Laboratory by electrochemical oxidation of BEDT-TTF in the presence of $Hg(SCN)_2$, KSCN, and 18-crown-6 ether using a mixture of 1,1,2-trichloroethane (TCE) and 10 vol.% of ethanol as a solvent.[16] Typical sample size was $2 \times 1 \times 0.5 mm^3$. Linear four electrical contacts were made by gold evaporated pads with 10 μm gold wire as electrical leads. AC(~13Hz) resistance measurement was made using a Lock-In Amplifier (EG&G PARC Model 5210). The high-field magnetoresistance measurements were carried out in quasi-static magnetic fields up to 23 T generated by a water-cooled Bitter-type magnet. The sample attached to a rotatable sample holder was immersed in liquid ^3He during the measurements, using a ^3He cryostat designed to operate in high field magnets at Francis Bitter National Magnet Laboratory at The Massachusetts Institute of Technology.

Crystal and Electronic Structures

The crystal structure of $(BEDT\text{-}TTF)_2KHg(SCN)_4$ at room temperature is shown in Fig. 1.[16] BEDT-TTF molecules form a conducting sheet in the *ac*-plane which is sandwiched by the insulating layers of polymeric anion sheets of $KHg(SCN)_4$. The anion-sheet thickness (6.8 Å) is much larger than other BEDT-TTF compounds because of its multi-layered structure, so that a strong two-dimensionality due to weak couplings between donor sheets was expected. Accordingly, the conductivity anisotropy (σ_c/σ_b) as large as 2000 at 1.2 K was reported, where σ_c and σ_b are the conductivity along and across the conducting plane, respectively.[11] The crystal structure belongs to triclinic with space group *P1*, and lattice parameters a=10.082 Å, b=20.565 Å, c=9.933 Å, α=103.70°, β=90.91°, γ=93.06°, V=1997 Å^3, Z=2 at 298 K, and a=9.948 Å, b=20.505 Å, c=9.833 Å, α=103.34°, β=90.53°, γ=92.80°, V=1949 Å^3, Z=2 at 104 K.[16]

The layered structure consisting of insulating anion sheets of $KHg(SCN)_4$ and conducting donor sheets of BEDT-TTF molecules yields two-dimensional anisotropy in the electronic structure. The 2D tight-binding band calculation using only the highest occupied molecular orbitals (HOMO's) obtained by the extended Hückel approximation was done by Mori *et al.*[17] Figure 2 shows the calculated electronic band structure and the Fermi surface (FS) of $(BEDT\text{-}TTF)_2KHg(SCN)_4$ at 298 K. The 2D Brillouin zone (BZ) has a closed FS, corresponding to about 19% of the first BZ ($3.94 \times 10^{15} cm^{-2}$ at 298 K), and a pair of open FS's along the c direction.

Shubnikov-de Haas Effect

The first magnetotransport measurements of $(BEDT\text{-}TTF)_2KHg(SCN)_4$ were carried out by Osada *et al.*[11] Figure 3 shows the transverse magnetoresistance under pulsed magnetic fields up to 40 T. Some remarkable and unusual features were found; a negative slope above 10 T, a

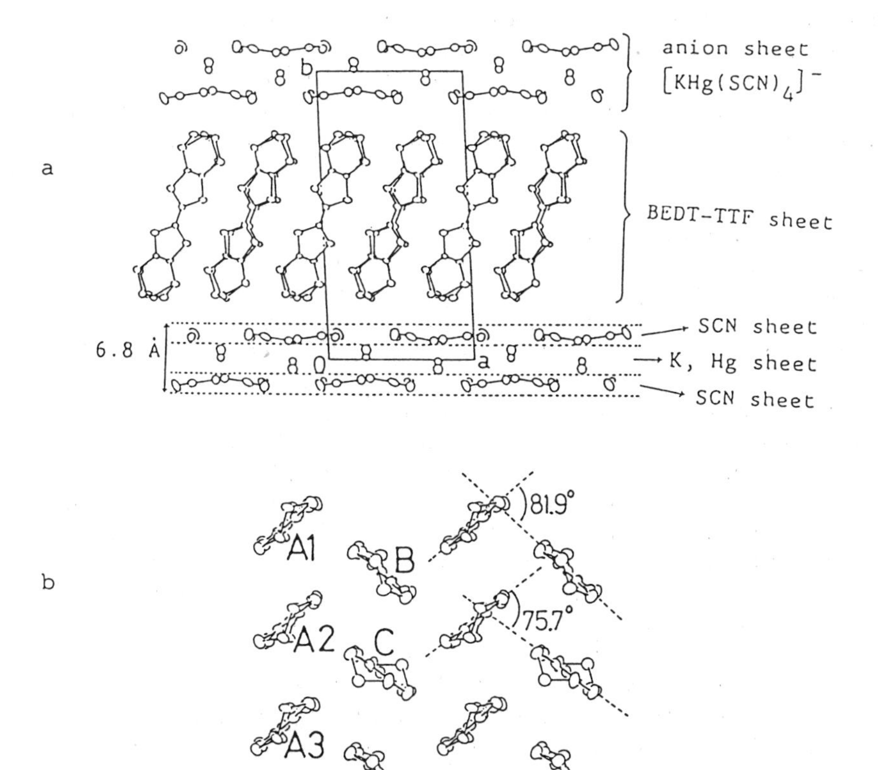

Fig. 1 Crystal structure of $(BEDT\text{-}TTF)_2KHg(SCN)_4$ at room temperature, from Oshima *et al.*[16] (a)along the *c* axis and (b) molecular stacking of BEDT-TTF, viewed along the molecular long axis.

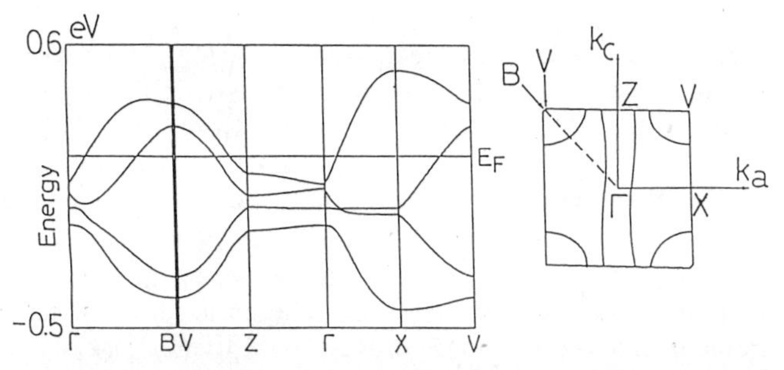

Fig. 2 Energy band structure and the Fermi surface of $(BEDT\text{-}TTF)_2KHg(SCN)_4$ calculated by Mori *et al.*[17]

sharp kink structure at 22.5 T, and a large enhancement of the Shubnikov-de Haas oscillations above the kink structure. These features of background magnetoresistance, except the SdH oscillations , are not well understood. The period of SdH oscillations $\Delta(1/H)=0.0015\ T^{-1}$ gives the cross section of FS: $S_{FS}=6.5\times10^{14}\ cm^{-2}$. This area corresponds to 16% of the first BZ. This value is in reasonable agreement with the cross-sectional area of the closed elliptical FS, i.e. the shaded area in Fig. 2, which amounts to 19% of the first BZ. The temperature dependence of the oscillation amplitude reflects the Landau-level spacing. Numerical fitting of the conventional formula of the SdH effect gave the cyclotron mass $m_c/m_0=1.4$. The Dingle temperature was deduced from the field dependence of the oscillation amplitude. Employing the theoretical formula of the SdH effect in the 2D system, they obtained the Dingle temperature $T_D=4.0$ K and the relaxation time $\tau=0.3$ ps.[11]

Figure 4(a) shows the magnetoresistance of our sample at 0.6 K under DC magnetic fields up to 23 T applied perpendicular to the conducting *ac* plane.[12] From the figure we can see clear SdH oscillations at fields above ~10 T. The overall field dependence of the background magnetoresistance is similar to the result in Fig. 3.[11] However, we notice a distinct difference in the oscillation waveform. That is, each oscillation has a double peak structure. A straightforward Fourier transform of the data gives us an anomalously large second harmonic at 1350 T, in addition to the fundamental frequency at 674 T as shown in Fig. 4(b). The period of SdH oscillations $\Delta(1/H)=0.00148\ T^{-1}$ or fundamental SdH frequency, $F=674$ T, corresponds to the cross-section of FS $S_{FS}=6.44\times10^{14}\ cm^{-2}$, i.e. 16% of the first BZ($4.04\times10^{15}\ cm^{-2}$ at 104 K), consistent with the value reported by Osada *et al.*[11] What is the origin of the anomalously large amplitude of the second harmonic? Large and anharmonic SdH oscillations were reported by Kang *et al.*[4] in the high-T_c state of β-$(BEDT\text{-}TTF)_2I_3$. They attributed these to both the 2D character of the electronic motion and high purity of the sample. In the present case, an oscillation at 1350 T, corresponding to twice as large a FS, cannot be a simple second harmonic, since it has even larger amplitude than the fundamental and we don't see ample harmonics higher than the second. Actually, a closer look into the SdH oscillation waveform reveals that the oscillations consist of uneven double peaks denoted by A and B in Fig. 5(b). These waveforms look very similar to those observed in the case of spin-split quantum oscillations[18] as will be discussed below. A crucial difference, however, is that such "**spin-splitting**" has normally been directly observed only near the quantum limit.

When the oscillation consists of two contributions from spin-split Landau levels, a simple Fourier transform gives us a misleading result. We should rather apply a technique, called best recursive fit (BRF),[19] in which we fit data consisting of a sum of exponentially damped sinusoids plus noise to a linearly recursive sequence. The BRF analysis of the data from 14 to 23 T tells us the presence of frequencies other than the fundamental and second harmonic. There appear to be two second harmonic frequencies (1352 T and 1373 T) separated by $\Delta F/F\approx0.015$. Such a long beat would show up only as a broadening of the fundamental. The beating frequency gives us directly the amplitude of warping of the cylindrical FS. The ratio directly measures the anisotropy of transfer integrals: $F/\Delta F=h^2k_F{}^2/2t_bm_{ac}=(k_Fa)(k_Fc)t_{ac}/t_b$ where t_{ac} and t_b are transfer integrals along and across the 2D plane, respectively. Assuming, for example, a cylindrical FS where $\pi k_F{}^2$ is equal to the experimental value ($6.44\times10^{14}\ cm^{-2}$) and taking a and c to be lattice

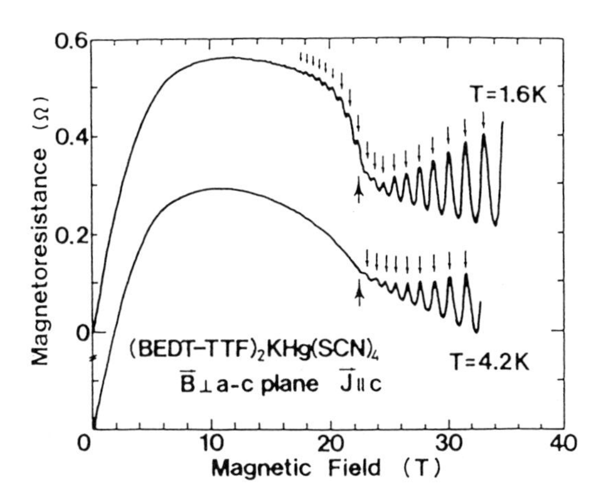

Fig. 3 The magnetoresistance of $(BEDT\text{-}TTF)_2KHg(SCN)_4$, under pulse fields applied perpendicular to the conducting plane, from Osada *et al.*[11]

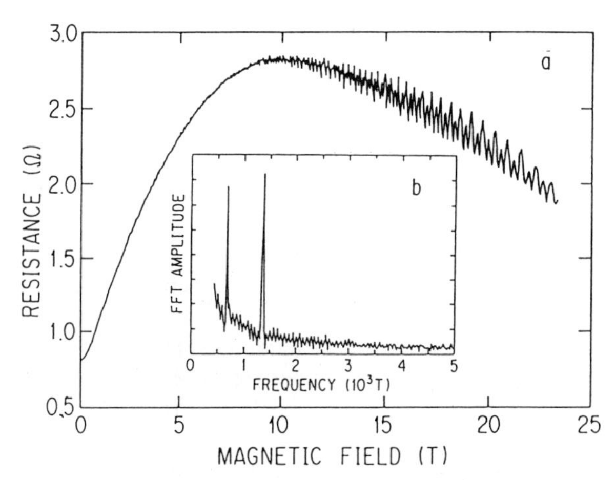

Fig. 4 (a)The magnetoresistance of $(BEDT\text{-}TTF)_2KHg(SCN)_4$, under the DC magnetic fields applied normal to the conducting plane (θ=0°) and (b)The Fourier transform of the data of (a), from Tokumoto *et al.*[12]

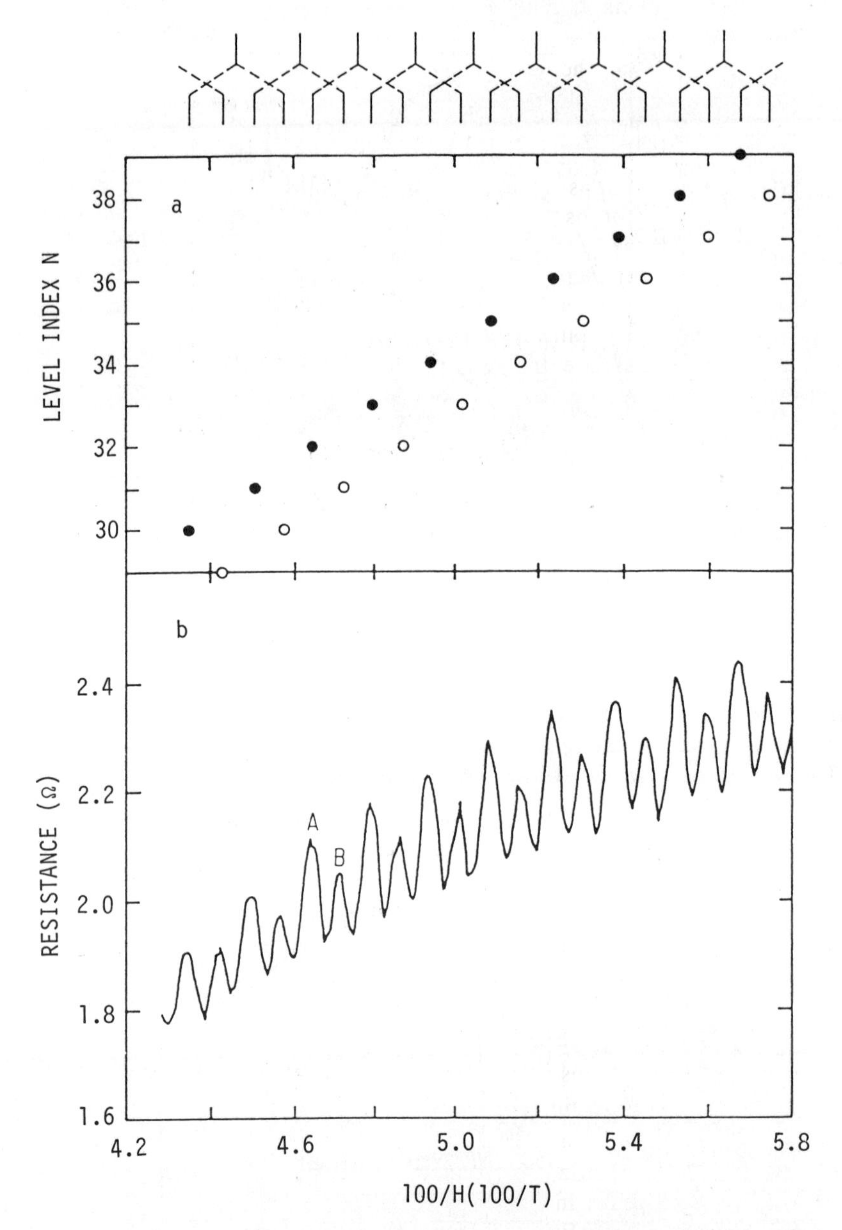

Fig. 5 (a)Plot of level index N vs. SdH peaks in reciprocal field. Each alternative peaks were assigned to spin-up(•) and spin-down(o) levels as shown schematically at the top of the frame and (b)The SdH peaks of $(BEDT\text{-}TTF)_2KHg(SCN)_4$, as a function of reciprocal field (θ=0°), from Tokumoto *et al.*[12]

parameters, leads to $t_{ac}/t_c=133$. On the other hand, if we take a typical intermolecular spacing of 3.7 Å instead of lattice parameters, we get $t_{ac}/t_c=950$. This value is roughly seven times larger than that obtained for β_H-(BEDT-TTF)$_2$I$_3$.[4]

The spin-splitting of the energy levels in a magnetic field can in favorable circumstances be directly observed as a splitting of the oscillations. The circumstances are most favorable (1) for oscillations when the components due to spin-up and spin-down levels are out of phase with each other, and (2) for oscillations of low quantum number, i.e. close to the quantum limit, and (3) for oscillations of as sharp a line shape as possible.[18] In fact, in the case of the needle in Zn, for example, dHvA oscillations of d^2M/dH^2,[20] and galvanomagnetic oscillations in magnetic breakdown conditions[21] show this direct splitting as we approach the quantum limit. However, any observable splitting of the oscillations is rapidly masked as H decreases, since the phase separation between up and down spins is independent of H, while the phase smearing due to the other causes varies as $1/H$.[18]

Let us check if these conditions are satisfied in the present case. The spin degeneracy of the conduction electrons is lifted in a magnetic field, leading to an energy difference between spin-up and spin-down electrons given by

$$\delta=(1/2)g\beta_0 H=(1/2)g(m^*/m_0)\beta H=(1/2)g(m^*/m_0)\Delta, \quad (1)$$

where Δ is the Landau level separation, i.e. $\Delta=h\omega_c=(eh/m^*c)H=\beta H$. For free electrons, g=2 and the spin splitting coincides with the Landau level separation. In real metals, however, spin-orbit coupling and many-body interactions can modify the g-factor and m^* considerably.

The values of $1/H$ for which the Landau tubes just part company with the FS are given by

$$F/\Delta=n+\gamma\pm(1/2)S, \quad (2)$$

where Δ is the Landau level separation in $1/H$, γ is a phase constant which is normally close to 1/2, and $S=(1/2)g(m^*/m_0)$ is spin splitting parameter. Figure 5(a) shows the twenty level indices vs. SdH peak positions in reciprocal field. By fitting peaks A and B separately we obtained values of phase constant and spin-splitting parameters as $\gamma=0.31$ and S=1.33, so that the value of spin-split parameter is close to the value $S=(1/2)g(m^*/m_0)\approx1.4$ estimated using g value ($g\approx2.014$) obtained by ESR measurement[22,23] and the cyclotron mass $m_c\approx1.4m_0$.[11] The resulting spin-split energy level scheme is illustrated at the top of Fig. 5(a). Horizontal position of each level corresponds to the value of $1/H$ for SdH peaks in Fig. 5(b). It is to be noted that this configuration obviously satisfies the first condition, and provides one of the most favorable circumstance for direct observation of spin-splitting in the sense that each level is almost evenly spaced in the energy diagram.

Now what about the second condition? In our experiment, level index tells us that more than 30 Landau tubes still remain inside the FS even at the highest field of 23 T. Thus the magnetic field is not even close to the quantum limit, and it is this crucial point which makes the present observation new and different from the cases reported so far.

Finally, we should examine the third condition, i.e. whether it is appropriate to observe the spin-splitting in the present experimental

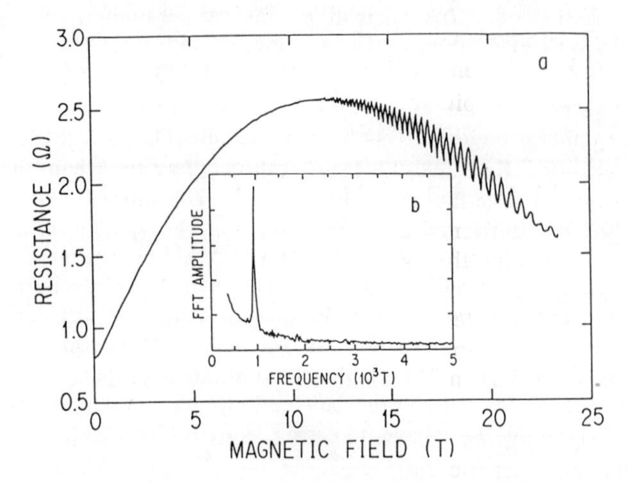

Fig. 6 (a)The magnetoresistance of $(BEDT\text{-}TTF)_2KHg(SCN)_4$, under the fields applied about 40° off the direction normal to the conducting plane ($\theta \approx 40°$) and (b)The Fourier transform of the data of Fig. (a), from Tokumoto *et al.*[12]

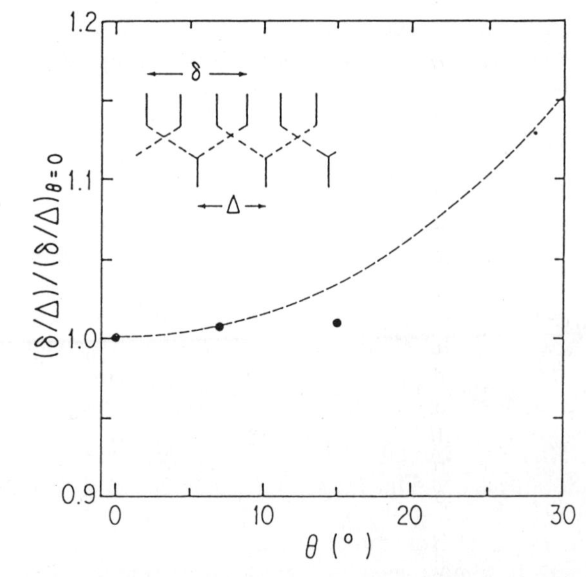

Fig. 7 Angular dependence of relative position(δ/Δ) of spin-split peaks, where Δ ($=h\omega_c$) is the spacing between adjacent Landau levels and δ is spin-splitting of each level, from Tokumoto *et al.*[12] The abscissa is the magnetic field direction(θ) measured from a perpendicular to the conducting plane. Dashed line shows $1/\cos\theta$.

situation. The spin-splitting would be seen only if the dephasing due to finite temperature and scattering is less than the spacing of energy levels due to spin up and down components. We estimate the energy difference between Landau levels at H=16 T from $\Delta=\hbar\omega_c$~0.9 meV. From the level scheme, the spacing between adjacent spin-split levels is $\Delta/2$~0.45 meV. The broadening due to finite temperature ($T\approx0.6$ K) and due to the Dingle temperature of the present sample ($T_D\approx0.5$ K)[24] is calculated from $k_B(T+T_D)$~0.3 meV. We notice that the broadening of Landau levels due to finite temperature and finite relaxation time of the carriers is barely less than the energy spacing between levels. This explains why the previous report[11] did not show spin-splitting for a sample with T_D=4.0 K measured at T=1.6 K. Also, as the sample is rotated in the field, the Landau levels are squeezed closer together, and higher effective fields are needed to observe spin-splitting. This tells us why splitting was easily lost when the field was tilted from normal to the conducting plane.

It would be interesting to see what happens to the splitting when we rotate the sample in magnetic field and change the angle (θ) between the field direction and the normal to the conducting layer. Figure 6(a) shows the magnetoresistance at 0.6 K under magnetic fields up to 23 T for $\theta\approx40°$. Note that we do not see the double peak structure any longer. The Fourier transform of the data has only one frequency at 939 T and no higher harmonics as shown in Fig. 6(b). The fundamental frequency was found to change roughly proportional to the inverse of $\cos\theta$. We found that the splitting disappears when the sample was tilted so that $\theta > 15°$.

A simple expectation of the effect of field rotation on spin-split pattern would be that δ/Δ changes inversely proportional to $\cos\theta$, if the separation of the Landau levels (Δ) due to orbital motion depends on the field component normal to the 2D plane, $H_\perp=H_{total}\cos\theta$, while spin-splitting ($\delta$) depend on H_{total} and the anisotropy of g-factor is negligibly small as shown by ESR measurement.[23] Figure 7 shows that, contrary to the above expectation, the angular dependence of relative position(δ/Δ) of spin-split peaks is very small and almost constant as a function of θ at least up to θ=15°. What causes the discrepancy? Since the fundamental SdH frequency is changing as $1/\cos\theta$, we suspect that the spin-splitting part(δ) is not constant, i.e. g value should be also changing as we tilt the magnetic field in contrast to the result given by ESR. At present, we don't have a good explanation for the discrepancy, but the following points about g value[18] are worthy of further examination. The main mechanisms responsible for departures from the free electron value g=2 are spin-orbit coupling and electron-electron interactions. Spin-orbit coupling is particularly important in small pieces of FS which are associated with electrons undergoing Bragg reflection at zone boundaries, while electron-electron many-body interactions are usually more important in larger pieces of FS. If spin-orbit coupling is the only relevant mechanism the g value obtained by spin resonance should be the same as the value deduced from SdH oscillations, but not if it is electron-electron interaction, because electron-electron interaction, although relevant to the energy separation of spin-up and spin-down states, produces an "effective" magnetic field which is always in the same direction as the instantaneous spin magnetic moment, and so can exert no couple on it which will affect the resonance frequency. Also, spin resonance can measure only an average of g over the FS and it is not always obvious how to compare this average with strongly orientation dependent g

values deduced from SdH oscillations which average over a particular orbit on the FS.

Recently Sasaki *et al.* found an anomalous temperature dependence of electrical resistance at 8 K.[25] Also, Kinoshita *et al.* found a corresponding anomaly in the temperature dependence of ESR signal.[26] These results seem to suggest that the ground state of $(BEDT\text{-}TTF)_2KHg(SCN)_4$ is not a simple metal, but rather has some magnetic character reminiscent of the Spin Density Wave. It is not unlikely if a pair of open FS's in Fig. 2 could nest with each other as in the case of $(TMTSF)_2X$ salts, where TMTSF is tetramethyl-tetraselenafulvalene. The nature of the ground state of $(BEDT\text{-}TTF)_2KHg(SCN)_4$ has not been clarified yet. Since an isostructural $(BEDT\text{-}TTF)_2HH_4Hg(SCN)_4$ was found to show superconductivity with an onset at 1.15 K,[27] the absence of superconductivity in $(BEDT\text{-}TTF)_2KHg(SCN)_4$ may be closely related to the presence of the phase transition at 8 K.

θ-$(BEDT\text{-}TTF)_2I_3$

The title compound, θ-$(BEDT\text{-}TTF)_2I_3$, is a superconductor with T_c=3.6 K under ambient pressure.[28-30] Recently a new type of oscillatory angular dependence of magnetoresistance were reported by Kajita *et al.*[31] Interestingly, this angle-dependent oscillation of magnetoresistance was observed at relatively high temperatures (< 6 K) and low magnetic field (> 3 T). A similar oscillation was observed by Kartsovnik *et al.* in β-$(BEDT\text{-}TTF)_2IBr_2$.[7] Yamaji proposed that such oscillatory behavior in the magnetoresistance can be explained in terms of the angular dependence of the area of semiclassical electron orbits under the magnetic field in the reciprocal lattice space.[32] Since this explanation assumed quantization of electron orbit, the observation of distinct angular dependent oscillation in θ-$(BEDT\text{-}TTF)_2I_3$ at relatively low magnetic field and high temperature has been regarded as contradicting the unsuccessful trial of observation of quantum oscillations in the same system. On the other hand, there has been a controversy about the electronic band structure or the Fermi surface of this material. Therefore, experimental observation of quantum oscillations has been desired in order to determine the Fermi surface and reconstruct the electronic band structure of this material.

Here we present the first observation of quantum oscillation due to de Haas-van Alphen effect in θ-$(BEDT\text{-}TTF)_2I_3$. The Fermi surface is shown to be uniquely determined based on the observed dH-vA oscillation frequencies. In addition, the observed oscillation shows a typical "saw-tooth" variation of magnetization which has been well-known as a characteristic of the 2D electron system, but is observed directly for the first time.

Experimental

Single crystals of θ-$(BEDT\text{-}TTF)_2I_3$ were grown at The University of Tokyo by the electrochemical crystal growth method described before.[28,33] Our magnetization measurements were made by the small-sample force magnetometer method.[34] A single crystal with dimension of $1.70\times0.95\times0.24$ mm^3 with weight of 0.66 mg was used for magnetization measurement. The sample was immersed in liquid ^3He during the

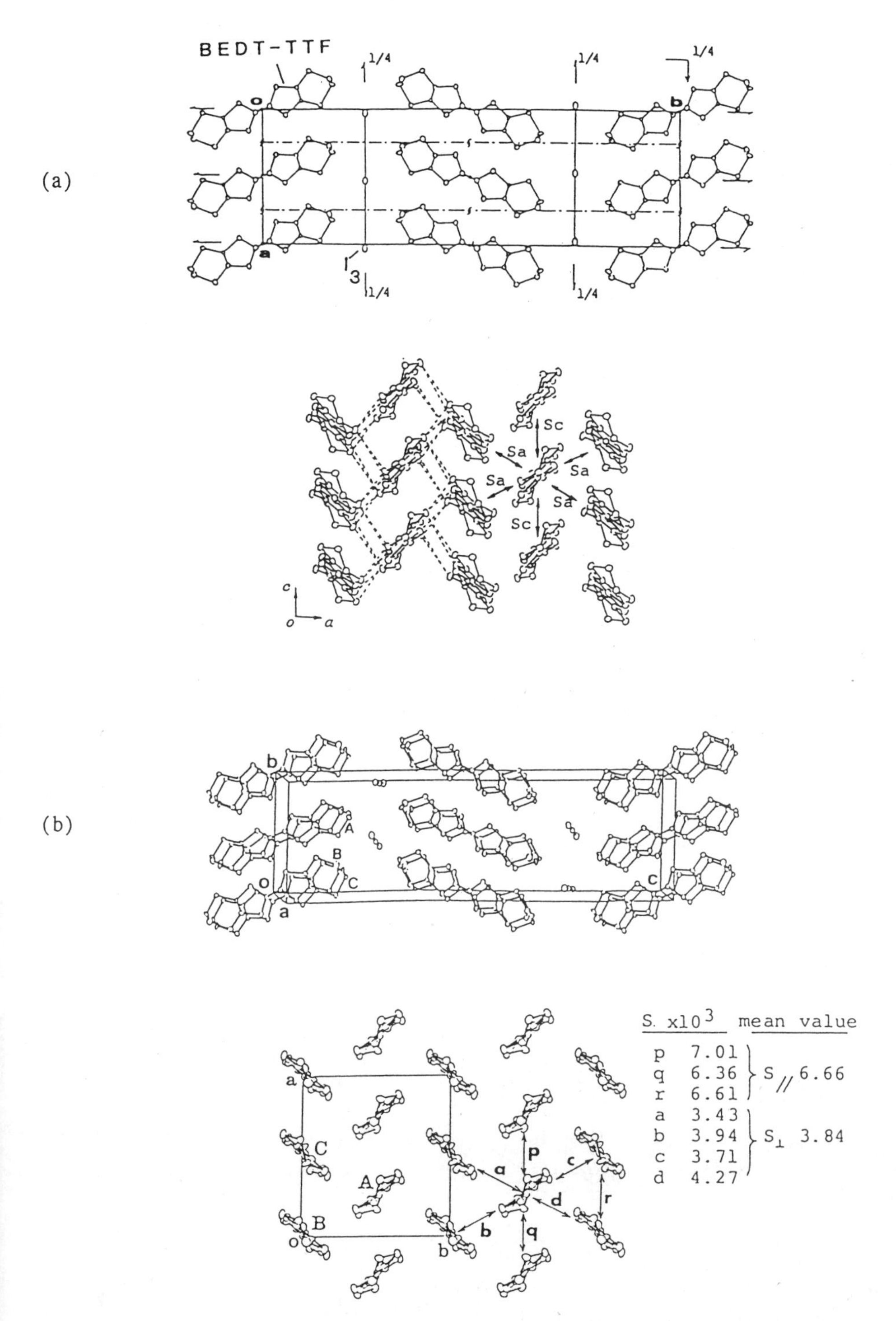

Fig. 8 Crystal structure of θ-$(BEDT\text{-}TTF)_2I_3$. (a)the orthorhombic average structure from Kobayashi *et al.*[29] (b)the monoclinic structure from Kobayashi *et al.*[35,36]

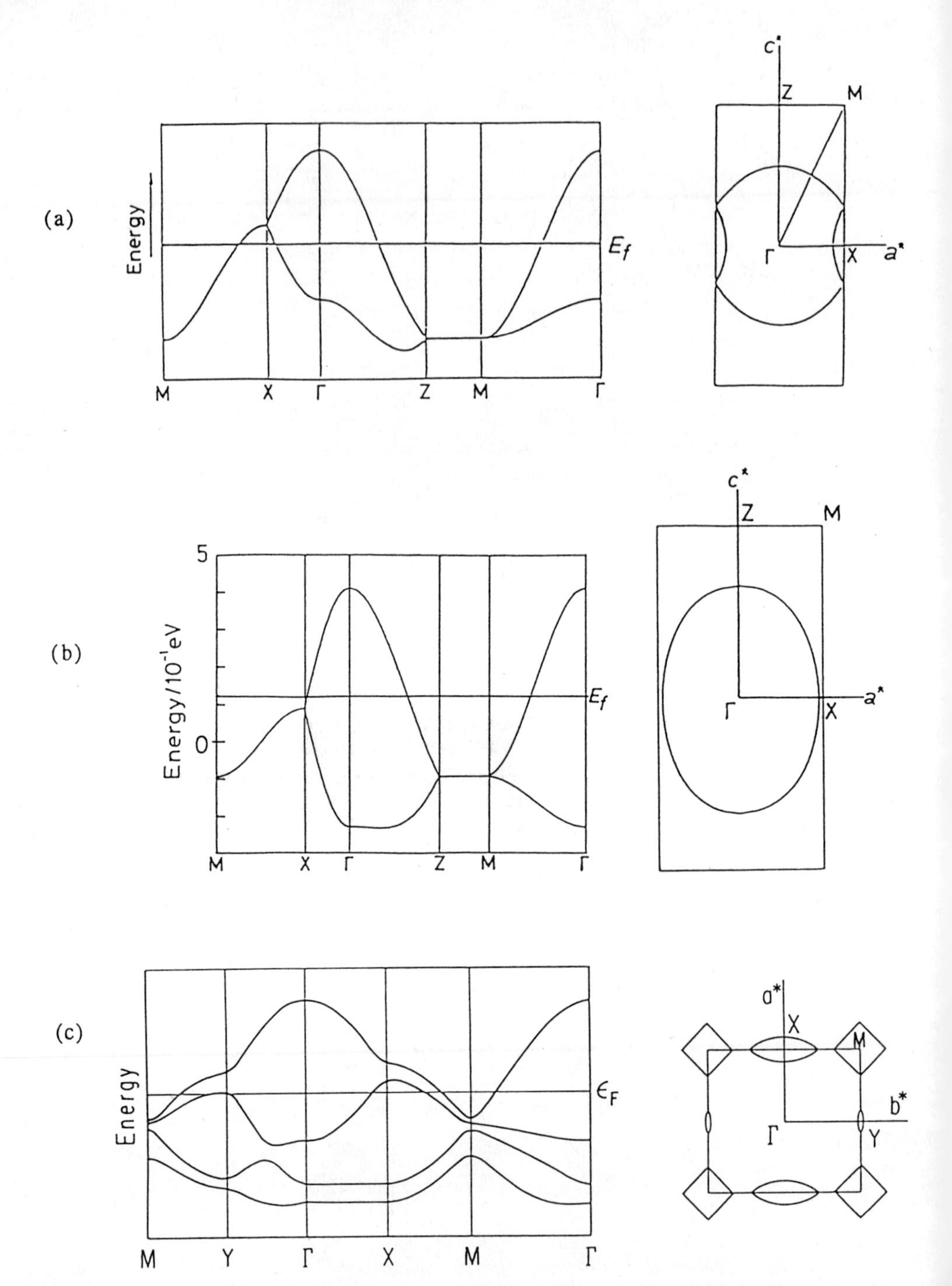

Fig. 9 Energy band structure and the Fermi surface of θ-$(BEDT\text{-}TTF)_2I_3$ (a)calculated by Kobayashi *et al.*[29] based on the orthorhombic average structure shown in Fig. 8(a) (b)estimated by Tamura *et al.*[33] from the reflectance spectra at 16 K, (c)calculated by Kobayashi *et al.*[36] based on the monoclinic structure shown in Fig. 8(b)[35]

measurements, using a ^{3}He cryostat designed to operate in high field magnets at Francis Bitter National Magnet Laboratory at The Massachusetts Institute of Technology.

Crystal and Electronic Structures

The crystal structure of θ-(BEDT-TTF)$_2$I$_3$ belongs to the monoclinic system, while most of the crystals are twinned to form a pseudo-orthorhombic lattice.[29,35] Figure 8(a) shows the orthorhombic average structure of θ-(BEDT-TTF)$_2$I$_3$ at room temperature with the lattice constants of a_o=10.076 Å, b_o=33.853 Å and c_o=4.964 Å, space group Pnma and Z=2.[29] Figure 8(b) shows the monoclinic structure with P2$_1$/c symmetry and the lattice constants of $a_m=2c_o$, $b_m=a_o$ and $c_m=b_o-c_o$, β=98.39°, where a_o, b_o, and c_o are those of the orthorhombic cell.[35] In the following, we use the notation of the crystal axes for the orthorhombic average structure: a=10.076 Å, b=33.853 Å and c=4.964 Å at room temperature.[29] Thus the conducting 2D layers spreads along the a-c plane, which is the most developed crystal face.

The energy band structure and the Fermi surface of θ-(BEDT-TTF)$_2$I$_3$ are shown in Fig. 9. Figure 9(a) shows the energy band calculated by Kobayashi *et al.*[29] based on the orthorhombic average structure by the tight binding approximation assuming that the transfer integrals are proportional to the calculated overlap integrals between HOMO's of BEDT-TTF molecules denoted as S_a, S_b and S_c in Fig. 8. Figure 9(b) shows the tight-binding band and the Fermi surface estimated by Tamura *et al.*[33] from the reflectance spectra at 16 K. A recent tight-binding band structure calculation by Kobayashi *et al.*[36] based on an improved monoclinic crystal structure shown in Fig. 8(b)[35] gives a new energy spectrum with the Fermi surface as shown in Fig. 9(c).

Angle-dependent Oscillation of Magnetoresistance

Here we briefly introduce a new type of oscillatory angular dependence of magnetoresistance reported by Kajita *et al.*[31] Figure 10(a) shows the magnetoresistance of θ-(BEDT-TTF)$_2$I$_3$ as a function of angle (θ) between a constant magnetic field and the normal to the conducting plane. The angle-dependent oscillation of magnetoresistance was observed at temperatures below 6 K and in the magnetic field above 3 T. It was found that this oscillation have a periodicity in $\tan\theta$, as shown in Fig. 10(b), and the period depend neither on temperature nor on magnetic field strength. A similar angle-dependent magnetoresistance was observed in β-(BEDT-TTF)$_2$IBr$_2$ by Kartsovnik *et al.*[7] and in (BEDT-TTF)$_2$KHg(SCN)$_4$ by Osada *et al.*[11]

The most plausible mechanism of the angle-dependent oscillations is the following model proposed by Yamaji.[32] The energy spectrum of the quasi 2D system with a weakly warped cylindrical FS under magnetic fields is a set of Landau subbands with the dispersion along the field direction. When the magnetic field is tilted from the normal to the 2D plane, the width of the Landau subband around the Fermi level oscillates against the tilted angle. The oscillatory behavior in the width of the Landau subband correlates with the angle dependence of the distribution width of the area of semiclassical closed orbits under the magnetic field in the reciprocal space. For simplicity, let us assume the following band energy:[32]

$$\varepsilon_k = h^2(k_a{}^2+k_c{}^2)/2m_{ac} - 2t_b\cos(bk_b), \qquad (3)$$

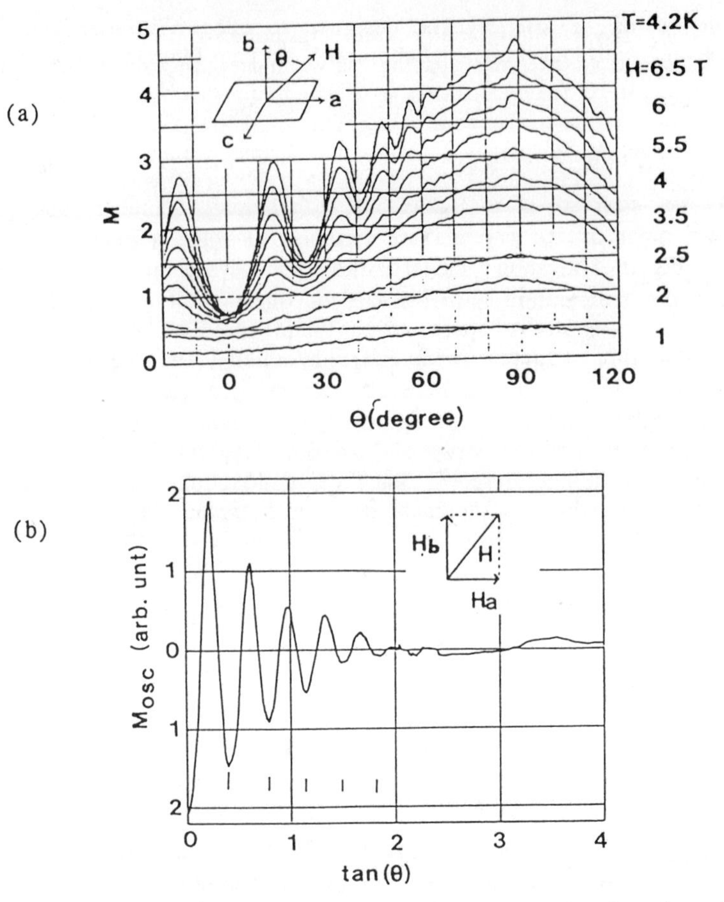

Fig. 10 (a)Magnetoresistance, M, of θ-(BEDT-TTF)$_2$I$_3$ plotted against the magnetic field direction θ and (b)the oscillatory part of the magnetoresistance, M_{osc} versus $\tan\theta$, after Kajita *et al.*[31]

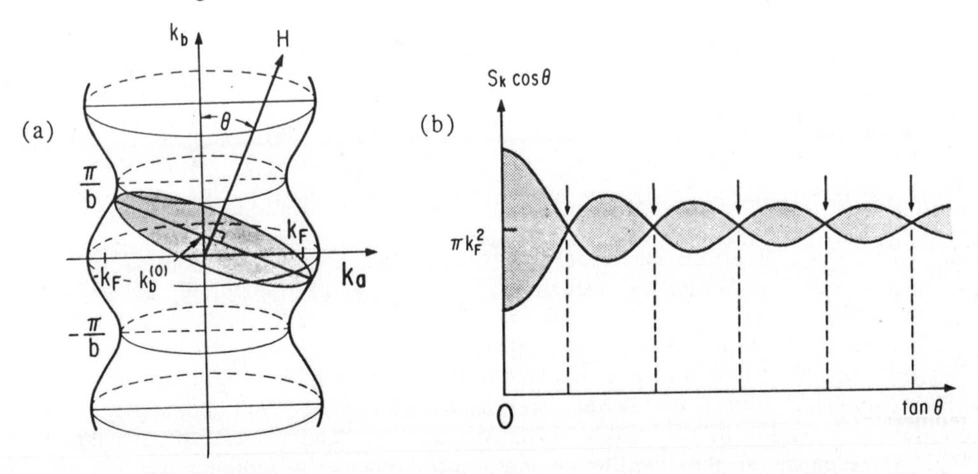

Fig. 11 (a)Semiclassical electron orbit on a weakly warped cylindrical FS perpendicular to the magnetic field and (b)The distribution of the area of closed orbit S_k when magnetic field is tilted by θ from the normal to the 2D plane, after Yamaji.[32]

where k_a and k_c are the conducting-plane components of the crystal wave vector $\boldsymbol{k}$, and b and k_b are the spacing between neighboring conducting planes and the component of the wave vector perpendicular to the planes. m_{ac} is the effective electron mass in the conducting sheet. Here an isotropic dispersion within the conducting ac-plane is assumed. Transfer energy t_b is assumed to be much smaller than the Fermi energy $\varepsilon_F = h^2 k_F{}^2 / 2m_{ac}$ where k_F is the Fermi wave vector for $t_b=0$. The corresponding Fermi surface in the extended Brillouin zone is shown in Fig. 11(a). The area of semiclassical electron orbit perpendicular to the magnetic field, intersecting the k_b-axis at $k_b(0)$, was calculated to the first order approximation of t_b/ε_F as

$$S_k(k_b)\cos\theta = \pi k_F{}^2 + 4\pi(m_{ac}/h^2)t_b\cos(bk_b)\, J_0(bk_b\tan\theta), \tag{4}$$

where J_0 is the Bessel function. Since $J_0(z) \approx (2/\pi z)^{1/2}\cos(z-\pi/4)$ for $z>1$, the $k_b(0)$-dependent term in eq. (4) vanishes periodically for magic angle θ satisfying

$$bk_b\tan\theta = \pi(n-1/4), \tag{5}$$

with integer n. Thus, for these values of magic angle, all the orbits on the Fermi surface have an identical value of orbital area, and the width of Landau subbands due to dispersion along k_b vanishes as shown in Fig. 11(b). In this situation, the extreme quantum limit applies in the b direction, while the quasi-classical regime for motion applies in the a-c plane.[37]

de Haas-van Alphen Effect

Fig. 12 shows magnetization of θ-(BEDT-TTF)$_2$I$_3$ as a function of magnetic field up to 23 T applied perpendicular to the conducting plane, i.e. $H // b$. Oscillations due to de Haas-van Alphen effect appear above 14 T at 0.5 K. We noticed that diamagnetic signal due to superconductivity was not observed at low magnetic field. This does not, however, exclude the possibility of superconductivity of the present sample by itself, since the sensitivity of our method vanishes as the magnetic field tend to zero. The recent study of low field magnetization and Meissner effect on θ-(BEDT-TTF)$_2$I$_3$ revealed that superconductivity of this material is a bulk effect in some samples, although the superconducting volume fractions of other crystals are significantly small.[38] Actually, the present sample was found to be one of the samples which belong to the latter category by a measurement using SQUID magnetometer.[38]

Figure 13 shows the Fourier-transformed spectrum of the data in Fig. 12. The Fourier spectrum shows the fundamental dHvA frequency of 4170 T and its higher harmonics (8340 T, 12510 T and 16680 T) up to the fourth. The observed fundamental frequency gives the Fermi surface cross section of 3.98×10^{15} cm^{-2}, amounting to 50.4% of the first BZ (7.89×10^{15} cm^{-2}) calculated on the basis of the orthorhombic lattice parameter at room temperature. If we assume an isotropic (i.e. circular) Fermi surface, this area leads to the Fermi wave number $k_F=0.356$ Å^{-1}. The present result is consistent with the picture of simple closed orbit for one carrier per unit cell as expected from stoichiometry. Apparently, it is consistent with the closed Fermi surface in Fig. 9(b) estimated by Tamura *et al.* from the reflectance spectra at 16 K based on the nearly free electron model and the tight binding

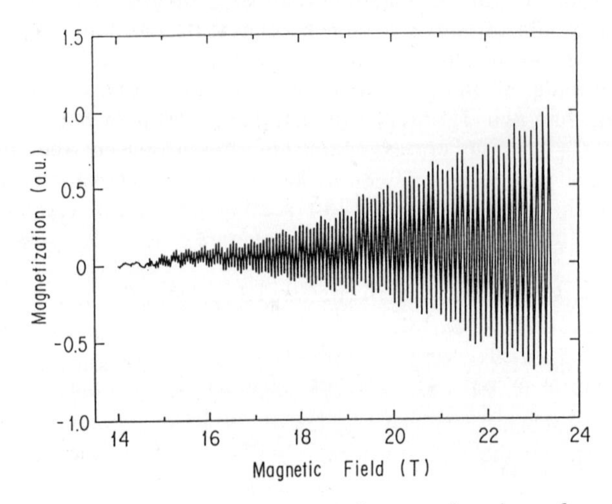

Fig. 12 Magnetization of θ-(BEDT-TTF)$_2$I$_3$ as a function of magnetic field, perpendicular to the 2D conducting plane, up to 23.4 T at T=0.5 K, after Tokumoto *et al.*[15]

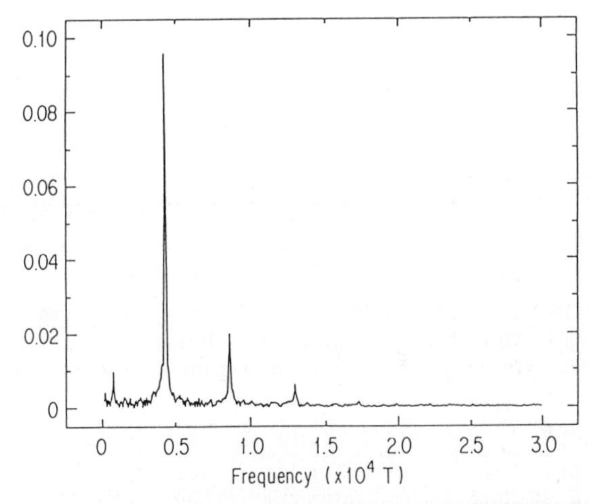

Fig. 13 Fourier-transformed spectrum of the data in Fig. 12, after Tokumoto *et al.*[15]

model,[33] but not so with the recent tight binding band structure calculation as shown in Fig. 9(c)[36] based on a new crystal structure in Fig. 8(b).[35]

In Fig. 13 we can see the presence of higher harmonics up to the fourth order. In the usual description of dH-vA oscillation for an isotropic Fermi surface, the amplitude of the pth harmonic is of the order of $(H/2pH_0)^{1/2}$ where H_0 is the fundamental field proportional to an extremal area of the Fermi surface cross section, $H_0 = A_0 h/2\pi e$.[18] The observed amplitude of higher harmonics is much larger, and is comparable to that observed by Kang *et al.* in the high-T_c state of β-(BEDT-TTF)$_2$I$_3$.[4] For an anisotropic system with different effective mass, m_b, along the field and $m_b << m_{ac}$, where m_{ac} is the effective electron mass in the conducting ac-plane perpendicular to the field and assumed to be isotropic within the conducting plane, the amplitude of the oscillations depends on the curvature of the Fermi surface and the harmonics are enhanced by the factor $(m_b/m_{ac})^{1/2}$.[4] It is already known, from superconducting critical field, H_{c2}, and conductivity anisotropies, that the ratio of the transfer integrals t_{ac}/t_b is of order of 30 in this compound.[39]

Figure 14 shows the details of the oscillation of magnetization as a function of inverse magnetic field, showing the periodicity and the "**saw-tooth**" variation. The "saw-tooth" variation of the magnetization of a 2D electron system at absolute zero is given by calculation of $-\partial U/\partial H$, where U is the total electronic energy in a magnetic field H.[18,40] The discontinuous change of the magnetization occurs as each successive Landau level passes through the Fermi level. It is remarkable that we can observe this "saw-tooth" variation of magnetization, considering the fact that we still have more than 170 Landau tubes inside the Fermi surface.

To our knowledge, this is the first time that such clear "saw-tooth" de Haas-van Alphen oscillation waveform, known as characteristic of 2D electron system, have been directly observed. There are many physical systems in which the electrons are constrained to move in two dimensions, including electrons held by image forces to the surface of liquid helium, various semiconductor device arrangements, such as the inversion layer in a silicon MOSFET[41] and heterostructure such as GaAs-Al$_x$Ga$_{1-x}$As[42,43] and graphite intercalation compound,[44] some of which do show quantum oscillations. In principle, magnetization of ideal 2D systems should give strong saw-tooth oscillations as a function of magnetic field, but in practice both amplitude and waveform can be easily lost by various phase smearing effects.[18] The present observation demonstrates two important facts. First, this organic system, θ-(BEDT-TTF)$_2$I$_3$, is highly two-dimensional with very small dispersion along the direction perpendicular to the conducting plane. Secondly, in organic metals, high purity of the sample can realize an ideal situation, in which phase smearing due to finite electron relaxation time as well as sample inhomogeneity is extraordinary small. And it is basically this saw-tooth variation which explains such ample higher harmonics as shown in Fig. 13, quantitatively.

Another interesting and important feature of our result is the presence of a peak at 730 T in Fig. 13. This frequency corresponds to a slow oscillation of small amplitude on which the fast oscillation of the fundamental frequency is superposed in Fig. 12. In general, presence of a slow oscillation suggests the existence of either a beat of two nearby frequencies around the fundamental field or a small pocket of much lower frequency. In the former case, the beating could be explained by a quasi-2D cylindrical Fermi surface with a small warping along the axial direction of the cylinder. The beating frequency

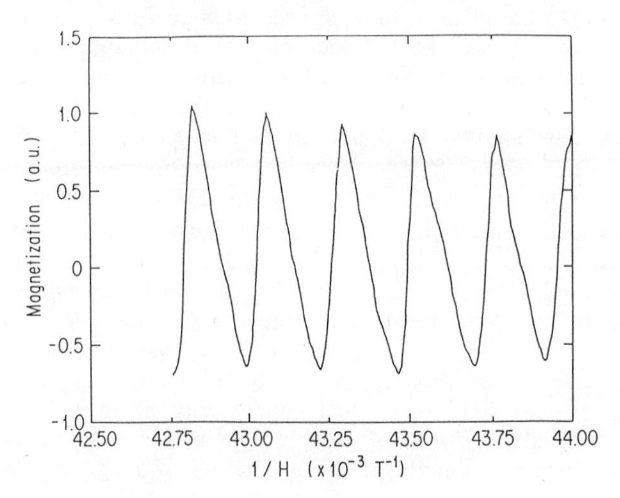

Fig. 14. Magnetization of θ-(BEDT-TTF)$_2$I$_3$ as a function of inverse magnetic field, after Tokumoto *et al.*[15]

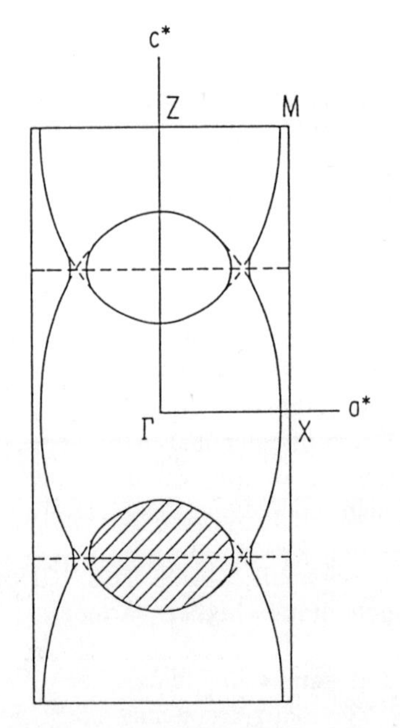

Fig. 15. Fermi surface topology consistent with two dH-vA oscillations at 730 T and 4170 T. Shaded area shows a hole pocket, corresponding to the former oscillation, while a large orbit across the gap corresponds to the latter one after magnetic breakdown above 14 T, after Tokumoto *et al.*[15]

gives us directly the amplitude of this warping. The area of orbit on the Fermi surface is obtained as

$$S_k(k_b)=\pi k_F{}^2+4\pi(m_{ac}/h^2)t_b\cos(bk_b). \tag{6}$$

The two extremal orbits give us two fundamental frequencies H_0+H_1 and H_0-H_1 where $H_1=2m_{ac}t_bh/e$, and the ratio H_0/H_1 directly measures the anisotropy of the transfer integrals:

$$H_0/H_1=h^2k_F{}^2/4m_{ac}t_b=k_F{}^2act_{ac}/2t_b \tag{7}$$

since $t_{ac}=h^2/2acm_{ac}$.
From the experimental values of the frequency ratio and relations $\pi k_F{}^2$ = 3.98×10^{15} cm^{-2}, a=10.076×10^{-8}cm, c=4.964×10^{-8}cm, we obtain $t_{ac}/t_b=2(H_0/H_1)/k_F{}^2ac=3.6$. This number is too small for this highly two-dimensional system. Therefore, it is not appropriate to assign this slow oscillation at 730 T to the beat of the fundamental dHvA oscillations. Actually, we notice some indication that this slow oscillation starts to appear at much lower field than the fundamental oscillations, indicating that it corresponds to a small pocket rather than a beat.

In the latter case, the slow frequency of 730 T corresponds to 8.8% of the first BZ(7.89×10^{15} cm^{-2}) calculated from the orthorhombic lattice parameter. This oscillation may correspond to a small pocket in the tight-binding band calculated by Kobayashi *et al.*[29,36] However the area of these pockets is much smaller than the observed value, i.e. the one around X in Fig. 9(a)[29] amounts to only 4%, and those around M or Z in Fig. 9(c)[36] correspond to about 3.5% and 3.2% of the first BZ (7.89×10^{15} cm^{-2}) for the average structure, respectively. Thus, the observed slow oscillation at 730 T cannot be explained by any of these pockets as calculated, suggesting that we need further modification.

Let us now consider the possibility of "**magnetic breakdown**". Assuming that the Fermi energy (ε_F) measured from the top of the conduction band is 0.2 eV, we can obtain the energy gap value of about 50 meV at the zone boundary from the band structure calculation shown in Fig. 9(c).[36] With the effective mass value: $m^*=3m_0$, we can get a rough estimate of the characteristic magnetic field for the magnetic breakdown as low as $H_0=m^*cE_g{}^2/h\varepsilon_Fe\sim 10$ T. This order of magnitude estimation suggests that magnetic breakdown is likely to occur in the present system in the measurement of the quantum oscillation above ~ 14 T. Then, combined with the observation of a single frequency at 730 T, we can almost uniquely construct a Fermi surface topology as shown in Fig. 15. Here, in order to enlarge the hole orbit (shaded area) up to 8.8% of the 1st BZ, a larger anisotropy of effective mass ratio (m^*_c/m^*_a=2.5), compared to the value (m^*_c/m^*_a=2.0) obtained by optical study[33] was employed. At higher fields, electrons do not see the presence of small energy gaps so that a connected large elliptic orbit may become dominantly detected.

On the other hand, the magnetotransport at low field, however, could be seriously affected by the presence of such small energy gaps. The magnetoresistance result[45] was considered as a "weak" field one and an estimation of the value of $\omega_c\tau$ (ω_c is the cyclotron frequency and τ is the carrier scattering time) was given as 0.06 which is much less than unity for T=1.4 K and H=9 T. If this estimation is correct, the value of $\omega_c\tau$ would be only 0.1 for T=0.5 K and H=15 T, since the resistivity saturates to a residual value below 1.4 K.[30] Then, the carrier scattering time of the present sample would be at least an order of magnitude longer than that of ref. 45 in order to satisfy the condition of $\omega_c\tau > 1$ at T=0.5 K and H=15 T. Detailed study of the temperature dependence of dHvA oscillation is under way, and its analysis in terms of the effective mass and Dingle temperature will be published separately.

CONCLUSION

We have presented some of the most attractive features observed in the recent Fermi surface study of two-dimensional organic metals $(BEDT\text{-}TTF)_2X$.

First, we reported the first direct observation of "spin-splitting" in the Shubnikov-de Haas oscillations in a quasi 2D organic conductor $(BEDT\text{-}TTF)_2KHg(SCN)_4$. The result suggests that organic metals can be anomalously clean system which enables such observation in a condition far from the quantum limit. The angular dependence of the splitting suggests that the g value anisotropy obtained by spin resonance may not be the same as the value deduced from SdH oscillations.

Secondly, we have presented the first observation of "saw-tooth" de Haas-van Alphen oscillations in a two-dimensional organic superconductor θ-$(BEDT\text{-}TTF)_2I_3$. The oscillation yields a fundamental frequency of 4170 T, corresponding to the closed Fermi surface area of 50% of the first Brillouin zone, as well as a slow oscillation at 730 T, corresponding to a pocket of 8.8% of BZ. The Fermi surface topology, consistent with both oscillations are proposed. The observation of the "saw-tooth" variation of magnetization demonstrates that this organic metal is a typical example of highly two-dimensional electron system with extraordinary high purity.

Although they have already provided very attractive physical properties, we must admit that many puzzling problems still remain to be solved. Since the Fermiology of organic metals started only two years ago, further elaborating studies including more sophisticated band structure calculations based on the crystal structures at very low temperatures would be required before we reach the goal of sufficient understanding of electronic structure of organic superconductors.

ACKNOWLEDGEMENTS

The present authors would like to thank Professor N. Miura of the Institute for Solid State Physics and Dr. T. Osada of Department of Pure and Applied Sciences, The University of Tokyo for kindly sending a preprint on $(BEDT\text{-}TTF)_2KHg(SCN)_4$ prior to publication. They also appreciate useful discussion with Professor G. Saito of Kyoto University, Dr. T. Mori of the Institute for Molecular Sciences, Dr. K. Yamaji of Electrotechnical Laboratory, Professors H. Kobayashi and K. Kajita of Toho University, and Professor N. Toyota and Mr. T. Sasaki of Institute for Materials Research, Tohoku University. They are indebted to Professor P. M. Chaikin of Princeton University for stimulating discussion.

They are grateful to Professor H. Kobayashi and Dr. R. Kato of Toho University, Dr. A. Kobayashi of The University of Tokyo and Professor K.

Yakushi of Institute for Molecular Science for their helpful advice on the preparation of the samples of θ-$(BEDT\text{-}TTF)_2I_3$ and encouragement.

They would like to thank the staff of the Francis Bitter National Magnet Laboratory (supported by the National Science Foundation) at the Massachusetts Institute of Technology for their assistance in these measurements. This work is supported by the National Science Foundation under grant DMR-88-18510. One of them (MT) acknowledges a substantial support from his colleagues at Electrotechnical Laboratory as well as the financial support of the Science and Technology Agency of Japan, which enabled his stay in Boston and accomplishment of this work.

REFERENCES

1 K. Oshima, T. Mori, H. Inokuchi, H. Urayama, H. Yamochi and G. Saito, Phys. Rev. B37, 938.(1988)
2 K. Murata, N. Toyota, Y. Honda, T. Sasaki, M. Tokumoto, H. Bando, H. Anzai, Y. Muto and T. Ishiguro, J. Phys. Soc. Jpn. 57, 1540.(1988)
3 N. Toyota, T. Sasaki, K. Murata, Y. Honda, M. Tokumoto, H. Bando, N. Kinoshita, H. Anzai, T. Ishiguro and Y. Muto, J. Phys. Soc. Jpn. 57, 2616.(1988).
4 W. Kang, G. Montambaux, J. R. Cooper, D. Jerome, P. Batail and C. Lenoir: Phys. Rev. Lett. 62, 2559.(1989)
5 M. V. Kartsovnik, P. A. Kononovich, V. N. Laukhin, S. I. Pesotskii and I. F. Schegolev: JETP Lett. 49, 519.(1988)
6 M. V. Kartsovnik, V. N. Laukhin V. I. Nizhankovskii and A. A. Ignat'ev, JETP Lett. 47, 363.(1988)
7 M. V. Kartsovnik, P. A. Kononovich, V. N. Laukhin and I. F. Schegolev: JETP Lett. 48, 541.(1988)
8 I. D. Parker, D. D. Pigram, R. H. Friend, M. Kurmoo and P. Day: Synth. Metals 27, A387.(1988)
9 F. L. Pratt, A. J. Fisher, H. Hayes, J. Singleton, S. J. R. M. Spermon, M. Kurmoo and P. Day: Phys. Rev. Lett. 61, 2721.(1988)
10 A. G. Swanson, J. S. Brooks, M. Tokumoto, A. Ugawa and K. Yakushi, these proceedings.
11 T. Osada, R. Yagi, A. Kawasumi, S. Kagoshima, N. Miura, M. Oshima, and G. Saito, Phys. Rev. B41, 5428.(1990)
12 M. Tokumoto, A. G. Swanson, J. S. Brooks, C. C. Agosta, S. T. Hannahs, N. Kinoshita, H. Anzai and J. R. Anderson, J. Phys. Soc. Jpn. 59, 2324 (1990).
13 T. Sasaki, N. Toyota, M. Tokumoto, N. Kinoshita and H. Anzai, to be published in Solid State Commun.
14 A. G. Swanson, J. S. Brooks, H. Anzai, N. Kinoshita, M. Tokumoto and K. Murata: Solid State Commun. 73, 353.(1990)
15 M. Tokumoto, A. G. Swanson, J. S. Brooks, M. Tamura, H. Tajima and H. Kuroda, to be published in Solid State Commun.
16 M. Oshima, H. Mori, G. Saito, and K. Oshima, Chem. Lett., 1159.(1989)
17 H. Mori, S. Tanaka, M. Oshima, G. Saito, T. Mori, Y. Maruyama and H. Inokuchi, Bull. Chem. Soc. Jpn. in press.
18 D. Shoenberg: *Magnetic Oscillations in Metals* (Cambridge Univ. Press, 1984).
19 J. R. Anderson, P. Heinmann, W. Bauer, R. Schipper and D. Stone: Inst. Phys. Conf. Ser. No. 39, p.81.(Inst. Phys. London, 1978)

20 W. J. O'Sullivan and J. E. Schirber, Phys. Rev. 162, 519 (1967).
21 R. W. Stark, Phys. Rev. 135, A1698 (1964); R. W. Stark and L. M. Falikov, *Progress in Low Temp. Phys.* 5, ed by C. J. Gorter, p235.(North-Holland, 1967)
22 M. Oshima, H. Mori, G. Saito, and K. Oshima, to be published in *Proc. 1st ISSP Symposium on the Physics and Chemistry of Organic Superconductors* (Springer-Verlag,1990)
23 N. Kinoshita, M. Tokumoto and H. Anzai, in preparation.
24 A. G. Swanson, M. Tokumoto, J. S. Brooks, C. C. Agosta, S. T. Hannahs, N. Kinoshita, H. Anzai, J. R. Anderson, in preparation.
25 T. Sasaki, N. Toyota, M. Tokumoto, N. Kinoshita and H. Anzai, to be published in Solid State Commun.
26 N. Kinoshita, M. Tokumoto and H. Anzai, submitted to J. Phys. Soc. Jpn.
27 H. H. Wang, K. D. Karlson, U. Geiser, W. K. Kwok, M. D. Vashon, J. E. Thompson, N. F. Larsen, G. D. McCabe, R. S. Hulscher and J. M. Williams, Physica C, 166, 57 (1990).
28 H. Kobayashi, R. Kato, A. Kobayashi, Y. Nishio, K. Kajita and W. Sasaki, Chem. Lett. 789 (1986).
29 H. Kobayashi, R. Kato, A. Kobayashi, Y. Nishio, K. Kajita and W. Sasaki, Chem. Lett. 833 (1986).
30 R. Kato, H. Kobayashi, A. Kobayashi, Y. Nishio, K. Kajita and W. Sasaki, Chem. Lett. 957 (1986).
31 K. Kajita, Y. Nishio, T. Takahashi, W. Sasaki, R. Kato, H. Kobayashi, A. Kobayashi, and Y. Iye, Solid State Commun., 70, 1189 (1989).
32 K. Yamaji, J. Phys. Soc. Jpn., 58, 1520 (1989); Yamaji, unpublished.
33 M. Tamura, K. Yakushi, H. Kuroda, A. Kobayashi, R. Kato and H. Kobayashi, J. Phys. Soc. Jpn. 57, 3239 (1988).
34 J. S. Brooks, M. J. Naughton, Y. P. Ma, P. M. Chaikin, and R. V. Chamberlin, Rev. Sci. Instrum. 58, 117 (1987).
35 A. Kobayashi, R. Kato, H. Kobayashi, S. Moriyama, Y. Nishio, K. Kajita and W. Sasaki, Chem. Lett., 2017 (1986)
36 H. Kobayashi, R. Kato, A. Kobayashi, T. Mori, H. Inokuchi, Y. Nishio, K. Kajita and W. Sasaki, Synthetic Metals, 27, A289 (1988); H. Kobayashi, unpublished.
37 P. M. Chaikin, T. Holstein and M. Ya. Azbel, Phil. Mag. B, 48, 457 (1983)
38 M. Tamura, H. Tajima, H. Kuroda and M. Tokumoto, J. Phys. Soc. Jpn. 59, 1753 (1990).
39 K. Kajita, Y. Nishio, S. Moriyama, W. Sasaki, R. Kato, H. Kobayashi and A. Kobayashi, Solid State Commun., 64, 1279 (1987).
40 C. Kittel, *Introduction to Solid State Physics* (John Wiley & Sons, 1986).
41 F. F. Fang and P. J. Stiles, Phys. Rev. B28, 6992 (1983)
42 J. P. Eisenstein, H. L. Stormer, V. Narayanamurti, A. Y. Cho, A. C. Gossard and C. W. Tu, Phys. Rev. Lett. 55, 875 (1985)
43 I. M. Templeton, J. Appl. Phys. 64, 3570 (1988)
44 R. S. Markiewicz, M. Meskoob and B. Maheswaran, Phys. Rev. B36, 7859 (1987)
45 K. Kajita, Y. Nishio, T. Takahashi, W. Sasaki, R. Kato, H. Kobayashi, and A. Kobayashi, Solid State Commun., 70, 1181 (1989).

TRANSPORT AND MAGNETIZATION STUDIES OF $\beta''(BEDT\text{-}TTF)_2AuBr_2$

A. G. Swanson and J. S. Brooks

Physics Department, Boston University, Boston, MA 02215, USA

M. Tokumoto

Electrotechnical Laboratory, Tsukuba, Ibaraki 305, Japan

A. Ugawa and K. Yakushi

Institute for Molecular Science, Myodai Ji-Cho, Okazaki 444, Japan

ABSTRACT

The de Haas van Alphen and Shubnikov de Haas frequencies of $\beta''(BEDT-TTF)_2AuBr_2$ indicate that the closed portion of the Fermi surface corresponds to 2.9% of the first Brillouin zone. The angular dependence of the data suggest a quasi three dimensional Fermi surface. The Hall measurements yield a hole like carrier concentration of $\sim 3 \times 10^{19}$ cm^{-3}, substantially less than the assumed carrier concentration of one carrier per unit cell.

INTRODUCTION

$\beta''(BEDT-TTF)_2AuBr_2$ is a layered organic metal, the structure of which contains stacks of the organic molecules separated by the stacks of $AuBr_2$ anions. The characteristic of the β'' phase is that the organic molecules within a stack are displaced relative to one another and the angle between the stacking axis and the plane of the (BEDT-TTF) molecules is 60°.[1] The band structure calculations yield a Fermi surface containing both open and closed portions.[2-3] The area of the closed portion is estimated to be ~ 2.9% of the first Brillouin zone, which is small compared to other compounds of $(BEDT\text{-}TTF)_2X$ family. Experimentally, the shape of a closed portion of the Fermi surface may be obtained by measuring the de Haas van Alphen (dHvA) and Shubnikov de Haas (SdH) frequencies.

The unit cell of $\beta''(BEDT-TTF)_2AuBr_2$ is triclinic with parameters a = 9.027, b = 16.2372, c = 5.712 Å, $\alpha = 97.60°$, $\beta = 102.94°$, $\gamma = 92.09°$, V = 813.7 Å^3, and z = 1 (Ref. 2). In this cell notation the conduction mainly occurs in the a-c plane, and the open planes of the Fermi surface are parallel to (001).

Organic Superconductivity, Edited by V.Z. Kresin and W.A. Little, Plenum Press, New York, 1990

The first SdH measurements on $\beta''(BEDT-TTF)_2AuBr_2$ were made by Pratt *et al.*[4] The authors reported complex SdH oscillations where four different frequencies were observed. These oscillations were observed in only one of their samples. In this paper we report both the dHvA and SdH frequencies of $\beta''(BEDT-TTF)_2AuBr_2$. Our results also indicate the presence of more than two frequencies, which can be explained in terms of the mixing of two frequencies. A hole like carrier concentration of $\sim 3.1 \times 10^{19} cm^{-3}$ is obtained from the Hall measurements.

EXPERIMENT

The crystals of $\beta''(BEDT-TTF)_2AuBr_2$ were synthesized at the Institute for Molecular Science, Okazaki, Japan. The crystals were roughly hexagonal in shape with a typical dimension of 2mm x 1mm x 0.2 mm. The magnetization measurements were performed with a small-sample torque magnetometer.[5] For transport measurements six gold pads were evaporated on the a-c plane of the crystal so that both magnetoresistance and Hall measurements could be performed on the same sample. The current direction was parallel to the c-axis. The resistance of the sample decreased monotonically from room temperaure down to 0.5 K. The room temperature resistivity of our sample was ~ 0.05 Ωcm and the resistivity at 60 mK was $\sim 10^{-6}$ Ωcm. The magnetization and the high temperature ($T > 1.2$ K) transport measurements were performed in a ^4He cryostat and the low temperature transport measurements were performed in a top loader dilution refrigerator in fields up to 23.9 tesla at the Francis Bitter National Magnet Laboratory in Cambridge, USA.

RESULTS: MAGNETIZATION AND MAGNETORESISTANCE

Fig. 1 shows the torque on the sample versus field at 2.3 K and 0.56 K. To the first order, the torque on the sample equals magnetization times field.[6] The 2.3 K curves are shifted up so that the difference between the curves may be seen clearly. The magnetic field was applied along the z-direction. At zero field the angle between the b-axis of the sample and the z-direction was $\sim 5°$. From the calibrations of our magnetometer we estimate that the change in angle due to torque never exceeded 2°. The tilt of the sample was such that a positive torque implied paramagnetic moment while a negative value of the torque implied diamagnetic moment. At both temperatures the oscillations are clearly visible on top of a smooth paramagnetic background. The size of the paramagnetic background is independent of temperature from 4.2 K down to 0.56 K.

The torque oscillations show paramagnetic jumps in the magnetization of the sample followed by smooth diamagnetic slopes. The torque measurements also show strong hysteresis, the size of which is field sweep rate independent. The size of the jumps and hysteresis increases with decreasing temperature. Similar paramagnetic jumps and hysteresis have been observed in another organic compound $(TMTSF)_2ClO_4$.[7] In this case the jumps are believed to represent field-induced phase transitions between different spin-density wave states. The spin density waves occur because of the one-dimensional nature of $(TMTSF)_2ClO_4$. The oscillations in $(TMTSF)_2ClO_4$ represent phase transitions, and their positions in field are temperature dependent. We rule out the possibility of spin-density waves in $\beta''(BEDT-TTF)_2AuBr_2$ because the positions of the oscillations are independent of temperature.

The hysteresis in magnetization and the paramagnetic jumps are probably due to the torque instability of the sample, an artifact of measurement technique. Shoenberg[8] points out that magnetic torques on a sample can produce perturbing effects and if the sample is not mounted rigidly, a torque instability may occur which changes the sample's angle of tilt discontinuously at a magnetic field. The detailed calculations[9] show that these torque instabilities lead to paramagnetic jumps with a hysteretic behavior of magnetization.

Fig. 2 shows the magnetoresistance at 70 mK and 400 mK. The 400 mK curves are shifted up in order to show the difference between the temperatures clearly. The direction of the magnetic field is perpendicular to the ac-plane. The oscillations are clearly observed. A slight hysteresis due to torque instability is observed up to 1.1 K.

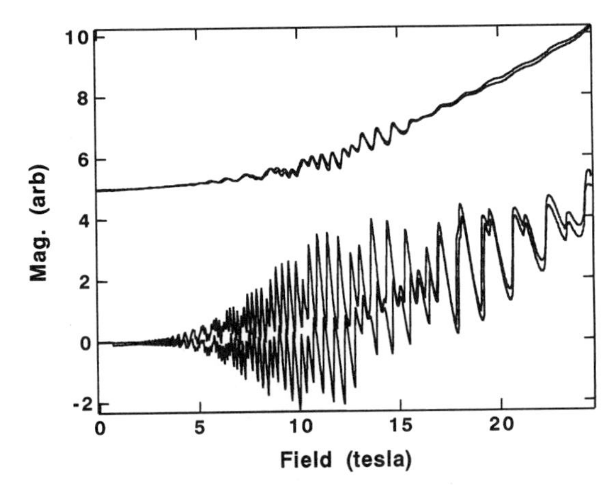

Figure 1. Magnetization vs. H for 0.56 K (bottom curve) and 2.3 K (top curve). The curves at 2.3 K are shifted up.

For transport measurements the sample can be mounted more rigidly, hence the size of the hysteresis loop decreases. Compared to the conventional metals this compound has a large magnetoresistance: at 23 tesla and 70 mK the resistance is within 60% of its room temperature value. The magnetoresistance has two components, one of which saturates at 16 tesla while the other one continues to rise. We expect that the non-saturating part of the resistance results from the open band in a-direction. Qualitiatively, our data matches the previously reported data,[4] although in Ref. 4 the saturating part of the resistance occurs at 5 tesla. This difference may be explained by the misalignment of the current pads.[10]

Fig. 3 and Fig. 4 show the Fourier transform of the magnetization and magnetoresistance data respectively. Both sets of measurements yield peaks in Fourier transform at 47, 99, 168, 218, 268 and 510 tesla $\pm$ 10 tesla, where the 99 and 510 tesla are the 1st harmonics of 47 and 268 tesla respectively. Also, we note that 168 and 218 tesla may be obtained from the linear combination of 47 and 268 tesla. The two independent frequencies obtained are 47 and 268 tesla. The 268 and 47 tesla frequency correspond to $\sim$ 2.9% and 0.6% of the First Brillouin zone respectively. Pratt *et al.* reported SdH

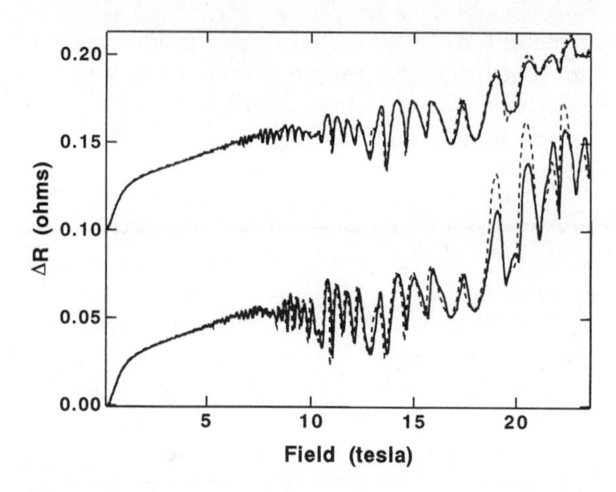

Figure 2. Magnetoresistance vs. H at 70 mK (bottom curve) and 400 mK (top curve). The solid lines represent field sweep up and the dashed lines represent field sweep down. The curves at 400 mK are shifted up.

frequencies of 140, 190, 220 tesla, and the presence of an unresolved slower oscillation. These frequencies differ from our frequencies by as much as 20% . The discrepancy between the measured frequencies may be due to misalignment of the samples with respect to the magnetic field. However, we estimate that the orientation of our samples could be off by a maximum of 5° which cannot explain the 20% difference.

The extremum area of the Fermi surface may be calculated by using a simple relation $S = \frac{2\pi e}{\hbar c}F$, where F is the fundamental frequency of dHvA or SdH oscillations. The following table summarizes the results of our SdH and dHvA frequencies. Here S_{BZ}, the area of the first Brillouin zone, is estimated to be $\sim$ 7.7×10^{15} cm^{-2}.

Table 1. Extremum areas

F_0(T)	S($\times 10^{14} cm^{-2}$)	S/S_{BZ}%
50	0.48	0.6
168	1.6	2.1
218	2.1	2.7
268	2.6	2.9

As noted above the 168 and 218 tesla frequencies may be obtained from the linear combination of 50 and 268 tesla frequencies.

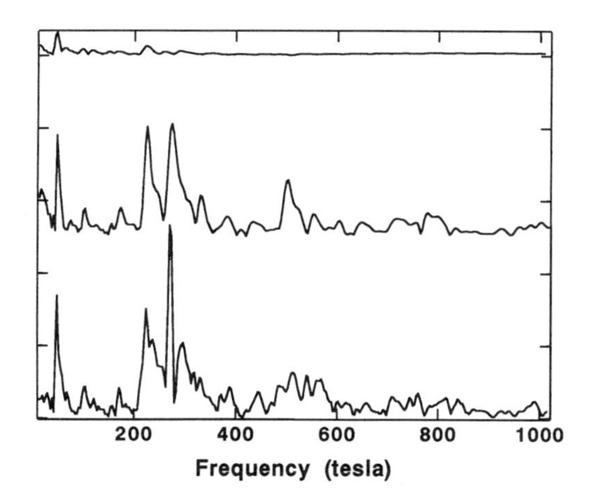

Figure 3. Fourier transform of the magnetization data at 3 K, 2.3 K, and 0.56 K. The top curve is for the highest temperature and the bottom curve is for the lowest temperature.

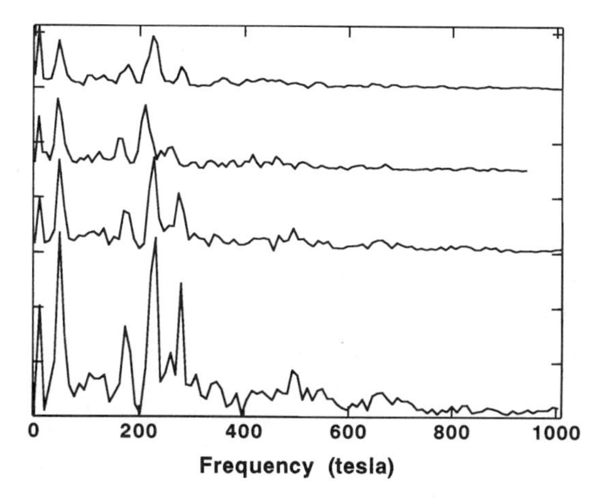

Figure 4. Fourier transform of the resistance data at 0.8 K, 0.36 K, 0.15 K, and 0.07 K. The top curve is for the highest temperature and the bottom curve is for the lowest temperature.

From the temperature dependence of the amplitude, we obtain an effective mass, $m = (0.5 \pm 0.1)\, m_o$ for the 50 tesla oscillations and $m = (1.9 \pm 0.5)\, m_o$ for 268 tesla oscillations. From the field dependence of the amplitude a Dingle temperature of $1.4\ K \pm 0.5\ K$ is obtained. These values of effective mass and Dingle temperature agree with the previously reported results.[4]

The band calculations indicate that the closed portion of the Fermi surface occupies 2.5% of the first Brillouin zone area which is close to the experimentally obtained value of 2.9% (F_0=268 tesla). However, the band calculations do not predict any small areas corresponding to the 50 tesla frequency. A beat like frequency, frequency $\leq$ 60 tesla, has previously been reported in many other $(BEDT\text{-}TTF)_2X$ compounds.[11] This beat frequency originates from the warping of the Fermi surface in z-direction. If we assume that the 50 tesla frequency is due to the warping of the closed cylindrical Fermi surface, the anisotropy ratios may then be estimated by:[12]

$$H_0/H_1 = k_F^2 act_{ac}/2t_b$$

where H_0 is the frequency corresponding to the cylindrical Fermi surface, 268 tesla, H_1 is the beat frequency, 50 tesla, and $\pi k_F^2 = 3.46 \times 10^{15} cm^{-2}$. These parameters lead to anisotropy ratio, $t_{ac}/t_b \sim 10$.

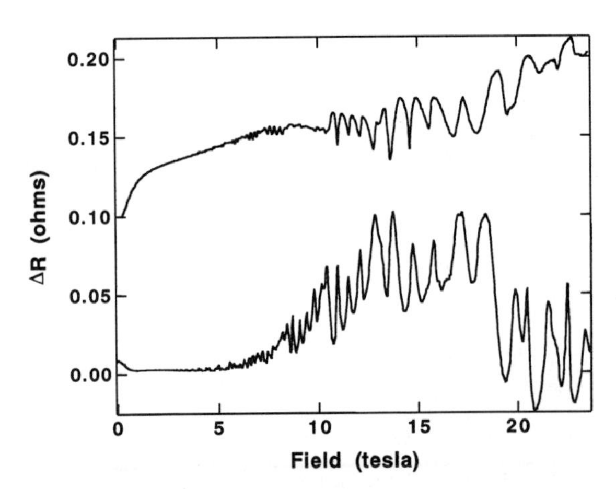

Figure 5. Magnetoresistance vs. H at 70 mK. The top curve is with H perpendicular to the conducting layers and the bottom curve is at $\theta = 25°$. The curve at 0° is shifted up.

Fig. 5 shows the magnetoresistance curve at 70 mK, the sample was rotated in the ab-plane with H at 25° to the normal of the ac-plane. Fig. 6 shows the Fourier transform of this data. The peaks of the Fourier transform for H at 25° are obtained at 46, 99, 220 and 268 $\pm$ 10 tesla. Our results indicate that therr is no change in SdH frequencies due to a change in angle of 25° within our experimental resolution. These

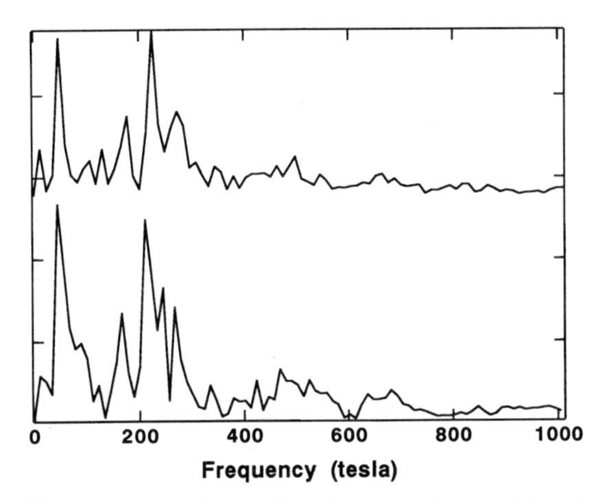

Figure 6. The top curve is the Fourier transform of the data with $H_\perp$ and the bottom curve is the Fourier transform of the data with H at 25° to the normal of the layers.

results show that the closed portion of the Fermi surface is quasi three dimensional. A lack of angular dependence of SdH frequency is not unique to $\beta''(BEDT-TTF)_2AuBr_2$ and has been previously observed in $\beta(BEDT\text{-}TTF)_2AuI_2$.[13]

RESULTS: HALL MEASUREMENTS

The Hall measurements were made by gold painting voltage contacts perpendicular to the current and measuring a potential drop across those contacts. Since the contacts were not perfectly aligned we measured the magnetoresistance simultaneously with Hall resistance. In order to obtain the Hall voltage the direction of H was reversed, taking care to keep the temperature constant. By reversing H magnetoresistance remains unchanged but the Hall voltage reverses sign, so the difference between the two voltages results in twice the Hall voltage, V_H. Most of the measurements were done using ac phase detection technique, however one set of measurements was made using a dc technique in order to determine the sign of the charge carriers. The Hall curves at high and low temperatures are shown in Fig. 7 and Fig. 8. From dc measurements, the Hall coefficient R_H, defined as 1/nec, is found to be positive. R_H is obtained from the slope of the Hall voltage vs. H, at low fields. This positive sign suggests the presence of hole like carriers at low fields as expected from the charge-transfer in this compound. The carrier concentration calculated from the relation, $V_H/H = \frac{1}{nec}$, n = 3.1×10^{19} cm^{-3}. This concentration is 2.6% of the total number of charge carriers calculated from the size of the crystals assuming one carrier per unit cell. It is interesting to note that the ratio of n obtained from the Hall voltage to the n estimated from one carrier per unit cell is equivalent to the ratio of the closed portion of the Fermi surface to the first Brillouin zone. At low temperatures the oscillations swamp the signal and it is hard to determine the overall shape of the curve. The frequencies of these oscillations match those of SdH. At high temperatures, the shape of the Hall voltage is not swamped by oscillations and V_H reaches a maximum value in the field, as shown

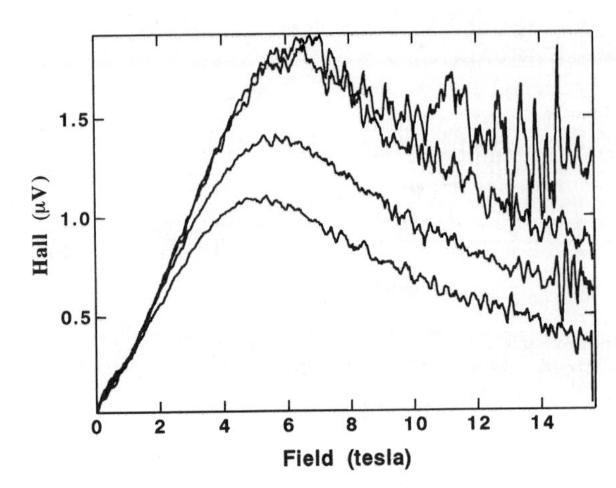

Figure 7. Hall voltage at 2.0, 2.5, 3.5, and 4.2 K. The top curve is at 2.0 K and the bottom curve is at 4.2 K. I = 0.1 mA.

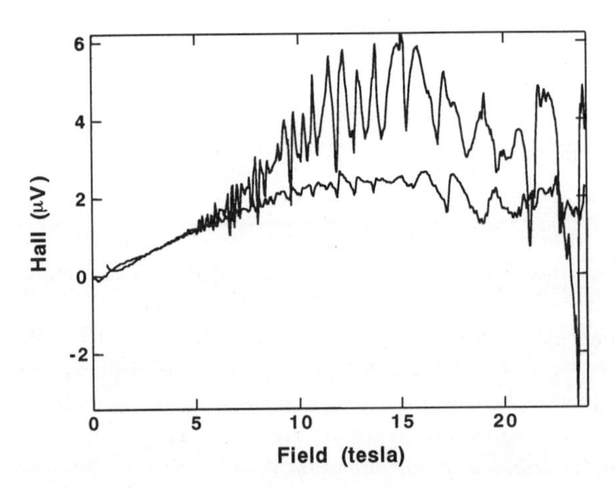

Figure 8. Hall voltage at 0.56, 0.92, and 1.2 K. The top curve is at 0.56 K and the bottom curve is at 1.20 K. I =0.5 mA.

in Fig. 7. This type of V_H is characteristic of a two band model with different number of electron and hole carriers of different mobilities.[14]

CONCLUSIONS

We have obtained the shape of the closed portion of the Fermi surface from the complicated SdH and dHvA frequencies. The closed portion of the Fermi surface corresponds to $\sim 2.9\%$ of the First Brillouin zone area and is warped along the b-direction with an anisotropy ratio of ~ 10. The angular dependence of SdH frequencies indicate the quasi three-dimensionality of the closed Fermi surface. The Hall measurements yield hole like carrier concentration of $3.1\times10^{19}\mathrm{cm}^{-3}$. The ratio of the carrier concentration obtained from the Hall measurements to the concentration estimated from one carrier per unit cell is similar to the ratio of the areas of the Fermi surface to the First Brillouin zone.

ACKNOWLEDGEMENTS

We are grateful to Dr. C. Agosta for his participation in some of the measurements. We thank Professor Robert Markiewicz for many useful discussions. We would like to thank the staff of Francis Bitter National Magnet Laboratory (supported by the National Science Foundation) at the Massachusetts Institute of technology for their assistance in these measurements. This work is supported by the National Science Foundation under grant DMR-88-18510. One of us (MT) acknowledges a substantial support from his colleagues at Electrotechnical Laboratory, including Dr. N. Kinoshita, Dr. H. Anzai and Dr. K. Tanaka as well as the financial support of the Science and Technology Agency of Japan, which enabled his stay in Boston.

REFERENCES

1. M. Kurmoo, D. R. Tolman, P. Day, I. D. Parker, R. R. Friend, A. M. Stringer and J. A. K. Howard, Solid State Commun., 61, 459 (1987).

2. T. Mori, F. Sakai, G. Saito and H. Inokuchi, Chem Lett., 1986, 1037 (1986).

3. K. Kajita, Y. Nishio, S. Moriyama, W. Sasaki, R. Koto, H. Kobayashi, and A. Kobayashi, Solid State Commun. 60, 811 (1986).

4. F. L. Pratt, A. J. Fisher, W. Hayes, J. Singleton, S.J.R.M. Spermon, M. Kurmoo, P. Day, Phys. Rev. Lett. 61, 2721 (1988).

5. J. S. Brooks, M. J. Naughton, Y. P. Ma, P. M. Chaikin, R. V. Chamberlin, Rev. Sci. Instrum. 58, 117 (1987).

6. J. D. Jackson, *Classical Electrodynamics*, John Wiley & Sons, Inc., (1975).

7. M. J. Naughton, J. S. Books, L. Y. Chaing, R. V. Chamberlin, P. M. Chaikin, Phys. Rev. Lett.55, 969 (1985).

8. D. Shoenberg, *Magnetic Oscillations in Metals*, Cambridge, (1984).

9. R. S. Markiewicz, Y. P. Ma, J. D. Hettinger and J. S. Brooks, unpublished.

10. A. B. Pippard, *Magnetoresistance in Metals*, Cambridge, University Press, (1989).

11. A. G. Swanson, J. S. Brooks, H. Anzai, N. Kinoshita, M. Tokumoto, K. Murata, Solid State Commun. 73, 353 (1990).

12. W. Kang, G. Montambaux, J. R. Cooper, D. Jérome, P.Batail, and C. Lenoir, Phys. Rev. Lett. 62, 2559 (1989).

13. I. D. Parker, D. D. Pigram, R. H. Friend, M. Kurmoo, and P. Day, Synthetic Metals 27, A387 (1988).

14. N. W. Ashcroft, N. D. Mermin, *Solid State Physics*, Saunders College, Philadelphia, (1976).

Magneto-resistance of $(TMTSF)_2ClO_4$ in Pulsed Magnetic Fields

Charles C. Agosta[a], S. Foner[a], J. S. Brooks[a,b],
W. G. Clark[c], and P. M. Chaikin[d]

[a]Francis Bitter National Magnet Lab, MIT
[b]Boston University
[c]University of California at Los Angeles
[d]Princeton University

Introduction

$(TMTSF)_2ClO_4$ was the first ambient pressure organic superconductor. Although its superconducting transition temperature has been surpassed many times, it remains an interesting compound because of its quasi one dimensional nature and its rich phase diagram[1] which includes superconductivity, metallic, field-induced spin density wave (FISDW) and re-entrant phases, as shown in Fig. 1. Our work has focussed on the magneto-resistance in the high field region of the phase diagram where it is suggested that $(TMTSF)_2ClO_4$ becomes re-entrant and we have paid special attention to the fast oscillations present in the data. To collect the data at the high fields necessary for this study we used a pulsed magnetic field with a novel high frequency lock-in amplifier. Our measurements reached a peak field of 41 tesla and temperatures down to 1.6K.

Experimental Details

The pulsed magnetic field was generated in a wire-wound multilayer copper conductor magnet with a 2.2cm I.D. When cooled to liquid nitrogen temperatures it gave a slightly damped sinusoidal pulse with a ~12ms pulse duration. A field of 41T was generated with 80kJ at 3kV. These pulsed fields are accompanied by the following drawbacks: there is a limited time to signal average the data due the short pulse length, a large background voltage is induced into the signal leads from the quickly changing magnetic flux, and by other noise created from switching, etc.

Organic Superconductivity, Edited by V.Z. Kresin and
W.A. Little, Plenum Press, New York, 1990

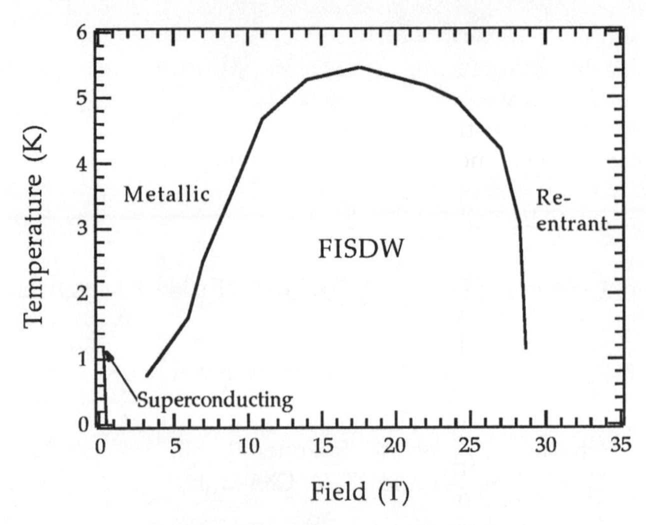

Fig. 1. The magnetic phase diagram of $(TMTSF)_2ClO_4$. (from Ref. 1).

Some of these extraneous effects can be attenuated by the use of AC detection where the signal is modulated on a carrier which allows us to use narrow band electronics optimized for a particular frequency. Unfortunately, off-the-shelf lock-in amplifiers do not have the output bandwidth necessary to resolve the details of our signal given the 12ms pulse length. Therefore, we built a lock-in tailored to our needs which was a modification of an NMR receiver developed at UCLA. A balanced transformer-coupled input helped attenuate common mode signals which may be large in the harsh magnetic and electrical environment of the pulsed field. After a balanced preamp stage the signal was fed to two double balanced mixers where it was mixed with the reference signal and a quadrature reference. This allowed us to monitor both the in-phase and out-of-phase signals simultaneously. The lock-in also included the circuitry to produce a balanced constant current between 10uA and 1600uA p-p which supplied the current to the sample. We operated the lock-in amplifier at 1.91MHz with a time constant between 3 and 100us using a standard four probe resistance measurement with a constant RF current. The RF lock-in system virtually eliminated the dB/dt background and allowed us to resolve a signal of 1uV with a 100us time constant.

One of the difficulties of working with $(TMTSF)_2ClO_4$ is that it has an ordering transition at 24K. To insure the compound is in the relaxed state it must be cooled through this transition temperature at a rate less than 0.1K/min., although this rate is not very well established. As other groups[2,3] have done we slow cooled over a ±10K region to guarantee a long annealing time at less than 0.1K per minute. This could take up to 14 hours so a new super-insulated Dewar, made mostly of G10 epoxy tube with detachable tails, was designed to maximize the helium hold time in the available space while avoiding the use of any conducting materials in the bore of the magnet. The new Dewar holds helium for about 24 hours if the tails are kept in liquid nitrogen as is the normal case since the pulsed magnet is run submerged in liquid nitrogen.

The experimental probe fit inside a thin glass Dewar insert which was needed to isolate the sample from the ^{4}He bath during the slow cool. Due to the magnet bore (2.2cm) we were limited to an I.D. of 1cm inside the plastic Dewar. In turn this limited the glass insert to an O.D. of 7.5mm and the probe on which the sample was mounted to a maximum diameter of 3mm. In addition to the sample, the probe held a heater and a carbon glass thermometer. These where connected to a analog integrating-differentiating temperature controller and a lock-in amplifier with a constant current source to form a temperature measurement and control loop. Together with a MacIntosh system running Labview we could type in any appropriate cooling rate and the system would accurately cool the sample while keeping track of both the temperature as a function of time and the sample resistance as a function of temperature. This system enabled us to cool at rates down to 2mK per minute and for up to 24 hours or more if the experiment could cope with some temperature fluctuations during a helium transfer.

Our sample was approximately 2mm X 0.7mm X 0.05mm. It was mounted on four 1mil gold wires with the conducting plane perpendicular to the field and had a resistance of 0.78Ω at room temperature. In a normal sequence the outer Dewar was cooled to 4.2K with liquid helium while the sample sat in vacuum and cooled only to 170K. A small amount of exchange gas was then added to the inner glass Dewar to bring the sample space down to 5-8K. With this amount of exchange gas the heater could be used to raise the temperature of the sample to 35-40K where the slow cool could start. After a 6-14 hour slow cool the glass Dewar was back filled with liquid helium to provide good thermal contact to the sample. Fig. 2 shows the even slow cool and a knee in the sample resistance at the ordering transition. If the sample was just in an exchange gas it would heat considerably during the field pulse. Therefore, in all the runs below the sample was immersed in liquid helium. The small amount of liquid helium in the glass insert could be pumped on subsequently to obtain temperatures down to 1.6K.

Results

Fig. 3 shows the magneto-resistance of the $(TMTSF)_2ClO_4$ at 4.2K up to 34.5T. Two breaks in the curve at 10.5T and 24.5 T show the transition to and from the FISDW and a rapid rise is seen in the re-entrant region. Superimposed on this are the rapid oscillations, which are most evident in the re-entrant region.

Fig. 4 shows data taken at 1.6K with the previous 4.2K data overlaid to show the change in magnitude of the resistance. Again sharp transitions are seen at the boundaries of the FISDW region (6.5 and 26.5T) and fast oscillations are evident especially in the re-entrant phase.

The FISDW state is fairly well understood although none of the current theories [4,5] can explain all of the features we see especially in regard to the shape of the re-entrant phase boundary. On the other hand the fast oscillations do not agree well with any theory and remain a mystery. One would think at first glance that they are Schubnikov-de Haas (SdH) oscillations but a number of anomalies differentiate them from normal closed orbit Fermi surface effects.

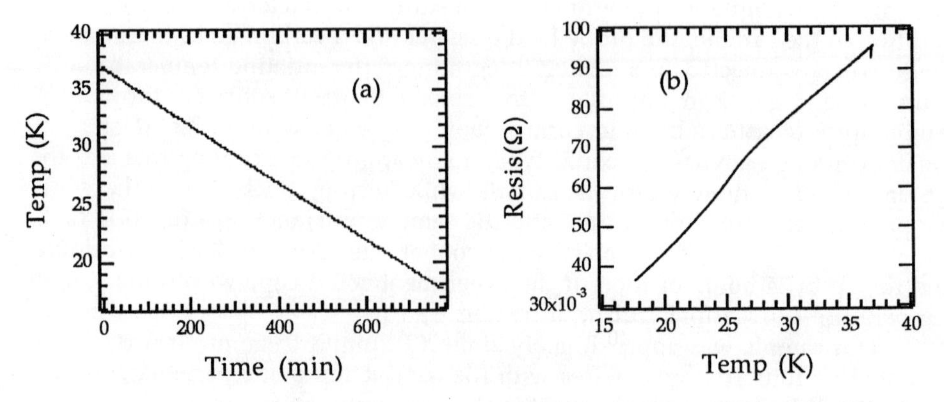

Fig. 2. The slow cool. (a) The even slow cool from 37K to 17K at 25mK/min. (b) The knee at ~25K where the ordering transition takes place.

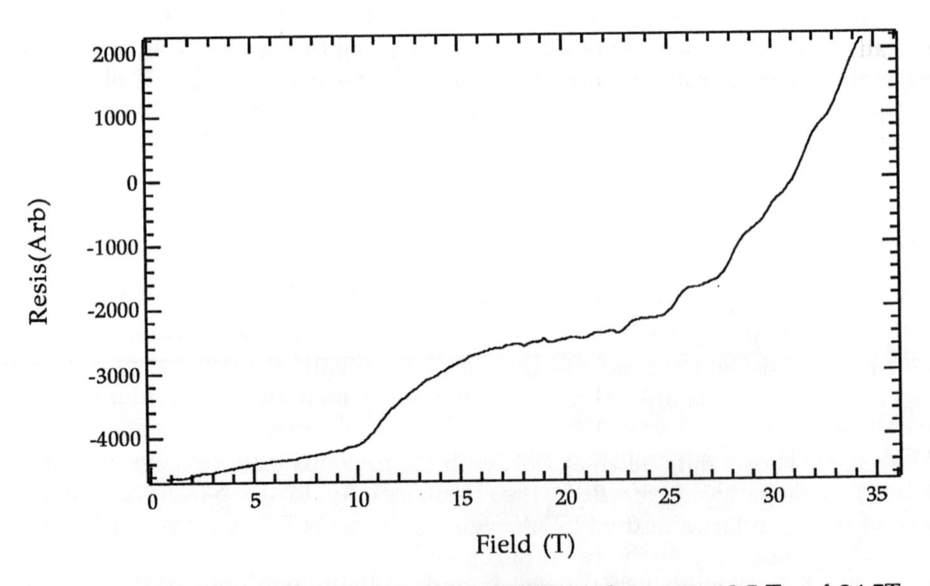

Fig. 3. Magneto-resistance at 4.2K. The FISDW is between 10.5 T and 24.5T. Rapid oscillations can be seen starting at about 20T.

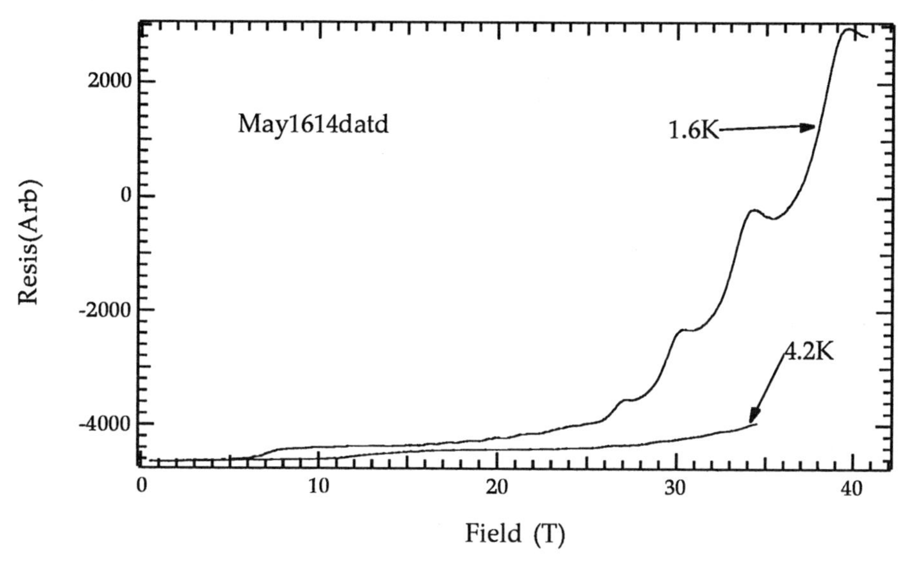

Fig. 4. The data at 1.6K with the previous 4.2K data superimposed. Notice the sharp rise in the magneto-resistance after the re-entrant transition at 26.5T and the large oscillations superimposed on the rise.

The most obvious argument against SdH is the lack of any closed orbit in the calculations of the $(TMTSF)_2ClO_4$ Fermi surface. It is in fact the quasi one dimensional nature of the Fermi surface that leads to the FISDW. Yet the oscillations do follow a 1/B periodicity, like SdH, and have been seen at a single frequency before, during and after the FISDW as the field is swept. In our data we see the oscillations at a single frequency through the FISDW state into the re-entrant phase but with a large increase in the amplitude as the phase boundary is crossed (see Fig. 5). This is consistent with the data of Osada et al[6] who show that unlike SdH these oscillations decrease in amplitude when the temperature is decreased below ~3K inside the FISDW state, while increasing in a normal fashion as the temperature is decreased outside the FISDW state. Contrary to other experiments though, Osada sees oscillations both at a frequency of 250T and 500T. In our data we only see the 250T frequency which corresponds to ~3% of the Fermi surface if they are SdH oscillations.

A subtle feature at ~15T, reported earlier[6], was also seen in our data. Shown in Fig. 6, it can be seen as a slight bend in the background of the magneto-resistance. It is observed in other measurements as a change in sign of the magnetization[7] and a minimum in the specific heat[8]. Its origin is not understood.

An additional feature is seen if we plot the peaks of the fast oscillations verses an arbitrary index (see Fig. 7). The graph shows a phase shift in the oscillations as the re-entrant boundary is crossed. This effect is not evident in Osada's data or in data taken by Naughton[9] up to 34T. Although this could be an anomaly of the instrumentation, previous data taken with the same system has been consistent with work done with Bitter magnets up to 31T. Furthermore, because of the high field at which the re-entrant phase begins previous experiments may not have had the base line necessary to see the phase shift clearly.

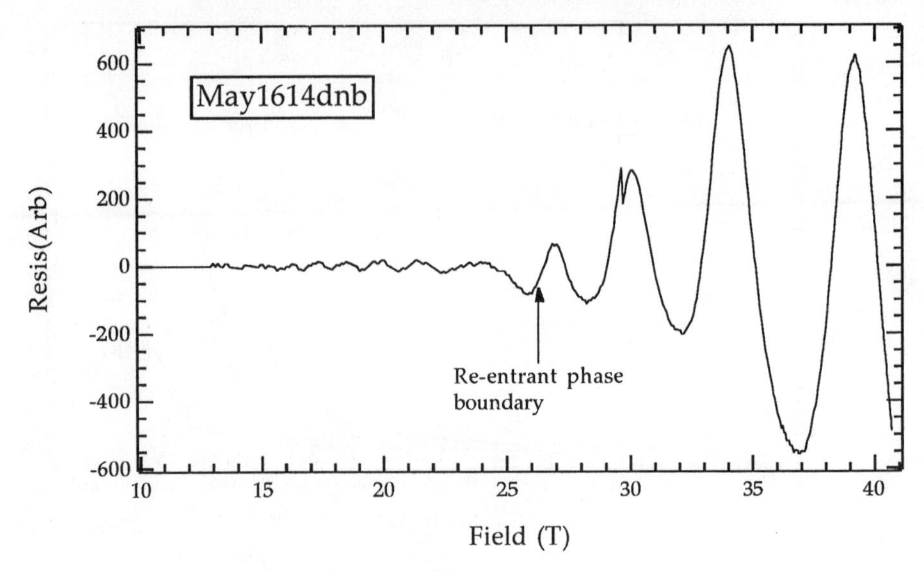

Fig. 5. The magneto-resistance at 1.6K with the background removed to show the oscillations. Notice the sharp rise in amplitude at the re-entrant transition.

If there were a closed orbit to produce the fast oscillations Chamberlin[10] has shown that the anomalous temperature dependence of the oscillations inside the FISDW state and the phase lag at the re-entrant line could be explained by competition for carriers between the quantum Hall transport seen in the FISDW state and the Fermi surface orbit. This would also explain the flat background of the magneto-resistance inside the FISDW phase, but more extensive data is needed to evaluate this theory.

Summary

We have built and tested an RF system for measuring magneto-resistance in pulsed fields. The RF detection system provides good noise attenuation even in the harsh environment of the pulsed magnetic field. Included in this apparatus is a convenient system for accurately slow cooling samples at a continuously variable rate down to 2mK/min.. In preliminary runs we have made measurements of $(TMTSF)_2ClO_4$ up to 41T and at 1.6K. Our data is consistent with the previous work on this compound at lower fields, and we have some evidence for a phase shift in the fast oscillations which has not been observed before. This phase shift was predicted by Chamberlin in a theory which also accounts for many of the other anomalies in the fast oscillations.

Our plans are to continue these measurements at a temperatures between 4.2K and 1.6K to map out the phase diagram more completely in the re-entrant region and eventually extend the range of temperature and field and also to examine the Hall effect.

One of the authors CCA acknowledges IBM for support. This work is supported by the NSF.

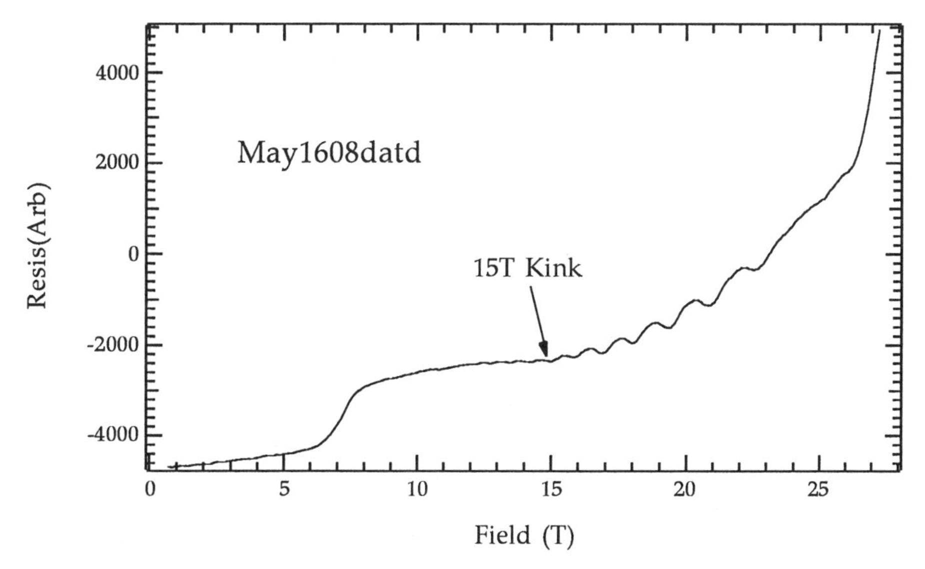

Fig. 6. A kink is seen in the magneto-resistance at 15T consistent with both magnetization data and specific heat data. Notice also the clear rapid oscillations here inside the FISDW state.

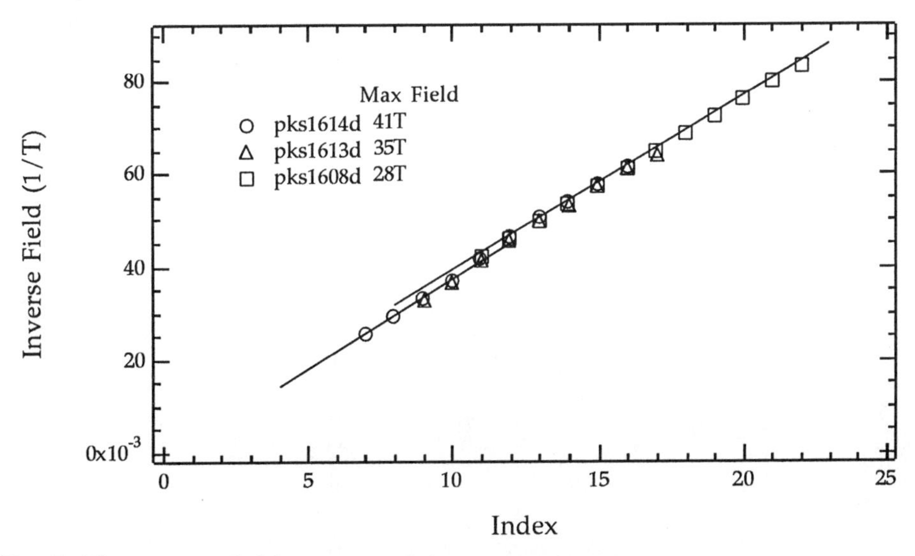

Fig. 7. The inverse field position of the peaks plotted verses an arbitrary index. The different symbols represent different runs with varying maximum fields and therefore show a consistency in the data despite the different pulses.

REFERENCES

1. G. Montambaux, M.J. Naughton, R. V. Chamberlin, X. Yan, P. M. Chaikin and M. Ya. Azbel , Phys. Rev. B, 39, 1, 885, 1989.

2. J. P. Ulmet, A. Khmou, P. Auban and L. Bachere, Solid State Comm., 58, 10, 753, 1986.

3. R. V. Chamberlin, M.J. Naughton, X. Yan, L. Y. Chiang, S. Y. Hsu and P. M. Chaikin, Phys. Rev. Let., 60, 12, 1189, 1988.

4. A. G. Lebed and Per Bak, Phys. Rev. B, 40, 16, 11433, 1989.

5. V. M. Yakovenko, Sov. Phys. JEPT, 66, 335, 1987.

6. T. Osada, N. Miura and G. Saito. Solid State Comm., 60, 5, 441, 1986.

7. X. Yan, M.J. Naughton, R. V. Chamberlin, S. Y. Hsu, L. Y. Chiang, J. S. Brooks and P. M. Chaikin, Phys. Rev. B, 36, 3, 1799, 1987.

8. N. A. Fortune, J. S. Brooks, M. J. Graf, G. Montambaux, L. Y. Chiang, P. M. Chaikin, Jos A. A. J. Perenboom, D. Althof, Phys. Rev. Let., 64, 2054, 1990.

9. M.J. Naughton, To be published.

10. R. V. Chamberlin, Private communication.

BAND STRUCTURE CALCULATION AND TUNNELING MEASUREMENTS IN $(BEDT\text{-}TTF)_2X$ ($X= I_3$, IAuI)

M. Weger*, A. Nowack+, D. Schweitzer@, J. Kubler#,
J. M. Van Bentum¢, and C.S. Sommers$

* Hebrew University, Jerusalem + Koln University

@ Stuttgart University # Technische Hochschulle Darmstadt

¢ Catholic University, Nijmegen $ Faculte des Sciences, Orsay

INTRODUCTION

At about the same time when Little proposed room-temperature superconductivity in organic polymers, it was suggested that the high-T_c of the more conventional A-15's is associated with their one-dimensional electronic band structure[1]. When TTF-TCNQ was discovered in 1973, it was suggested that the electron-phonon coupling in this 1-D organic molecular crystal is responsible for the metal-to-insulator (Peierls) transition at 52 K[2], and reducing λ will cause a crossover to a superconducting state[3]. Since then, the electronic structure, the conduction mechanism, and the superconducting mechanism were subject to controversy. Therefore, it is of some importance to establish whether the electronic band structure, and conduction mechanism, are similar to those in more conventional metals, and whether the superconductivity mechanism is the normal BCS phonon-mediated interaction

AN AB-INITIO BAND CALCULATION FOR $\beta\text{-}(BEDT\text{-}TTF)_2I_3$

The extended-Huckel method has been used extensively for the description of the electronic band structure of organic metals for more than a decade[4]. In this method, only one molecular orbital is considered - the highest occupied (HOMO) or lowest unoccupied (LUMO) one, and banding of this orbital is treated by the tight-binding approximation. Calculations of this type have been performed for ET salts since the work of Mori[5]. These calculations account for the 2-D band structure, the recent dHvA and Shubnikov data described by Tokumoto in this conference, as well as other phenomena described in the talks of Saito, Uemura, Ishiguro, Eldridge, Whangbo, and Kasowski.

As the organic molecule gets larger, the molecular levels get denser, and eventually their separation becomes less than the bandwidth (Fig. 1). As a result, different molecular levels overlap. This effect can be treated by the tight-binding approximation, as in the work of Whangbo et al (this conference), but also by other "standard" methods employed for calculating band structures of normal metals. We employed the ASA method of Williams[6] for the calculation of the band structure of $\beta\text{-}(BEDT\text{-}TTF)_2I_3$. Such a calculation is very difficult because of the large number of atoms per unit cell (55); the low symmetry ($P\bar{1}$) and particularly because the atomic spheres occupy only about 1/3 of the total volume.

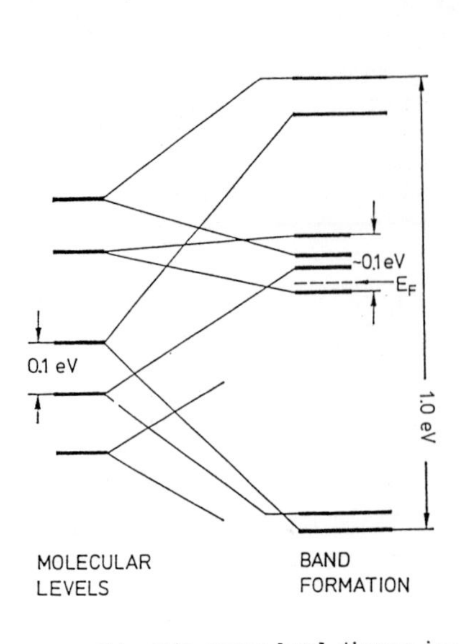

Schematic energy level diagram indicates how molecular levels broaden to form bands.

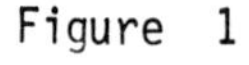

Figure 1

The separation between molecular levels is smaller than bandwidth.

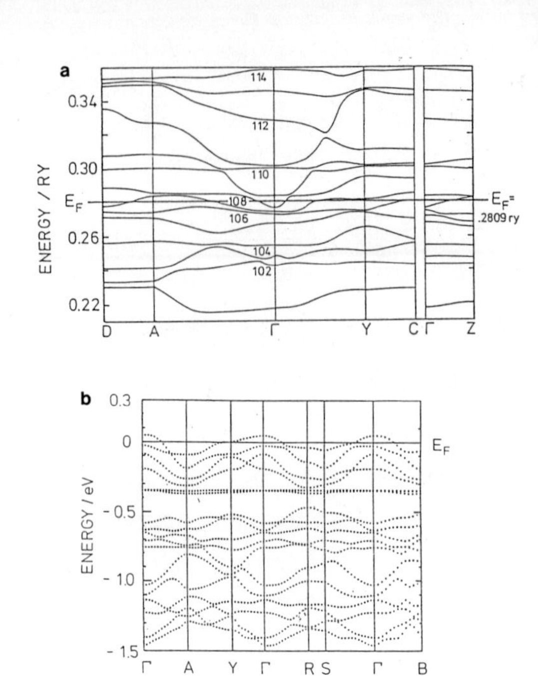

Figure 2

E vs. k curves for β-$(ET)_2I_3$. Top: IBM 3090. Bottom: Cray II.

We carried out some preliminary calculations; the volume between the atomic spheres is filled with "empty" spheres, so that the total volume of the atomic and empty spheres equals the volume of the unit cell. Thus, there is some overlap between the spheres. It turns out that when the positions, and radii, of the empty spheres are not precise, the calculation does not converge at all. We were able to obtain convergence after a meticulous placing and adjustment of the radii of the spheres[7]. Still, the calculation on an IBM 3090 takes a very long time, and could be carried out only on weekends and holidays. A significant improvement can be achieved by using a Cray II[8]. In Fig. 2 we show the band structure as calculated by the IBM (a) and Cray (b). In (a), the charges on all carbons and sulphurs are set to be equal, to reduce the size of the required CPU space and computing time. In (b), the charges on crystalographically (and chemically) inequivalent carbons are allowed to adjust themselves self-consistently. (b) gives the correct density of states, but not correct spin densities (as determined by NMR). Nor is the Fermi surface accurate, as determined by the dHvA method. Calculation (b) suggests a large energy gap about 50 meV above the Fermi level. To test for this gap, we performed a vacuum tunneling experiment in Nijmegen (Fig. 3). If there is a gap close to E_F, there should be rectification, as in ordinary semiconductors. The I-V curves are found to be precisely symmetric, (for positive and negative voltages) up to 2 V, thus we feel that there cannot be a gap close to E_F ; note that Fig. 2a does not show such a gap. Our conclusion is that the accuracy of our energy levels is not yet adequate; since the bands are so densely spaced, errors of $\sim$0.1 eV can distort the band structure very much. We are not yet in a position to say whether the ASA method is suitable to achieve the high accuracy required. We may note the similarity to "normal" metals like V_3Si, Nb_3Sn for which initial band calculations were rather inaccurate, and it took more than a decade to achieve adequate accuracy[9].

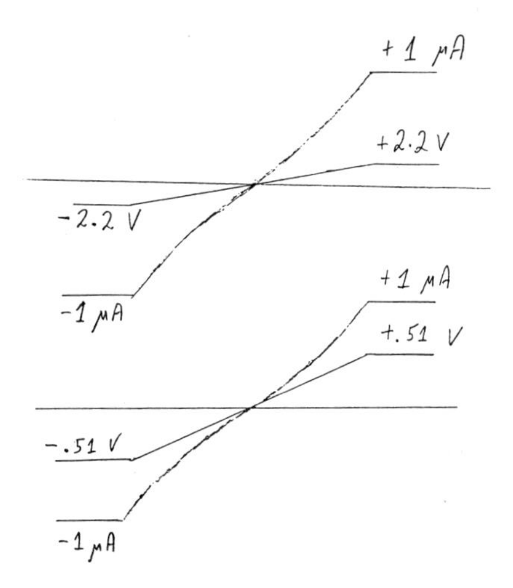

Figure 3
Vacuum Tunneling on β-ET. No gap is seen close to E_F.

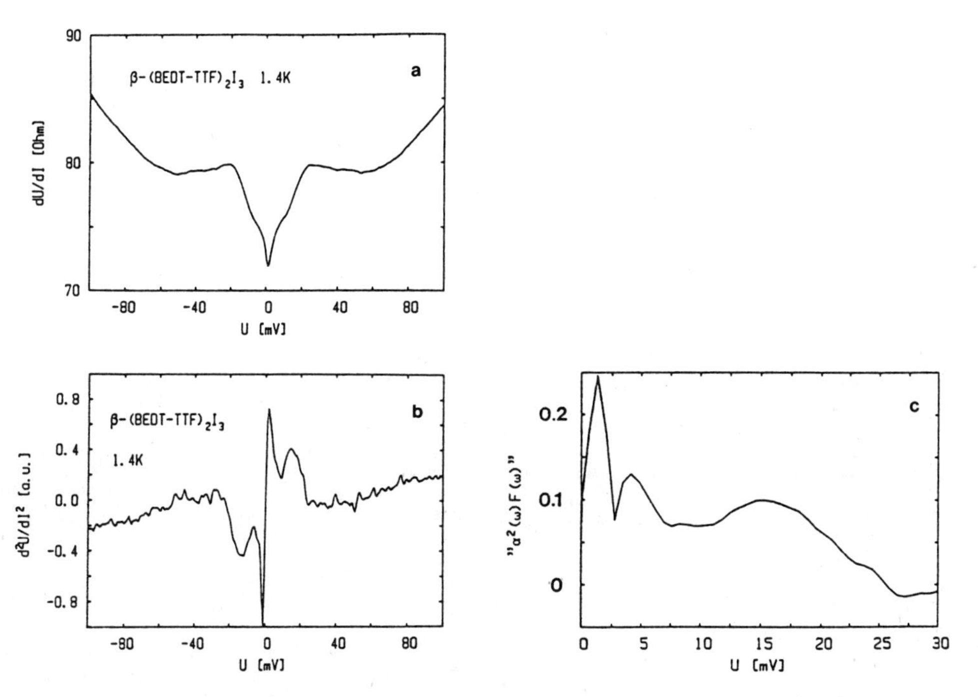

Figure 4
R & dR/dI vs. V curves for β-ET

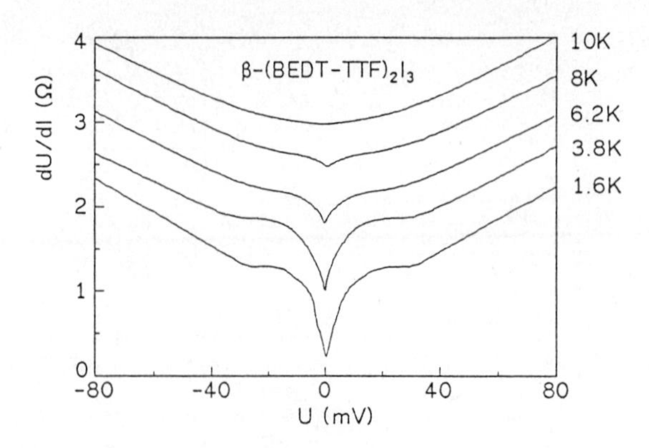

Figure 5
Rvs. V curves for β-ET at various Temperatures.

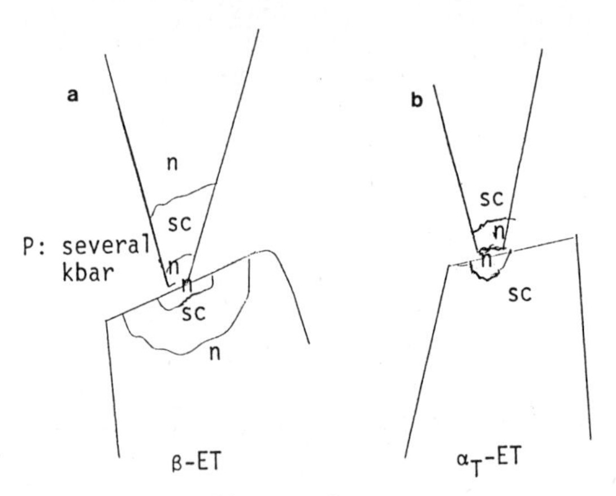

Figure 6
Configuration for point-contact spectroscopy. Contact pressure has a drastic effect.

In any case, we demonstrated that a band calculation by conventional methods is at all possible; moreover, the two-dimensional nature is apparent already at the present stage, and the shape of the Fermi surface is at least approximately correct.

POINT CONTACT SPECTROSCOPY AND TUNNELING - REVIEW

Point contact spectroscopy and tunneling measurements have been carried out on ET salts with I_3 and IAuI since 1986[10,11]. In Fig. 4 we show R vs. V and dR/dI vs. V curves for β-$(BEDT\text{-}TTF)_2I_3$. These curves were taken at a temperature above T_c of this material (T_c = 1.35 K) and therefore attributed to the normal state[11]. In Fig. 5 we show R vs. V curves up to 10 K[12]. It is seen that the structure disappears at 8 - 8.5 K, which is the maximum value of T_c under pressure. This behavior was verified on a large number of samples. Thus, we attribute the structure to the superconducting state under pressure. Since the crystals are pressed against each other, and the contact radius is about a few hundred Angstrom, even a force of a milligram gives rise to a pressure of a few kbar, as can be verified by the breaking of the surface area throughout the experiment. Thus, we have a structure as illustrated in Fig. 6. The pressure at the contact area is estimated to be a few kbar, therefore T_c is small and at about $T \simeq 1.5$ K the material is normal. At some distance from the contact area, the pressure falls to about 1-2 kbar, and T_c is about 8 K. At a larger distance, the pressure is small and T_c has the bulk value of about 1.35 K, thus for $T \simeq 1.5$ K the material is normal. Thus we have four n-s interfaces, and the voltage per interface is the overall voltage divided by 4. Thus, the sharp structure occurs at about 0.25 meV and a broad structure occurs at about 3.75-4 meV (Fig. 4). The presence of 4 interfaces causes considerable smearing, therefore it is difficult to draw definitive conclusions from these data.

It is possible to avoid the pressure at the interface by vacuum tunneling (or tunneling in helium gas). Measurements on ET IAuI by this method were performed by Hawley et al[10] and by us[11]. In Fig. 7a we show a dI/dV vs. V. curve reconstructed from the I-V curves of Hawley et al, for a crystal-to-gold needle interface. In Fig. 7b we show a dI/dV vs. V curve reported by us for a crystal-to-crystal interface. Fig. 7a suggests a gap of about 2.5 meV (as suggested by Hawley et al from an I^2 vs. V^2 curve). Fig. 7b suggests a gap of about 0.7 meV. At an early stage it was suggested that perhaps in the needle-to-crystal tunneling the needle touches a grain of the crystal, and the tunneling is between this grain and the bulk of the crystal, thus it is s-to-s tunneling and Δ has to be halved to 1.25 meV; or that in the crystal-to-crystal tunneling one surface is not superconducting, but normal, therefore Δ should be doubled (to about 1.4 meV). However, it seems that both data are good and such excuses are fortuitous. Moreover, the structure of Fig. 7a is measured between half-maximum points, while between the maxima the width corresponds to about $\Delta \simeq 4$ meV; while in the data of Fig. 7b, if we measure the distance between half-maximum points, the width corresponds to $\Delta \simeq 0.35$ meV. Thus, the difference between 7a and 7b is about a factor of 7 (for half-intensity points) or 6 (for the maxima of dI/dV), and such a large discrepancy cannot be accounted for by attributing it to faulty measurements.

We note that the shape of Fig. 7b agrees very well with the BTK theory[13], and the value of $\Delta \simeq 0.7$ meV corresponds very well to the BCS value for $T_c \simeq 4.1$ K ($2\Delta/k_BT_c \simeq 4$, which agrees with the BCS value to within the accuracy of the experiment). In contrast, Fig. 7a gives a value 4 to 6 times the BCS value, as pointed out already by Hawley et al. Values of $2\Delta/k_BT_c$ between 2 and 10 were reported in ref. 11 (Table I). There it was suggested that the vacuum tunneling data for ET-IAuI are the most trustworthy; however, on the basis of more experiments we feel that all the values reported in Table I are credible, and it is not possible to establish a unique value of $2\Delta/k_BT_c$.

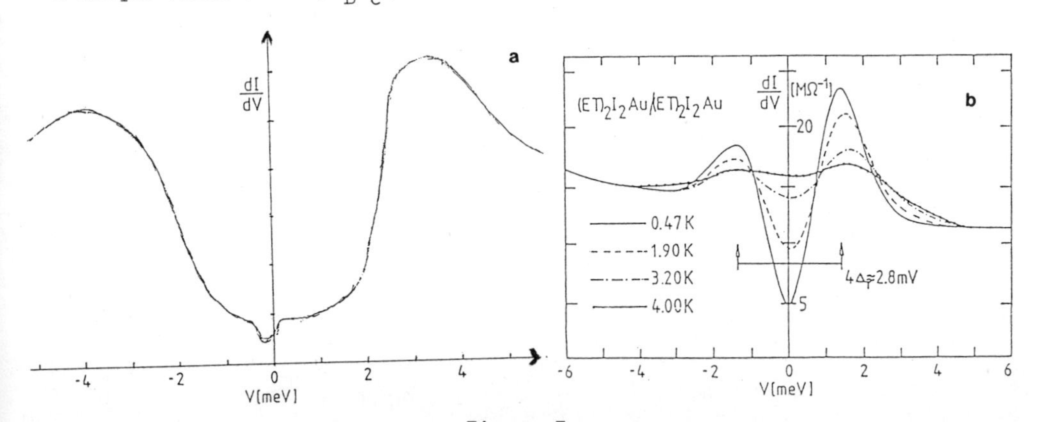

Figure 7
dI/dV vs. V tunneling curves for ET-IAuI.
Left: Hawley et al. (reconstructed). Right: Nowack, Poppe, et al.

POINT CONTACT SPECTROSCOPY - NEW DATA

To avoid the ambiguity of 4 n-s interfaces (Fig. 6a) we performed experiments on α_T-(BEDT-TTF)$_2I_3$, for which T_c = 8-8.5 K at ambient pressure; thus there are only 2 n-s interfaces (Fig. 6b). R vs. V and dR/dI vs. V data are shown in Fig. 8. The α_T crystals have a slight mosaic structure, but the angle between the individual crystalites is only a few degrees[14]. The crystals are pressed against each other with an orientation in which the current flows along the a-b planes.

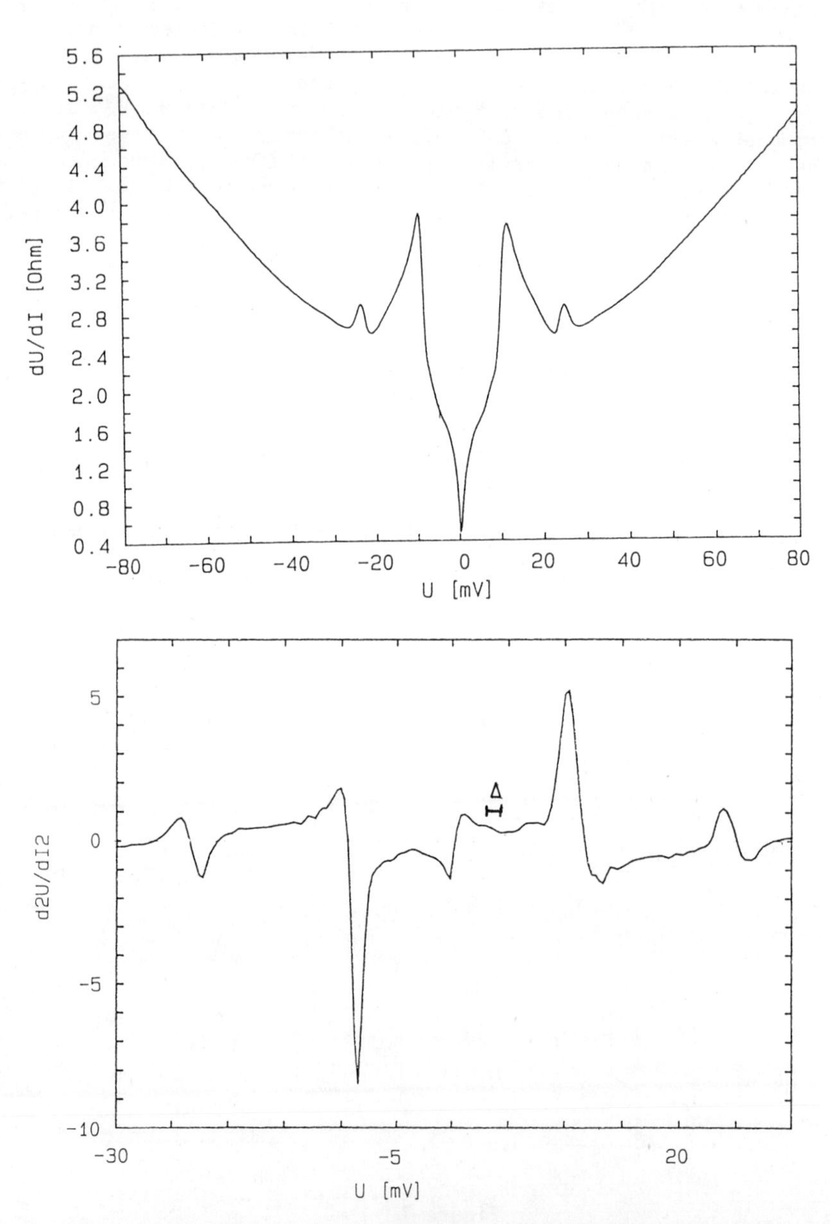

Figure 8
R vs. V and dR/dI vs. V curves for α_T-(ET)$_2I_3$ at T = 1.5 K.

Note that the R vs. V curve indicates a very sharp minimum at V=0; the width of this minimum is less than 0.5 mV. Near V = ±10 mV, there is a sharp rise in R ro a maximum, followed by a slight fall. Near ±20 mV, there is another sharp structure. Between these three structures, the curve of dR/dI vs. V is quite flat, without discernible structure.

We wish to point out that the data of Fig. 8 are DEFINITIVE. We performed measurements over a period of several years with more than a dozen crystals, with several measurements being taken for each pair of crystals. The data are reproducible, and did not change as the quality of the crystals improved over the last years. The very sharp rise of R with V near ±10 mV is by itself an indication of the high quality of the data. (Note in contrast Fig. 4, which shows how this structure is smeared out when the crystals are not so "good").

Since we have an s-n-s structure, the measured voltage is twice the voltage of each n-s interface; therefore the "giant" structure occurs at a voltage of about 5 mV across an n-s interface. We denote the structure near V=0 by $\delta^{(o)}$; the structure near 5 mV by $\delta^{(1)}$, and the structure near 10 mV by $\delta^{(2)}$. We denote the voltage at which R is maximum (for point contact spectroscopy) or minimum (for vacuum tunneling) by M; i.e. $\delta^{(o)}{}_M$, $\delta^{(1)}{}_M$, $\delta^{(2)}{}_M$. We denote the voltage at the half-maximum point by ½, i.e. $\delta^{(o)}{}_{1/2}$, $\delta^{(1)}{}_{1/2}$, $\delta^{(2)}{}_{1/2}$. This is nearly the same voltage as that when dR/dI is maximum, i.e. the second derivative vanishes, or the extrapolation of the I^2 vs. V^2 curves (Hawley et al[10]). The data for all samples, old and new, are summarised in Table I.

Table I

	T K	$\delta_M^{(o)}$	$\delta_{1/2}^{(o)}$	$\delta_M^{(1)}$	$\delta_{1/2}^{(1)}$ (mV)	$\delta_M^{(2)}$	$\delta_{1/2}^{(2)}$	T_c K
β-$(BEDT\text{-}TTF)_2I_3$ (point contact)	0.08	0.5	0.21					1.35
	1.4	0.5	0.3	5.5	3.75			?
$(BEDT\text{-}TTF)_2IAuI$ (Poppe)	0.47	0.7	0.35					4.1
(Hawley et al) (vacuum tunneling)	2.4			4	2.5			4.1
α_T-$(BEDT\text{-}TTF)_2I_3$ (point contact)	1.4		0.25	5.2	4.5	11.5	12.5	8.5

Note that all ET crystals (β-ET I_3, ET IAuI, α_TET) possess nearly the same value of δ. Also, the discrepancy between the Argonne and Julich vacuum tunneling data is accounted for - the Argonne group measured $\delta^{(1)}$, while the Julich group measured $\delta^{(o)}$.

Note that δ is independent of T_c; thus the quantity $2\delta/k_BT_c$ doesn't seem to be meaningful, in accord with the data of Table I of ref. 11.

It is instructive to compare this with the Raman data on ET crystals[15] (Fig. 9). The Raman data indicate a sharp line at about 4 meV. This line is extremely sharp for the α-phase, and slightly split for the α_T phase. Since Raman data are due to a q=0 mode, this is an optical phonon. Because of its low frequency, it is probably a libron. We suggested[11,7] that due to the 2-D band structure, a libron around the c-axis possesses a linear coupling with the electrons. The line at 15 meV is a stretching vibration of I_3, which probably does not interact too strongly with the BEDT-TTF molecule. The lines at higher energies are harmonics of this stretching mode. Thus, the phonon seen by the Raman experiments is amazingly close to the giant structure seen in the I-V curves at 5 mV.

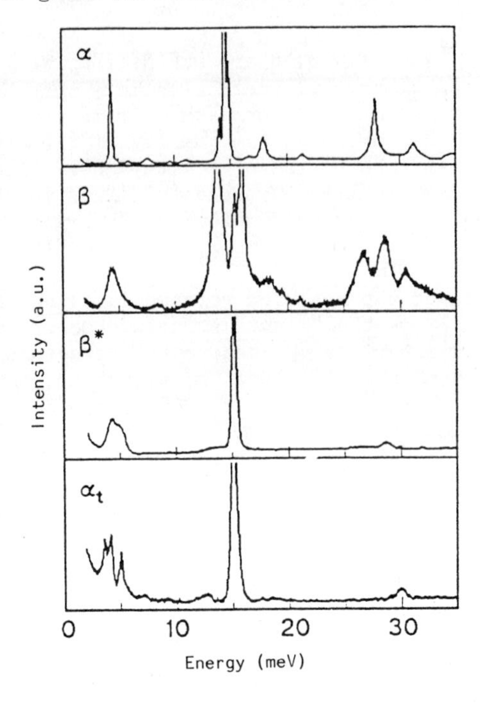

Figure 9

Raman data for various $(BEDT\text{-}TTF)_2I_3$ salts.

POSSIBLE INTERPRETATION OF THE I/V CURVES

(a) Attempt at a Conventional Interpretation

A conventional interpretation of the R vs. V data is based on the BTK theory[13]. This theory accounts for the maximum in R at $eV = \Delta$ (Δ being the superconducting gap parameter) for a good n-s interface, due to Andreev reflection, while for a "bad" interface (i.e. with an oxide, or other insulating layer) normal tunneling gives rise to a minimum in R at this voltage (following the McMillan-Rowell theory[16]). Harmonics of the gap are frequently seen in High-T_c superconductors as well[17].

The minimum in R at zero voltage can be attributed to a Josephson effect, or more likely to a proximity effect induced in the normal region by the superconductor across the interface. Thus, for this interpretation, $\delta^{(1)} \equiv \Delta$.

This interpretation poses the following dilemmas:
(1) Δ is independent of T_c. For all superconductors, with T_c ranging from 1.35 K (possibly) to 8.5 K, Δ is about 4-5 meV.
(2) $2\Delta/k_BT_c$ is unreasonably large. This was already pointed out by Hawley et al[10]. We tried to account for this large value (about 4 to 6 times the BCS value) by means of soft phonons, with the aid of D. Rainer, using the Bergmann-Rainer algorithm[18], without success. We tried a soft phonon with an energy of 1 meV, with a huge coupling to the electrons[11], and were not able to get even close to these values of $2\Delta/k_BT_c$.
(3) We have to assume that the very close coincidence between $\delta^{(1)}$ and ω_{ph} is a pure accident. In superconductivity theory there is <u>no</u> tendency whatsoever of Δ to "stick" to ω_{ph}[19].

We may try a two band model, like that of Kresin (this conference), where $\delta^{(o)}$ represents the gap of one band, and $\delta^{(1)}$ of the other. However, the differences (a factor of 16) seem to be too large for such a model.

(b) Unconventional Interpretation

We may assume that the close agreement between $\delta^{(1)}$ and ω_{ph} is not an accident, i.e.

$$\delta^{(1)} \equiv \omega_{ph}.$$

Since $\delta^{(1)}{}_M$ is nearly temperature independent, while $\delta^{(1)}{}_{\frac{1}{2}}$ falls with increasing temperature (and can be roughly fitted to the temperature dependence of a BCS gap, as pointed out by Hawley et al), it makes more sense to identify $\delta^{(1)}{}_M$ with ω_{ph}.

Thus, there is <u>no</u> structure in the I/V curves at the BCS gap Δ (about 1.4 - 1.6 meV); the region of possible values of Δ for weak and strong coupling is indicated in Fig. 8b , and it is seen that there is no structure there. (A similar absence of IR absorption at the gap energy was reported by Eldridge et al at this conference).

This interpretation poses the following dilemmas:

(1) It is not consistent with the conventional McMillan-Rowell theory. According to it, the "Big" structure is at $eV/2 = \Delta$, as discovered by Giaever.

(2) T_c is unexpectedly large for $\omega_{ph} \simeq 4$ meV. Even if μ^* is very small, a value of $T_c = 8.5$ K or up to $T_c = 12.5$ K (for some ET salts, as discussed by Williams at this conference), requires <u>a very large</u> value of λ. According to Kresin et al[20], $T_c = 0.18 <\omega^2>^{\frac{1}{2}} \sqrt{\lambda}/(1+2.6\mu^*)$.

For $\mu^* = 0.2$, $\omega = 4$ meV, $T_c = 12.5$ K, we need: $\lambda \simeq 3.5$.

At first sight, these dilemmas appear overwhelming. However, a novel theory, suggested originally for the high T_c superconductors, gives $\Delta(\varepsilon)$ vs. ε curves strikingly similar to the observed dR/dI vs. V curve[21]. We sketch such a theoretical curve in Fig. 10. This theory is based on a novel solution of the Migdal-Eliashberg ladder diagram equation for the case of an Einstein spectrum[22].

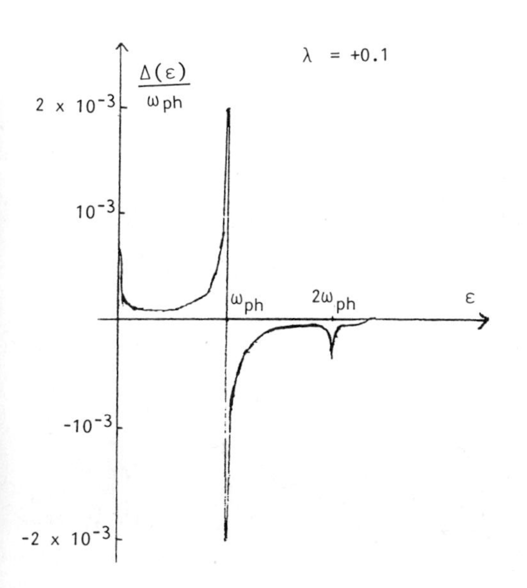

Figure 10

A novel calculation of the gap function $\Delta(\varepsilon)$ vs. ε for an Einstein spectrum. There is structure at $\varepsilon \simeq 0$, ω_{ph} , $2\,\omega_{ph}$ but not at $\varepsilon = \Delta(0)$.

In any case, the fact that $\delta^{(1)}$ and ω_{ph} coincide (at least approximately), and there is virtually no structure in the I/V curves at other energies, is a strong indication that the superconductivity is due to phonons, and specifically the 4 meV phonon. This is in accord with the prediction made in 1973 when TTF-TCNQ was discovered[3], and subsequent work[23]. This is also in line with the work of Whangbo et al, and Ishiguro, presented at this conference, about the role of the inter-molecular phonons, their softness, and T_c .

REFERENCES

1) M. Weger, Rev. Mod. Phys. 36(1964)175 ; Solid State Physics 28(1973)1.
2) J.P. Ferraris, D.O. Cowan, V.V. Walatka, J.E. Perlstein, JACS 95(1973) 948; L.B. Coleman, M.J. Cohen, D.J. Sandeman, F.G. Yamagishi, A.F. Garito, A.J. Heeger, Solid State Com. 12(1973)1125.
3) H. Gutfreund, B. Horovitz, M. Weger, J. Phys. C.7(1974)383; Phys. Rev. B12(1975)3174.
4) A.J. Berlinsky, J.F. Carolan, L. Weiler Solid State Com. 15(1974)795; M. Weger, Solid State Comm. 19(1976)1149.
5) T. Mori, A. Kobayashi, Y. Sasaki, H. Kobayashi, G. Saito, H. Inokuchi, Chem. Lett. 1984,957.
6) A.R. Williams, J. Kubler, C.D. Gelatt Phys. Rev. B19(1979)6094.
7) J. Kubler, M. Weger, C.B. Sommers, Solid State Comm. 62(1987)801.
8) J. Kubler & C.B. Sommers, Proc. ICSM Tokyo 1989.
9) T. Jarlborg, A.A. Manuel, M. Peter, Phys. Rev. B27(1983)4210.
10) M.E. Hawley, K.E. Gray, B.D. Terris, H.H. Wang, K.D. Carlson, J.M. Williams, Phys. Rev. Lett. 57(1986)629.
11) A. Nowack, M. Weger, D. Schweitzer, H.J. Keller, Solid State Comm. 60 (1986)199; A. Nowack, U. Poppe, M. Weger, D. Schweitzer, H. Schwenck, Z. Phys. B 68(1987)41.
12) A. Nowack, Thesis, Koln University, 1990.
13) G.E. Blonder, M. Tinkham, T.M. Klopwijk, Phys. Rev. B 25(1982)4515.
14) D. Schweitzer, Proc. ICSM Tokyo 1989.
15) R. Swietlik, D. Schweitzer, H.J. Keller, Phys. Rev. B36(1987)6881.
16) W.L. McMillan, J.M. Rowell, in "Superconductivity", R.D. Parks Ed. Marcel Dekker 1969.
17) A.I. Akimenko, N.M. Ponomarenko, V.A. Gudimenko, I.K. Yanson, P. Samueli, P. Kus, Sov. J. Low Temp. Phys. 12(1989)
18) G. Bergmann & D. Rainer, Z. Phys. 263(1973)59.
19) P. Allen & R. Dynes, Phys. Rev. B12(1975)905.
20) V. Kresin, H. Gutfreund, W.A. Little, Solid State Comm. 51(1984)339.
21) M. Weger, R. Englman, B. Halperin, NATO ASI on HTSC, Bad Windsheim, Aug. 1989, R. Kossowsky & S. Methfessel Edts; in "Recent Prog. in Many Body Theory", Y. Avishai Ed. Plenum Press 1990.
22) R. Englman, M. Weger, B. Halperin, Physica C162-164(1989)1339;Solid State Comm. 71(1989)17.
23) B. Horovitz, H. Gutfreund, M. Weger, Mol. Cryst. Liq. Cryst. 79(1982) 235.

PHASE DIAGRAM OF ANISOTROPIC TWO–DIMENSIONAL QUARTER–FILLED BAND OF INTERACTING ELECTRONS

S. Mazumdar[(1)], H. Q. Lin[(2)], and D. K. Campbell[(2)]

[(1)]Department of Physics, University of Arizona
Tucson, AZ 85721, USA

[(2)]CNLS, Los Alamos National Lab.
Los Alamos, NM 87545, USA

I. INTRODUCTION

Since its discovery(1), the magnetic field induced spin density wave (FISDW) transition observed in superconducting TMTSF salts has continued to attract considerable interest. This phenomenon has been studied most widely in the ambient pressure superconductor $(TMTSF)_2ClO_4$, where for moderate magnetic fields $H > H_f$(~5 Tesla) a cascade of transitions to semimetallic SDW phases was observed(2). Until very recently, it was generally accepted that the theoretical explanation for the FISDW lay in the gradual one dimensionalization of the Fermi surface in the presence of the magnetic field and the accompanying nesting instability(3-7). These theories do not account for the more recently discovered very high field transition (VFHT) for $H > H_v$. In contrast to the predictions of the above theories the SDW transition temperature T_{SDW} actually decreases for H > 17T and then goes to zero at H = 25–30T(8,9). At present, there exists only one theoretical attempt to explain the VFHT(10). Within this model(10) the material is at the one dimensional limit for very high fields, and the competition between the equally strong Peierls and Cooper channels in this limit destroys the FISDW. The system is then an *ordinary* metal for $H > H_v$.

We believe that the very high field phase is *not* an ordinary metal. Experimentally, it is characterized by a high resistivity, (weakly) activated conductivity and vanishing Hall signal(9). All of these suggest that the system is a semimetal, or perhaps even a narrow gap semiconductor. In addition the system is diamagnetic in this region, which also suggests that its nature is different from simple metallic.

In the present paper we present preliminary work on a new theoretical approach. We believe that the model presented here provides a better description of charge transfer solids in general, and FISDW behavior in particular, better. In contrast to much of the existing theoretical work (see, however, reference 5), we emphasize the direct Coulomb interaction between the carrier holes. More importantly, we consider a spatial broken

Organic Superconductivity, Edited by V.Z. Kresin and
W.A. Little, Plenum Press, New York, 1990

symmetry neglected in theories of FISDW so far, -- the bond order wave (BOW), -- which competes with the SDW as the one dimensional limit is approached (11). Finally, the actual band filling, or more precisely, the number of holes, ρ, per molecular site, plays an important role in our model. This is a consequence of explicit inclusion of both on-site and intersite Coulomb interactions(12,13). The very important role of ρ is missed in noninteracting models. While we agree with Yakovenko(10) that the very high field phase is near the one dimensional(1D) limit, we believe that the relevant competition is between the BOW and the SDW and not between superconductivity and SDW.

The motivation for the present work, however, goes beyond explaining FISDW. We intend to show that the same theoretical description applies to the normal states of all organic superconductors, as well as related materials with molecular cations that are similar to TMTSF structurally. Specifically, we have chosen $(TMTTF)_2X$ and $(BEDT-TTF)_2X$, since we believe that similarities at the molecular and structural levels of these materials warrant the same basic theoretical model, albeit with different magnitudes of the various parameters. For example, it is our belief that the pressure induced spin-Peierls(or BOW in our nomenclature) to SDW transition(14,15), that has been observed in $(TMTTF)_2PF_6$, is a manifestation of the same basic dimensionality crossover within the same model Hamiltonian.

II. THEORETICAL MODEL

The model that we consider is the single band quasi-2D extended Hubbard model,

$$\tilde{H} = \sum_{i\sigma}(t_x c^\dagger_{i\sigma}c_{i+x} + t_y c^\dagger_{i\sigma}c_{i+y} + H.c) + \sum_{i}(V_x n_i n_{i+x} + V_y n_i n_{i+y}) + U\sum_{i} n_{i\uparrow}n_{i\downarrow} \quad (1)$$

where $c^\dagger_{i\sigma}$ creates a hole with spin σ at the molecular site i, $n_{i\sigma} = c^\dagger_{i\sigma}c_{i\sigma}$ is a number operator, $n_i = \sum_{\sigma} n_{i\sigma}$, U is the on-site Coulomb repulsion between two holes occupying the same molecular site. V_x and V_y are the intrachain and interchain intersite nearest neighbor repulsions, and t_x and t_y are intrachain and interchain hopping integrals, respectively. We do not consider a realistic lattice at this preliminary stage, but assume a rectangular lattice with variable anisotropy ($0 < t_y/t_x < 1$, $0 < V_y/V_x < 1$). The stack axis is chosen to be the x-direction. The implicit parameter ρ is very important in our theory, particularly at moderate to large t_y/t_x. We consider $\rho = 1/2$ only corresponding to the one-quarter-filled band, although TMTSF salts may be characterized by weak incommensurability(16) arising due to band structure effects or incomplete charge transfer.

We have investigated numerically the phase diagram of the above theoretical model as a function of t_y/t_x and V_y/V_x. Before presenting our numerical results we discuss here the qualitative reasonings for our belief that Eq.(1) can reproduce the experimental behavior.

In the limit of $t_y = 0$ and $V_y = 0$, Eq.(1) has been widely studied as a model for $\rho = 1$ Mott insulators(11) as well as for conductors with arbitrary ρ(12,13). Because of the continuous nature of the broken symmetry in the SDW, the SDW never occurs as a distinct phase. Rather, the dominant broken symmetry here is an unconditional $2k_F$ BOW(11,17) (which, depending upon the parameters, may also be accompanied by a $4k_F$ instability(17)). For $V_x = V_y = 0$ and $t_y = 0$, it has been shown analytically that the charge-spin decoupling that occurs in the large U Hubbard model leads to the magnetic part of $\tilde{H}$ being described by an isotropic Heisenberg spin Hamiltonian (18). Similar results are expected to persist even for $V_x > 0$, although the expression for the Heisenberg exchange integral is expected to be different. It is known unambiguously that such a system exhibits a spin-Peierls (BOW) transition, and the SDW does not occur for any U, V_x. Experimental evidence for the spin-Peierls transition in charge transfer salts is common.(11)

The above situation is expected to change as t_y is increased from zero. Existing nonzero t_y results are only for the $\rho = 1$ limit, where the consensus is that the BOW is destroyed in the presence of Coulomb interaction(19,21). Although these results are for $t_y = t_x$, we believe that the disappearance of BOW (and appearance of SDW) can occur at t_y/t_x considerably less than 1. Furthermore, we believe that the spin-Peierls to SDW transition should not be limited to $\rho = 1$ only but also should be seen for $\rho = 0.5$, where the BOW vanishes for $t_y > t_{y_1}$. The special feature of $\rho = 0.5$ is that unlike $\rho = 1$, where the SDW gets progressively stronger with the increase of t_y, the $\rho = 0.5$ SDW is weakened for t_y greater than an upper critical value t_{y_2}. The reason for this is shown in Fig. 1. Due to the nonzero V_x and V_y, holes have a tendency to occupy alternate sites. For large enough t_y, the intrachain and interchain antiferromagnetic spin couplings can become comparable, and the resulting spin frustration (see Fig. 1) will destroy the SDW.

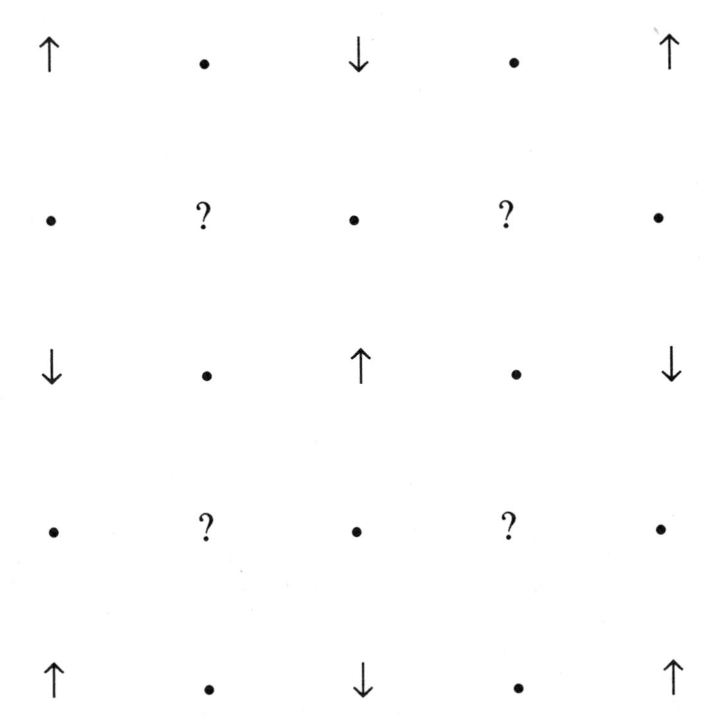

Fig. 1. Spin configurations for the 1/4-filled band. Note that because of the antiferromagnetic couplings between spins along $\vec{x}$, $\vec{y}$, and $\vec{x}+\vec{y}$ directions, spins at middle sites are frustrated.

Because of these considerations, we expect that there three distinct regions for t_y: $t_y < t_{y_1}$, $t_{y_1} < t_y < t_{y_2}$, and $t_y > t_{y_2}$. Generally speaking, TMTTF-based materials are in region 1, TMTSF salts in region 2, and BEDT-TTF salts in region 3. TMTTF and TMTSF materials can further be modified by pressure (which increases two dimensionality) or magnetic field (which decreases two dimensionality), and crossovers from one region to another have been observed. Note that for spin frustration to occur it is essential that V_x and V_y are explicitly included in the Hamiltonian, and that the proper role of $\rho = 0.5$ is taken into account. The importance of considering the role of V_x and ρ has been emphasized before(12,13).

III. NUMERICAL RESULTS

In this section we present some numerical results to give partial support to our qualitative considerations in the previous sections. We study the BOW and SDW phases within the Hamiltonian (1). The "structure factors" corresponding to the BOW and the SDW are defined as

$$\mathrm{BOW}(\vec{q}) = \frac{1}{N}\sum_{i,j} e^{i\vec{q}\cdot(\vec{r}_i - \vec{r}_j)} \left(\langle B_i B_j \rangle - \langle B_i \rangle \langle B_j \rangle \right) ,$$

where $B_i = \sum_\sigma (c_{i\sigma}^\dagger c_{i\sigma} + \mathrm{H.c})$ is the bond order parameter, and

$$\mathrm{SDW}(\vec{q}) = \frac{1}{N}\sum_{i,j} e^{i\vec{q}\cdot(\vec{r}_i - \vec{r}_j)} \langle (n_{i\uparrow} - n_{i\downarrow})(n_{j\uparrow} - n_{j\downarrow}) \rangle \ .$$

Both these quantities were calculated explicitly in the 1D limit and then for varying anisotropy. In the 1D limit, $t_y = 0$ and $V_y = 0$, the results were obtained by a world line quantum Monte Carlo simulation on lattice of size $N = 128$ atoms at temperature $T = 1/32$ in units of $t_x = 1$ (for a t_x of 0.2 eV this would be about 70K which is higher than the temperatures of interest in FISDW but which is reasonably close to the spin-Peierls temperatures in many organic systems). Only the results for $V_x = 0$ are shown here. For nonzero but moderate V_x (V_x less than or equal to 2t) the results are the same qualitatively (17). In Fig. 2 we plot $\mathrm{BOW}(\vec{q})$ as a function of the wave vector q for several values of U. In all cases a peak at $q = 2k_F$, where k_F is the Fermi wave vector within single-particle model, is clearly visible. Although the Coulomb repulsion suppresses the BOW here (unlike at $\rho = 1$, where an enhancement is seen for moderate U(11)), a logarithmic divergence as a function of N always exists, which implies that even for weak electron-lattice coupling a $2k_F$ BOW instability will occur, leading to a tetramerization in the present case.

Similar quantum Monte Carlo simulations for 2D lattices when ρ is not equal to 1 are difficult currently because of the well known "negative sign" problem. This problem becomes even more severe for large Coulomb interaction and low temperature. In 2D therefore we have done exact calculations for a 4×4 lattice. Notice that this already gives huge Hamiltonian matrices for Eq. (1). The exact ground state wave function was calculated using a Lanczös diagonalization procedure. As we are restricted to a single lattice, finite size scaling analysis is not possible and our conclusions are drawn from comparison with the results for the $U = V_x = V_y = 0$ case, for which the results are known analytically. Finally, the number of parameters that appear in Eq.(1) in 2D are rather large, so we assume $V_y/V_x = t_y/t_x$ in all our calculations. Since both the nearest neighbor hopping and Coulomb integrals are functions of intermolecular distances, this is not an unreasonable assumption. In any case, this restriction is not a limitation, and can be relaxed later.

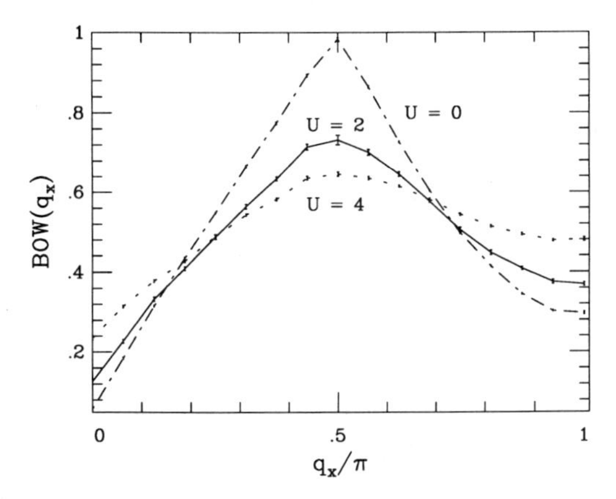

Fig. 2. Bond-Order-Wave Structure Factor for the 1D Hubbard Model.
N = 128, T = $t_x/32$, and U = 0, 2, 4.

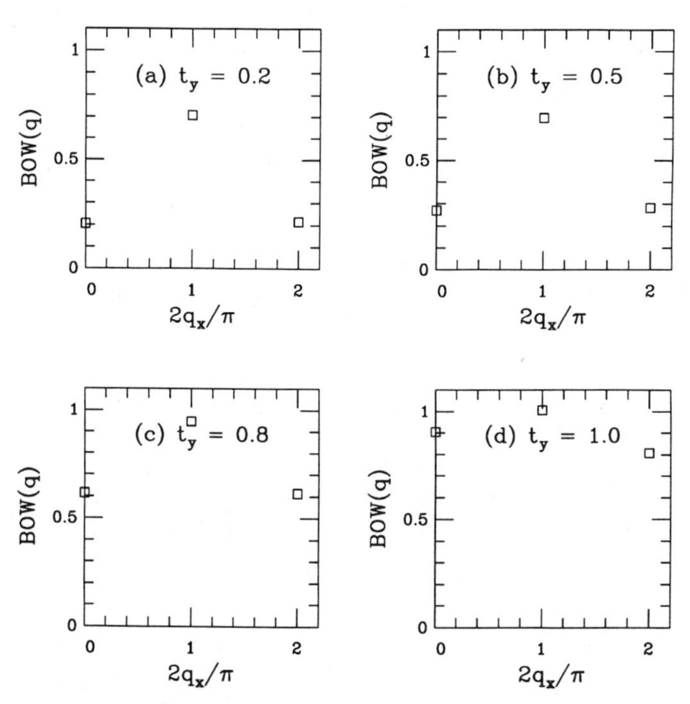

Fig. 3. Bond-Order-Wave Structure Factor BOW(q_x) as function of the anisotropy.(see the text)
U = 3, V_x = 1, and t_x = 1.

While the calculations were done for several different parameter values we present the results for $U = 3$, $V_x = 1$ (in units of $t_x = 1$), which are representative. Instead of showing the full set of results for the complete (q_x, q_y) subspace we present the results for $q_y = \pi$ only. This is because the correlation functions along (q_x,π) are the ones that are relevant for describing the spin-Peierls and SDW transitions in two dimensions. In Fig. 3 we have plotted $BOW(q_x)$ as a function of the anisotropy. It is clear that the BOW is progressively weakened as t_y and V_y are increased. This is what we expected, based on our earlier $\rho = 1$ results.(19) What is more interesting is the behavior of the

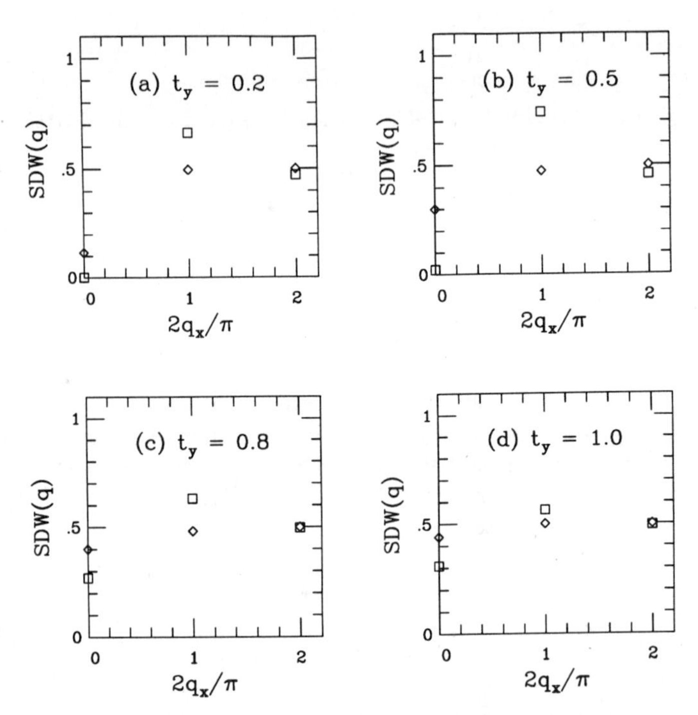

Fig. 4. Spin-Density-Wave Structure Factor $SDW(q_x)$ as function of the anisotropy.(see the text)
$U = 3$, $V_x = 1$, and $t_x = 1$;(squares)
$U = 0$, $V_x = 0$, and $t_x = 1$;(diamonds)

SDW, as seen in the plots of $SDW(q_x)$ for the same set of parameters in Fig. 4. Unlike the BOW, the SDW is enhanced initially, since the peak at $2k_F = \pi/2$ becomes more pronounced as t_y increases from $t_y = 0.2$ to $t_y = 0.5$. However, with further increase in t_y, the SDW amplitude *decreases,* until at $t_y = 1$, where its behavior becomes indistinguishable from single particle behavior, indicating a vanishing of the SDW due to the spin frustration discussed above. Qualitatively, this behavior is exactly what we claimed in section II., although the actual spin frustration here seems to become relevant only for relatively large t_y.

IV. COMPARISON WITH EXPERIMENTS

The rapid destruction of the BOW for even small nonzero t_y and the destruction of the SDW in the other limit of large t_y clearly suggests three distinct regions. Quantitative comparisons with results of simulations for such a small system would be meaningless, and any critical t_y obtained from our calculation would be too large. We therefore do not attempt quantitative comparisons at this preliminary stage. Our emphasis is rather on qualitatively explaining the behavior of all three families of materials, $(TMTTF)_2X$, $(TMTSF)_2X$, and $(BEDT{-}TTF)_2X$ within the same theoretical model.

Within our model ambient pressure $(TMTTF)_2X$ lies in the $t_y < t_{y_1}$ region. Since both bonding and Coulomb interactions are larger along the stacking axis than along the interchain direction, the lattice is expected to be more compressible along the latter. Pressure therefore increases t_y until the region $t_{y_1} < t_y < t_{y_2}$ is reached and the spin-Peierls to SDW transition occurs. Further increase in pressure increases T_{SDW}, which, however, reaches a maximum and then begins to decrease. Note that the peak in T_{SDW} is expected in our model and occurs in the region where spin frustration starts to be relevant. Experimentally, it is currently not clear whether there is a narrow region of coexistence between the spin-Peierls and the SDW phases. Due to the strong finite size effects in our numerical simulation, we are unable to resolve this issue theoretically.

The effect of magnetic fields on TMTSF is opposite to that of pressure in that the anisotropy increases with H. For $H < H_f$ and $t_y > t_{y_2}$, there is no SDW due to spin frustration. For $H > H_f$ the increase in anisotropy takes the system into the region $t_{y_1} < t_y < t_{y_2}$, where SDW is the characteristic broken symmetry. The current model is too simplistic to give the cascade of SDW transitions, but we believe that treating magnetic field effects as in standard theories(3,4,5) will permit us to incorporate FISDW effects in our model. However, the VFHT is *expected* in our model. In the extreme 1D limit $t_y < t_{y_1}$, the BOW phase occurs at low temperatures. This phase is then *not* a metal, and we expect semimetallic behavior (or even weakly semiconducting behavior), in agreement with the transport measurements(9). Note that the observed diamagnetism is also consistent with the spin-Peierls phase.

Finally in $(BEDT{-}TTF)_2X$ the anisotropy is considerably weaker and t_y is much larger than t_{y_2}. The absence of SDW in these materials is then expected from our model. Because of the smaller t_y in the sulphur-based materials, as compared to the selenium based materials, a tendency to charge density wave may be expected within model Hamiltonian Eq.(1), and this may explain the experimentally observed tendency to localization in these substances.

V. DISCUSSIONS

The present work is still in its initial stages and further work is in progress. Our principle motivation was to point out that a natural explanation for the VFHT follows once the competition between the BOW and the SDW as a function of effective dimensionality is taken into account. Several points are to be noted.

Firstly, we do expect a boundary between the low field metallic phase and the very high field phase. Presumably, this would be a line sharply rising with temperature T in the H-T plane from the SDW curve, separating a low T phase with activated conductivity and a high T phase with more metallic behavior. The high T phase should also exhibit different magnetic behavior. Currently, no phase boundary has been found, but search for this boundary must be extended to the high temperature region.

Secondly, the numerical results presented in this paper are for a rectangular lattice and for a perfectly commensurate electron density. The former is certainly not true for TMTSF salts where weak deviation from commensurability(16) is conceivable. Incorporation of the real lattice structure and weak incommensurability in our calculations would be simple. Currently the value of t_y at which spin frustration begins to occur is too large. We believe that more realistic estimates of t_{y_2} can be obtained once these features are incorporated in our calculations.

Finally, the magnetic field H is not explicitly included in our calculations, and as a result the cascade of transitions observed at moderate fields is missing in our treatment. Explicit inclusion of H will be attempted at a later stage.

ACKNOWLEDGEMENT

H. Q. Lin thanks P. Ford and staff members of Advanced Computing Laboratory at Los Alamos National Laboratory for their computational assistance. This work is partly supported by Los Alamos National Laboratory and the U.S. DOE Office of Energy Research.

REFERENCES

1. J. F. Kwak, J. E. Schirber, R. L. Greene, and E. M. Engler, Phys. Rev. Lett. **46**, 1296 (1981).
2. For a recent review, see X. Yan *et al.,* in Low-Dimensional Conductors and Superconductors, edited by D. Jerome and L. Caron, NATO Advanced Study Institute, Series B, Physics, Vol. 155 (Plenum, New York, 1987); pp. 211-220. See also, P. M. Chaikin *et al.,* Synthetic Metals, **27**, B163 (1988).
3. L. P. Gorkov and A. G. Lebed, J. Phys. (paris) Lett. **43**, L433, (1984).
4. G. Montambaux, M. Heritier, and P. Lederer, Phys. Rev. Lett. **55**, 2078 (1985).
5. K. Yamaji, J. Phys. Soc. Jpn., **56**, 1101 (1987) and references therein.
6. A. Virosztek *et al.,* Phys. Rev. B **34**, 3371 (1986).
7. M. Ya. Azbel *et al.,* Phys. Rev. A **39**, 1392 (1986).
8. T. Osada *et al.,* Solid State Commun. **60**, 441 (1986).
9. R.V. Chamberlin *et al.,* Phys. Rev. Lett. **60**, 1189 (1988).
10. V. M. Yakovenko, Zh. Eksp. Teor. Fiz. **93**, 627 (1987) [Sov. Phys. JETP 66, 355 (1987)].
11. For a recent review of the competition between and coexistence of various spatial broken symmetries in Coulomb correlated one dimensional systems, see D. Baeriswyl, D. K. Campbell, and S. Mazumdar, review article to be published in *Conducting Polymers,* edited by H. Kiess. (Springer-Verlag)
12. S. Mazumdar and A. N. Bloch, Phys. Rev. Lett. **50**, 207 (1983);
13 S. Mazumdar, S. N. Dixit, and A. N. Bloch, Phys. Rev. B **34**, 3683, (1986).

14. F. Creuzet *et al.,* Synthetic Metals, **19**, 289 (1987) and references therein.
15. L. G. Caron *et al.,* Synthetic Metals, **27**, B123, (1988).
16. S. Tomic, J. R. Cooper, D. Jerome, and K. Bechgaard, Phys. Rev. Lett. **62**, 462 (1989). This reference discusses non-ohmic conductivity in $(TMTSF)_2NO_3$.
17. J. E. Hirsch and D. J. Scalapino, Phys. Rev. B **29**, 5554 (1984).
18. D. J. Klein and W. A. Seitz, Phys. Rev. B **10**, 3217(1984).
19. S. Mazumdar, Phys. Rev. B **36**, 7190 (1988).
20. F. C. Zhang and P. Prelovsek, Phys. Rev. B **37**, 1569 (1988).
21. S. Tang and J. E. Hirsch, Phys. Rev. B **37**, 9546 (1988).

NATURE OF CHALCOGEN···CHALCOGEN CONTACT INTERACTIONS IN ORGANIC DONOR-MOLECULE SALTS

Juan J. Novoa[a] and Myung-Hwan Whangbo

Department of Chemistry
North Carolina State University
Raleigh, North Carolina 27695-8204

Jack M. Williams

Chemistry and Materials Science Divisions
Argonne National Laboratory
Argonne, Illinois 60439

The nature of chalcogen...chalcogen contact interactions in organic donor-molecule salts was examined by performing ab initio SCF-MO/MP2 calculations on $H_2X \cdots XH_2$ (X=O, S, Se,Te) and MM2 calculations on donor dimers $(TXF)_2$ (X=S, Se, Te) and $(BEDX\text{-}TTF)_2$ (X=O, S).

INTRODUCTION

With a number of monovalent anions organic donor molecules BEDT-TTF (**1**) and its analogs form charge-transfer salts that are semiconducting, metallic or superconducting.[1] These salts adopt a layered structure in which layers of donor molecules alternate with layers of anions. The crystal structures of these salts, and therefore their physical properties, are strongly governed by donor···donor and donor···anion interactions associated with the donor-molecule C-H bonds.[2,3] Within each layer of donor molecules, there exist also numerous

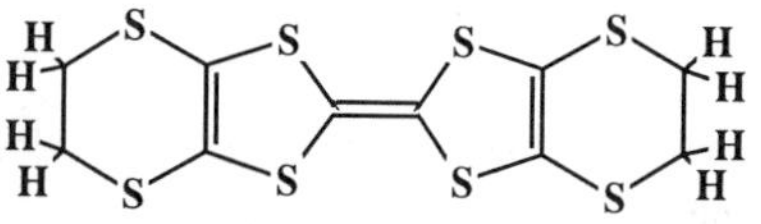

H H H H S S S S S S S S H H H H

1

[a]Permanent Address: Departmento de Quimica Fisica, Facultad de Quimica
Universidad de Barcelona, 08028 Barcelona, Spain

Organic Superconductivity, Edited by V.Z. Kresin and
W.A. Little, Plenum Press, New York, 1990

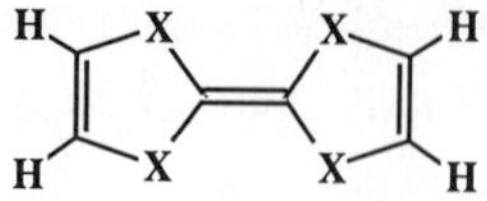

(X = O, Se, Te) 2

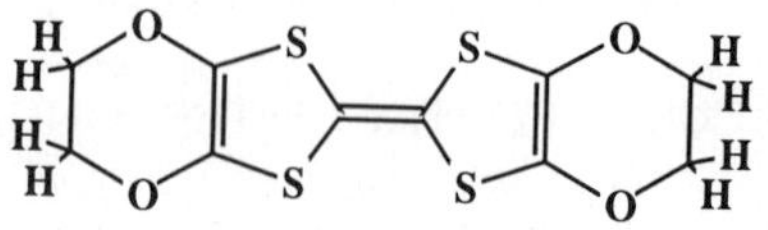

3

chalcogen···chalcogen contacts shorter than the van der Waals (VDW) radii sums. Whether an organic donor salt is a one- or two-dimensional metal depends largely upon these chalcogen···chalcogen contacts.[1] In our preliminary ab initio computational studies,[2,3] it was pointed out that, in contrast to the case of the C-H···donor and C-H···anion interactions, the chalcogen···chalcogen (X···X) contacts are essentially nonbonding in nature in the vicinity of their VDW X···X distances. In the present work, we examine the nature of the X···X contacts (X = O, S, Se, Te) by performing ab initio SCF-MO/MP2[4] calculations on $H_2X \cdots XH_2$ with extensive basis sets. To also examine the nature of the X···X interactions associated with real donor molecules, we also perform molecular mechanics[5] (i.e., MM2) calculations on $(TXF)_2$ (X = S, Se, Te) and $(BEDX\text{-}TTF)_2$ (X = O, S) dimers.

AB INITIO COMPUTATIONS

Basis Set

The ab initio SCF-MO/MP2 calculations of $H_2X \cdots XH_2$ (X=O, S, Se, Te) described in the present study were carried out by employing GAUSSIAN 86 program.[6] For the H, O and S atoms, the 6-311G(d, p), 6-311G(2d, 2p) and 6-311G(3df, 3pd) basis sets were used. For the Se and Te atoms, the core electrons were represented by the pseudo-potentials of Wadt and Hay[7] while the valence electrons were represented by their uncontracted (3s, 3p) orbitals so as to achieve a triple zeta quality as in the case of the H, O and S atoms. The (d, p), (2d, 2p) and (3df, 3pd) polarization sets of the Se and Te atoms are obtained as follows: We optimized the d-orbital exponents of the (d, p) polarization set by performing SCF-MO calculations on XH_2 (X = Se, Te). The optimized d-orbital exponents are 0.380 and 0.312 for X = Se and Te, respectively. Then the usual splitting procedure[8] was employed to obtain the d-orbital exponents of the (2d, 2p) and the (3df, 3pd) polarization sets. Thus the d-orbital exponents of Se are 0.760 and 0.190 for the (2d, 2p) set, and 0.760, 0.380 and 0.190

for the (3df, 3pd) set. Likewise, the d-orbital exponents of Te are 0.624 and 0.156 for the (2d, 2p) set, and 0.624, 0.312 and 0.156 for the (3df, 3pd) set. Finally, the f-orbital exponents of Se and Te were taken to be the same as that of S (i.e., 0.80). Thus the basis sets used for Se and Te are expected to have the 6-311G(d, p), 6-311G(2d, 2p) and 6-311G(3df, 3pd) qualities as far as their valence electrons are concerned. For simplicity, the basis sets of the Se and Te atoms described above will be referred to as 6-311G(d, p), 6-311G(2d, 2p) and 6-311G(3df, 3pd) sets.

The X···X (X = O, S, Se, Te) contacts of organic donor salts essentially give rise to VDW interactions. To test the adequacy of our basis sets in describing such interactions, we calculate the dipole moments and the polarizabilities of XH_2 (X = O, S, Se, Te) at the SCF-MO level by employing the geometries described in the next section. Results of our calculations are summarized in Table 1. The calculated dipole moments and the polarizabilities are in good agreement with the available experimental values, although the calculations were based upon the SCF-MO level. (As previous studies have shown,[9] inclusion of correlation should provide a closer agreement between theory and experiment.) Namely, the dipole moments and the polarizabilities of XH_2 are reasonably well described by our basis sets. Since dipole moment and polarizability terms are the two leading terms of a long-range intermolecular interaction, our basis sets are expected to provide an adequate description of the X···X interactions. It is estimated[10] that the interaction energy of a polar-

Table 1. Dipole moments (au) and polarizabilities (au) of H_2X (X = O, S, Se, Te) calculated at the SCF-MO level

Molecule	Basis set	Dipole moment	Polarizability
	6-311G (d, p)	0.853	5.562
	6-311G (2d, 2p)	0.786	6.568
H_2O	6-311G (3df, 3pd)	0.750	7.483
	experimental	0.724[a]	9.642[b]
	6-311G (d, p)	0.532	15.43
H_2S	6-311G (2d, 2p)	0.442	17.99
	6-311G (3df,3pd)	0.408	21.93
	experimental	0.382[a]	25.79[c]
	6-311G (d, p)	0.366	19.96
H_2Se	6-311G (2d,2p)	0.301	24.71
	6-311G (3d,3pd)	0.307	25.47
	experimental	0.247[a]	-
	6-311G (d, p)	0.102	29.05
H_2Te	6-311G (2d, 2p)	0.052	35.38
	6-311G (3df, 3pd)	0.051	36.05
	experimental	-	-

[a]Ref. 12 [b]Ref. 13 [c]Ref. 14

polar VDW interaction is calculated within 5% error when the dipole moment and the polarizability are computed within 2 and 10% errors, respectively.

Geometry

In our calculations on $H_2X \cdots XH_2$ the XH_2 (X = O, S, Se, Te) geometries were kept constant: For X = O, S, Se and Te, the X-H bond lengths are 0.958, 1.346, 1.470 and 1.654 Å, respectively, and the HXH angles are 104.45, 93.30, 91.00 and 90.60°, respectively. To probe the X···X contact interactions, we performed ab initio SCF-MO/MP2 calculations on $H_2X \cdots XH_2$ as a function of the X···X distance ($r_{X \ldots X}$) for the D_{2h}-arrangement shown in **4**. The angular dependence of the X···X interactions were examined by performing SCF-MO/MP2 calculations on $H_2X \cdots XH_2$ as a function of the bending angle θ for the C_s-arrangement shown in **5**.

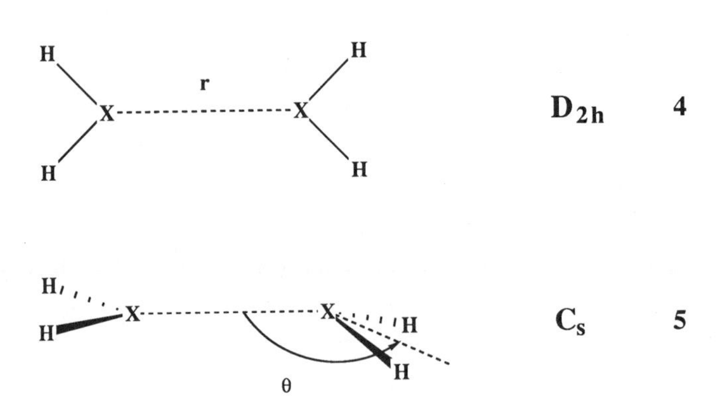

Nature of X···X interactions

Figure 1 shows a general feature of how the interaction energies ΔE of $H_2X \cdots XH_2$, calculated at the MP2 level using the 6-311G (d, p) basis set, vary as a function of $r_{X \ldots X}$ for the D_{2h}-arrangement **4**. Table 2 summarizes the optimum $r_{X \ldots X}$ and ΔE values (i.e., r_{opt} and ΔE_{opt}, respectively) calculated at the MP2 level using the 6-311G(d, p), 6-311G(2d, 2p) and 6-311G(3df, 3pd) basis sets. (For simplicity, we do not show the corresponding values obtained at the SCF-MO level. With all the three basis sets, the SCF-MO calculations show that the ΔE values of the D_{2h}-arrangement of $H_2X \cdots XH_2$ (X = O, S, Se, Te) are positive.) Table 2 also lists the ΔE values calculated at the VDW X···X distances (i.e., ΔE_{vdw} and r_{vdw}, respectively). Finally, Figure 2 shows how the ΔE values of $H_2X \cdots XH_2$ change as a

Table 2. r_{opt}, ΔE_{opt} and ΔE_{vdw} values calculated for $H_2X \cdots XH_2$ (X = O, S, Se, Te) by ab initio calculations at the MP2 level

Contact	Basis set	r_{opt} (Å)	ΔE_{opt} (kcal/mol)	r_{vdw} (Å)	ΔE_{vdw} (kcal/mol)
O···O	6-311G (d,P)	∞	0.00	3.00	3.42
	6-311G (2d, 2p)	∞	0.00	3.00	2.33
	6-311G (3df, 3pd)	∞	0.00	3.00	1.85
S···S	6-311G (d, p)	4.25	-0.03	3.60	1.37
	6-311G (2d, 2p)	4.28	-0.17	3.60	0.49
	6-311G (3df, 3pd)	4.05	-0.41	3.60	-0.10
Se···Se	6-311G (d, p)	4.53	-0.09	3.80	0.48
	6-311G (2d, 2p)	4.20	-0.46	3.80	0.20
	6-311G (3df, 3pd)	-	-	3.80	-0.19
Te···Te	6-311G (d, p)	5.02	-0.13	4.20	0.40
	6-311G (2d, 2p)	4.62	-0.79	4.20	-0.56
	6-311G (3df, 3pd)	-	-	4.20	-1.02

function of the bending angle θ (see **5**), in which the X···X distances were fixed at the r_{opt} values for X = S, Se and Te while the r_{vdw} value was used for the O···O distance of $H_2O \cdots OH_2$ (see below).

Table 2 reveals that, for the D_{2h}-arrangement **4**, $H_2O \cdots OH_2$ is not bound while the $H_2X \cdots XH_2$ (X = S, Se, Te) systems are bound. The binding energies of $H_2X \cdots XH_2$ (i.e., -ΔE_{opt}) increase in the order X = S < Se < Te, which follows the polarizability ordering of the X atoms. Within each $H_2X \cdots XH_2$ system, ΔE_{opt} becomes more attractive and r_{opt} generally

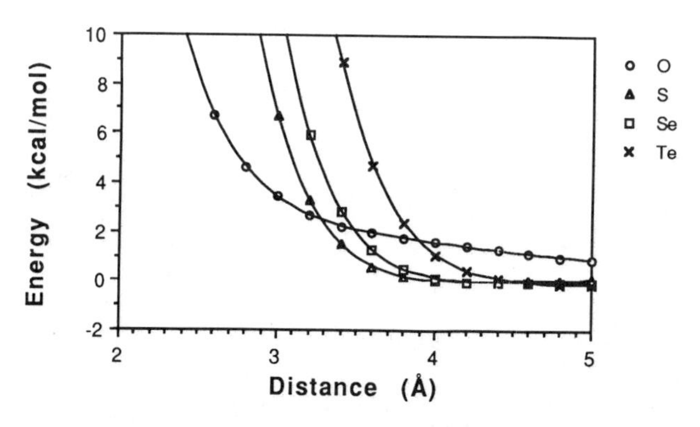

Figure 1. Interaction energies ΔE of $H_2X \cdots XH_2$ (X = O, S, Se, Te) calculated at theMP2 level with the 6-311G (d, p) basis set as a function of the X···X distance r for the coplanar arrangement **4**.

becomes shorter as the size of the polarization basis set increases. Since the more flexible basis set generally provides a better description of VDW interactions, the ΔE_{opt} and r_{opt} values computed with the 6-311G(3df, 3pd) basis set are expected to be most reliable. The dipole moments and polarizabilities of H_2O and H_2S calculated with this basis set are in excellent agreement with experiment even at the SCF-MO level (Table 1). Thus the interaction energies of $H_2X \cdots XH_2$ (X = O, S) calculated with the 6-311G(3df, 3pd) basis set at the MP2 level would be accurate enough so that the repulsive nature of the O···O interaction and the attractive nature of the S···S interaction are not expected to change with further increase in the basis set. Our calculations of the basis-set-superposition-error (BSSE) for $H_2S \cdots SH_2$ with the 6-311G(2d, 2p) basis set using the counterpoise method[11] show that the BSSE effect on ΔE_{opt} is negligible (smaller than 0.01 kcal/mol). We also calculated the ΔE value of $H_2S \cdots SH_2$ with the 6-311G(3df, 3pd) basis set and $r_{X...X}$ = 4.0 Å at the MP4 level. This gives the ΔE value (-0.34 kcal/mol) quite close to the corresponding ΔE value (-0.40 kcal/mol) obtained at the MP2 level. This suggests that the ΔE_{opt} values of $H_2X \cdots XH_2$ obtained at the MP2 level are close to the "converged" MPn value ($n \rightarrow \infty$).[4] It is noted from Figure 2 that the angular dependence of ΔE is weak. Thus the ΔE_{opt} values calculated with the D_{2h}-arrangement **4** are expected to be reasonable in describing the X···X interactions.

Table 2 shows that the r_{opt} values of $H_2X \cdots XH_2$ are significantly larger than their r_{vdw} values. At r_{vdw}, therefore, the X···X interactions less attractive (Figure 1), so that the ΔE_{vdw} values for $H_2X \cdots XH_2$ (X = S, Se, Te) become repulsive with the 6-311G(d, p) and 6-311G (2d, 2p) basis sets. The 6-311G(3df, 3pd) basis set gives a slightly attractive ΔE_{vdw} value for $H_2X \cdots XH_2$ (X = S, Se), but a substantially attractive ΔE_{vdw} value for $H_2Te \cdots TeH_2$. The ΔE_{vdw} values of $H_2X \cdots XH_2$ calculated with this basis set become more attractive in the order X = S < Se << Te, which follows the polarizability ordering of XH_2 (see Table 1). Other

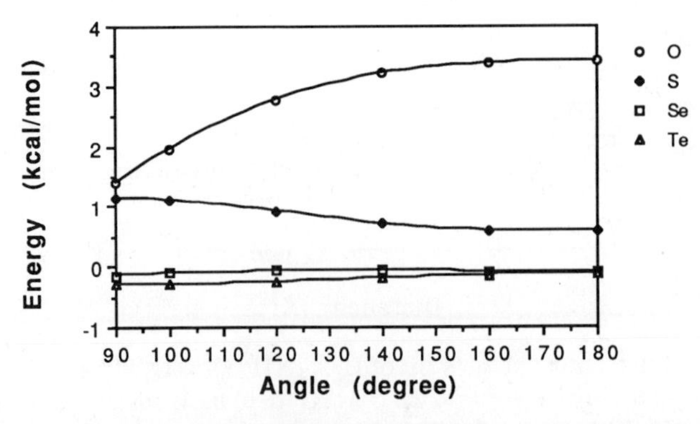

Figure 2. Interaction energies ΔE of $H_2X \cdots XH_2$ (X = O, S, Se, Te) calculated at the MP2 level with the 6-311G (d, p) basis set as a function of the bending angle θ for the arrangement **5**.

factors contributing to this trend may be that the dipole moment of TeH_2 is very small compared with that of XH_2 (X = S, Se) (See Table 1), and the net charge of Te in TeH_2 is postive while that of X in XH_2 (X = S, Se) is negative. Thus the X···X (X = S, Se) contact interactions are only slightly attractive in the vicinity of r_{vdw}. This is in contrast to the case of the C-H···donor and C-H···anion interactions, which are substantially attractive in the vicinity of their VDW contact distance.[3] Consequently, the X···X (X = S, Se) contact interactions between donor molecules are less important than the C-H···donor and C-H···anion interactions in the crystal packing of organic donor salts.

MOLECULAR MECHANICS COMPUTATIONS

We now examine the nature of the X···X (X = S, Se, Te) contact interactions on the basis of donor dimers $(TXF)_2$ (X = S, Se, Te) and $(BEDX\text{-}TTF)_2$ (X = O, S). Ab initio MP2 calculations of these systems with the basis sets used for $H_2X\cdots XH_2$ are beyond our reach at present, so we resort to MM2 calculations. To test the reliability of the MM2 method, we carried out MM2 calculations on $H_2X\cdots XH_2$ (X = O, S, Se, Te) with the D_{2h}-arrangement **4**. As shown in Figure 3 for $H_2S\cdots SH_2$, the potential energy curve obtained from the MM2 calculations is shifted to a longer distance compared with the curves derived from the ab initio MP2 calculations. The r_{opt}, ΔE_{opt} and ΔE_{vdw} values of the MM2 calculations are summarized in Table 3. It is clear from this Table that the qualitative conclusions of the ab initio MP2 calculations are well reproduced by the MM2 calculations (i.e., $r_{opt} > r_{vdw}$, $\Delta E_{opt} < 0$, and $\Delta E_{vdw} > 0$).

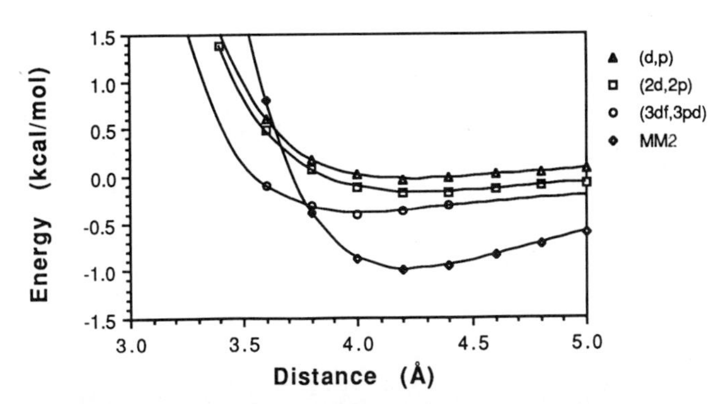

Figure 3. Interaction energies ΔE of $H_2S\cdots SH_2$ calculated by ab initio MP2 method with various basis sets and also by the MM2 method as a function of the S···S distance r.

Table 3. r_{opt}, ΔE_{opt}, and ΔE_{vdw} values calculated for $H_2X\cdots XH_2$ (X = O, S, Se, Te) by MM2 calculations

Contact	r_{opt} (Å)	ΔE_{opt} (kcal/mol)	r_{vdw} (Å)	ΔE_{vdw} (kcal/mol)
O···O	∞	0.00	3.00	3.28
S···S	4.25	-0.99	3.60	0.80
Se···Se	4.47	-1.35	3.80	1.34
Te···Te	4.80	-1.72	4.20	0.17

The interaction energies ΔE between donor molecules of $(TXF)_2$ (X = S, Se, Te) and $(BEDX\text{-}TTF)_2$ (X = O, S) were calculated for their coplanar "eclipsed" X···X arrangements (**6a** and **7a**, respectively) as a function of the shortest X···X distance. The optimum $r_{X...X}$ and ΔE values listed in Table 4 (r_{opt} and ΔE_{opt}, respectively) show that all the dimers are bound for the "eclipsed" X···X arrangement. As in the case of $H_2X\cdots XH_2$, $r_{opt} > r_{vdw}$ in $(TXF)_2$ and $(BEDX\text{-}TTF)_2$. The interaction energies of $(TXF)_2$ and $(BEDX\text{-}TTF)_2$ calculated as a function of coplanar donor-molecule sliding (i.e., **6a** → **6b** and **7a** → **7b**) reveal that both dimers become more stable (by ΔΔE, as listed in Table 4) when the donor molecules have the "staggered" X···X contacts (i.e., **6b** and **7b**). This is consistent with the general observation that organic donor salts rarely exhibit "eclipsed" X···X contact arrangements.

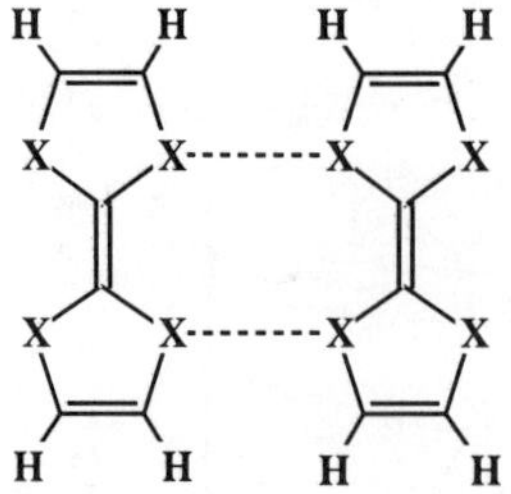

6a 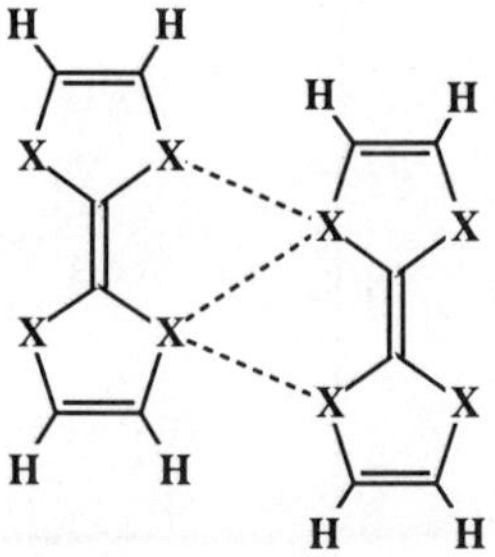 **6b**

Table 4. r_{opt}, ΔE_{opt}, and ΔE_{vdw} values calculated for $(TXF)_2$ (X = S, Se, Te) and $(BEDX\text{-}TTF)_2$ (X = O, S) by MM2 calculations

Dimer	r_{opt} (Å)	ΔE_{opt} (kcal/mol)	$\Delta\Delta E$ (kcal/mol)	ΔE_{vdw} (kcal/mol)
$(TTF)_2$	4.00	-6.71	2.25	-5.69
$(TSeF)_2$	4.21	-7.54	1.69	-4.71
$(TTeF)_2$	4.50	-8.74	1.92	-6.78
$(BEDO\text{-}TTF)_2$	3.72[a]	-10.9	2.60	-10.3
$(BEDT\text{-}TTF)_2$	3.75[b]	-12.6	2.89	-11.7

[a]S...S distance
[b]S···S contact between six-membered rings

Comparison of the ΔE_{opt} values for $(H_2S)_2$, $(TTF)_2$ and $(BEDT\text{-}TTF)_2$ (Tables 3 and 4) show that the interaction energy between donor molecules becomes more attractive as the number of the shortest X···X contacts increases. However, the magnitudes of the ΔE_{opt} values for $(TTF)_2$ and $(BEDT\text{-}TTF)_2$ are greater than the corresponding value for $(H_2S)_2$ by a factor of more than two and four, respectively. Figure 4 shows the potential energy curves

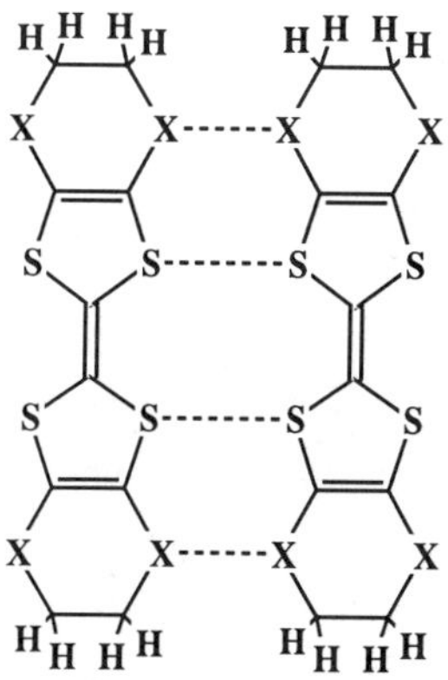

7a 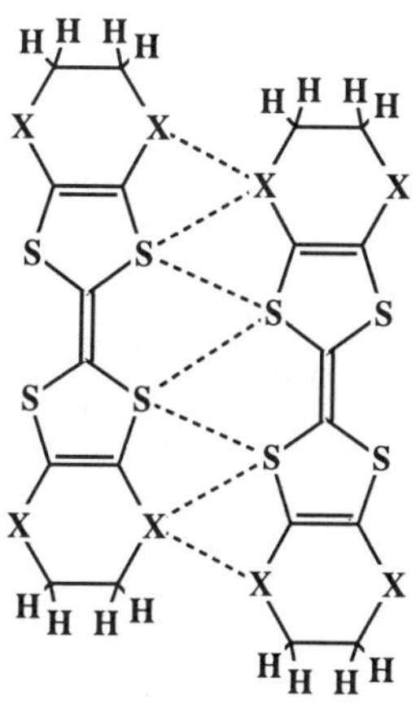 7b

of $(H_2S)_2$, $(TTF)_2$ and $(BEDT\text{-}TTF)_2$ calculated (for the eclipsed S···S arrangement) as a function of the shortest S···S contact distance. The potential energy minimum becomes deeper, and the r_{opt} value becomes shorter, on going from $(H_2S)_2$ to $(TTF)_2$ to $(BEDT\text{-}TTF)_2$. In addition, the S···S distance at which the interaction energy vanishes becomes

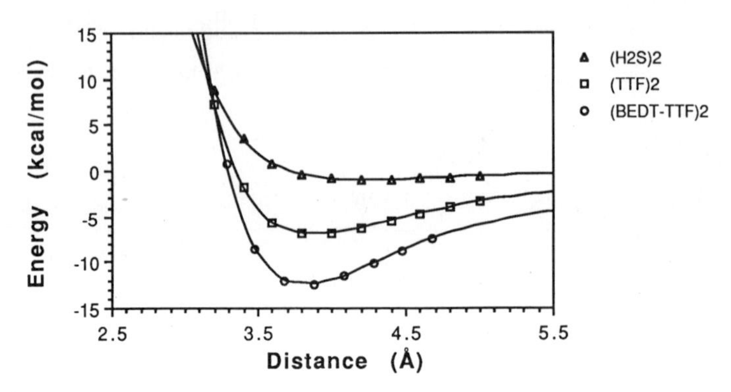

Figure 4. Interaction energies ΔE of $H_2S\cdots SH_2$, $(TTF)_2$ and $(BEDT\text{-}TTF)_2$ calculated by the MM2 method as a function of the shortest S···S distance r between donor molecules for the 'eclipsed' S···S arrangement.

shorter as we move from $(H_2S)_2$ to $(TTF)_2$ to $(BEDT\text{-}TTF)_2$. This reflects the fact that attractive VDW interactions arising from interdimer contacts other than the shortest S···S contacts are important in binding these donor molecules. This accounts for why the interaction energies of $(TXF)_2$ and $(BEDX\text{-}TTF)_2$ calculated at the VDW X···X distances (ΔE_{vdw} in Table 4) are attractive, in contrast to the case of $(H_2X)_2$.

CONCLUDING REMARKS

Our ab initio MP2 calculations on $H_2X\cdots XH_2$ (X = O, S, Se, Te) show that $H_2O\cdots OH_2$ is not bound, but $H_2X\cdots XH_2$ (X = S, Se, Te) are bound at the X···X distances longer than their VDW contact distances. In the vicinity of their VDW distances, the S···S and Se···Se interactions are slightly attractive, but the Te···Te interaction is substantially attractive. According to our MM2 calculations, the donor···donor interactions which occur through their X···X contacts become more attractive as the number of their X···X contacts increases. In $(TXF)_2$ (X = S, Se, Te) and $(BEDX\text{-}TTF)_2$ (X = O, S) attractive VDW interactions arising from contacts other than their shortest X···X contacts contribute significantly to the binding energy of these dimers.

ACKNOWLEDGMENTS

Work at North Carolina State University and Argonne National Laboratory is supported by the U. S. Department of Energy, Office of Basic Energy Sciences, Division of Materials Sciences, under Grant DE-FG05-86ER45259 and Contract W31-109-ENG-38, respectively. We express our appreciation for computing time on the ER-Cray Computer, made available by DOE, and that on VAX-8700 at Argonne National Laboratory provided by Chemistry Division. J. J. N. thanks the NATO and Ministerio de Educacion y Ciencia (Spain) for Fellowships, which made it possible to visit North Carolina State University, and also CICYT Grant PB86-0272.

REFERENCES

1. J. M. Williams, H. H. Wang, T. J. Emge, U. Geiser, M. A. Beno, P. C. W. Leung, K. D. Carlson, R. J. Thorn, A. J. Schultz, and M. -H. Whangbo, Prog. Inorg. Chem. 35:51 (1987).
2. M. -H. Whangbo, D. Jung, J. Ren, M. Evain, J. J. Novoa, F. Mota, S. Alvarez, J. M. Williams, M. A. Beno, A. M. Kini, H. H. Wang, and J. R. Ferraro, in: "The Physics and Chemistry of Organic Superconductors," G. Saito and S. Kagoshima, eds.,Springer-Verlag, in press.
3. (a) J. J. Novoa, M. -H. Whangbo, and J. M. Williams, Mol. Cryst. Liq. Cryst. 181:25 (1990).
   (b) M. -H. Whangbo, J. J. Novoa, and J. M. Williams, This volume.
4. C. Moller and M. S. Plesset, Phys. Rev. 46:618 (1934).
5. U. Burkett and N. L. Allinger, "Molecular Mechanics, ACS Monograph 177," American Chemical Society, Washington, DC (1982).
6. GAUSSIAN 86, M. J. Frisch, J. S. Brinkley, H. B. Schlegel, K. Raghavachari, C. F. Melius, R. L. Martin, J. J. P. Stewart, F. W. Bobrowicz, C. M. Rohlfing, L. R. Kahn, D. J. Defrees, R. Seeger, R. A. Whiteside, D. J. Fox, E. M. Fleuder, and J. A. Pople, Carnegie-Mellon Quantum Chemistry Publishing Unit, Pittsburgh, PA, 1984.
7. W. R. Wadt and P. J. Hay, J. Chem. Phys. 82:284 (1985).
8. T. H. Dunning and P. J. Hay, "Modern Theoretical Chemistry," Plenum, New York, 1976, Ch. 1.
9. H. -J. Werner and W. Meyer, Mol. Phys. 31:855 (1976).
10. J. H. van Lenthe, J. G. C. M. van Duijneveldt-van de Rijdt, and F. B. van Duijneveldt, in: "Ab Initio Methods in Quantum Chemistry II", K. P. Lawley, ed., Wiley, New York, 1987, p. 521.
11. S. F. Boys and F. Bernardi, Mol. Phys. 19:553 (1970).

12. F. K. Lovas, J. Phys. Chem. Ref. Data, 7:1445 (1978).
13. G. D. Zeiss and W. J. Meath, Mol. Phys. 33:1155 (1977).
14. M. P. Bogaard, A. D. Buckingham, R. K. Pierens, and A. H. White, Far. Trans. I, 74:3008 (1978).

IMPORTANCE OF C-H···DONOR AND C-H···ANION CONTACT INTERACTIONS FOR THE CRYSTAL PACKING, THE LATTICE SOFTNESS AND THE SUPERCONDUCTING TRANSITION TEMPERATURES OF ORGANIC CONDUCTING SALTS.

M. -H. Whangbo, J. J. Novoa and D. Jung

Department of Chemistry, North Carolina State University
Raleigh, North Carolina 27695-8204

J. M. Williams, A. M. Kini, H. H. Wang, U. Geiser,
M. A. Beno and K. D. Carlson

Chemistry and Materials Science Divisions, Argonne National Laboratory
Argonne, Illinois 60439

INTRODUCTION

The organic donor molecule BEDT-TTF (**1**) and its analogs **2-4** have yielded a number of ambient-pressure superconducting salts. What structural and electronic factors govern the magnitudes of their superconducting transition temperature T_c has been a topic of intense studies.[1-4] Examination of the band electronic structures of closely related superconducting salts shows[2,3] that the magnitudes of their T_c's are primarily determined by the softness of their crystal lattices. The crystal packing and the lattice softness of organic donor salts are strongly influenced by the donor···donor and donor···anion contact interactions involving the donor-molecule C-H bonds.[2,5,6] In the present work, we briefly

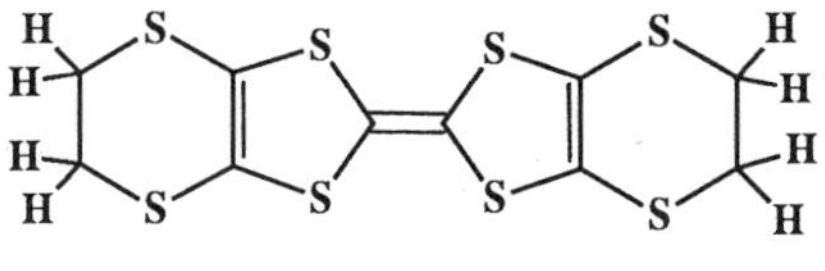

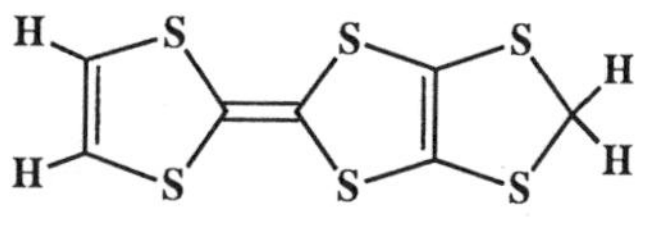

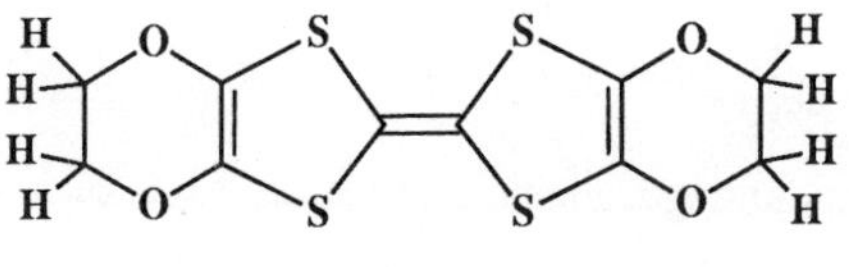

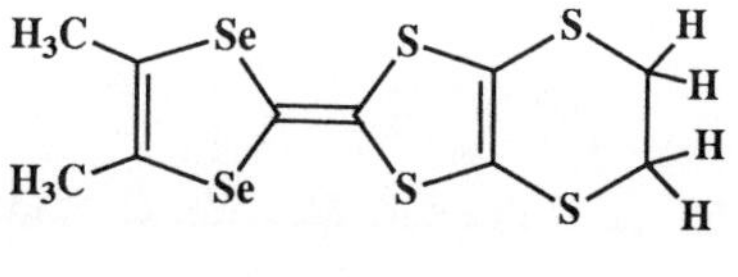

Organic Superconductivity, Edited by V.Z. Kresin and
W.A. Little, Plenum Press, New York, 1990

review the electronic structures of some representative organic salt superconductors and discuss the softness of their crystal lattices on the basis of the interaction energies calculated for the C-H···donor and C-H···anion contact interactions.

LATTICE SOFTNESS AND SUPERCONDUCTIVITY

Figure 1 shows how the T_c values of β-(BEDT-TTF)$_2$ X ($X^- = I_3^-$, AuI_2^-, IBr_2^-) vary as a function of applied pressure P.[7] For $X^- = AuI_2^-$ and IBr_2^-, the T_c of β-(BEDT-TTF)$_2$X decreases gradually with increasing P. For $P < 0.5$ kbar, the T_c of β-(BEDT-TTF)$_2I_3$ decreases as well with increasing P. At $P \cong 0.5$ kbar, however, the T_c of β-(BEDT-TTF)$_2I_3$ jumps to 8K but decreases as P increases further. [β-(BEDT-TTF)$_2I_3$ with $T_c \cong 8$K is commonly referred to as β*-(BEDT-TTF)$_2I_3$.[8]] To understand what causes the T_c jump in β-(BEDT-TTF)$_2I_3$ under pressure, we first examine whether or not the electronic structure of β-(BEDT-TTF)$_2I_3$ undergoes any drastic change by applied pressure. Figures 2a and 2b respectively show the dispersion relations of the two highest-occupied (HO) bands and the Fermi surface calculated[34] for β-(BEDT-TTF)$_2I_3$ on the basis of its crystal structure[9] determined at 9K under ambient pressure. The band dispersion relations and the Fermi surface of β*-(BEDT-TTF)$_2I_3$, calculated on the basis of its crystal structure[8] determined at 4.5K under 1.5 kbar, are shown in Figures 3a and 3b, respectively. Practically, the band electronic structures of the two salts are identical. Furthermore, the β-(BEDT-TTF)$_2$X ($X^- = AuI_2^-$, IBr_2^-) salts have the band dispersion relations and the Fermi surfaces very similar to those in Figures 2 and 3. Consequently, the pressure- and anion-dependence of the T_c in β-(BEDT-TTF)$_2$X do not originate from their electronic structures and therefore should be explained in terms of factors associated with their crystal structures.[2]

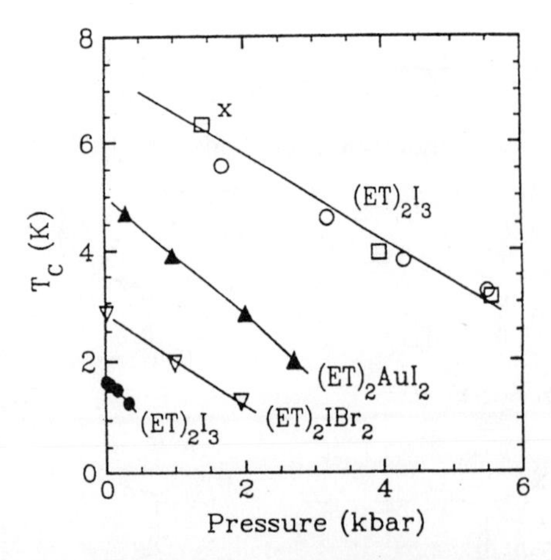

Figure 1. Pressure dependence of the superconducting transition temperatures T_c of β-(BEDT-TTF)$_2$X ($X^- = I_3^-$, AuI_2^-, IBr_2^-)

With the BCS theory of superconductivity,[10,11] Cooper pair formation is induced by electron-phonon interaction, and the extent of this interaction is measured by the electron-phonon coupling constant λ. Given a lattice with Debye temperature θ_D that has a phonon spectrum (i.e., phonon band) effective for electron-phonon coupling, the T_c is related to θ_D and λ as follows:[11]

$$T_c = \frac{\theta_D}{1.45} \exp\left[- \frac{1.04\,(1+\lambda)}{\lambda - \mu^*\left(1 + \lambda \frac{<\omega>}{\omega_o}\right)} \right] \qquad (1)$$

where $<\omega>$ and ω_o are the average and the maximum frequency of the phonon band, respectively. The Coulomb pseudopotential μ^* and the frequency ratio $<\omega>/\omega_o$ are on the

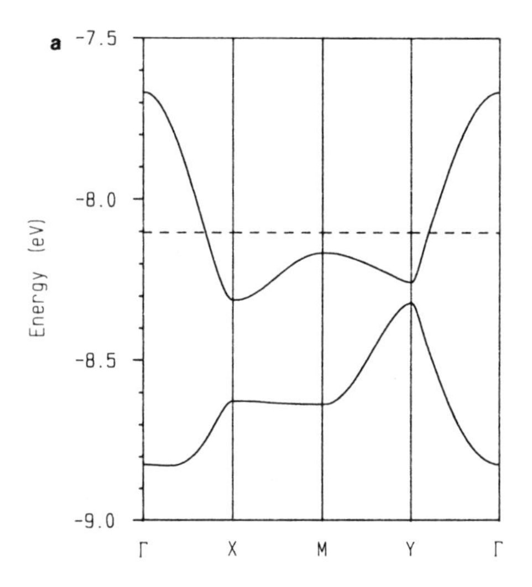

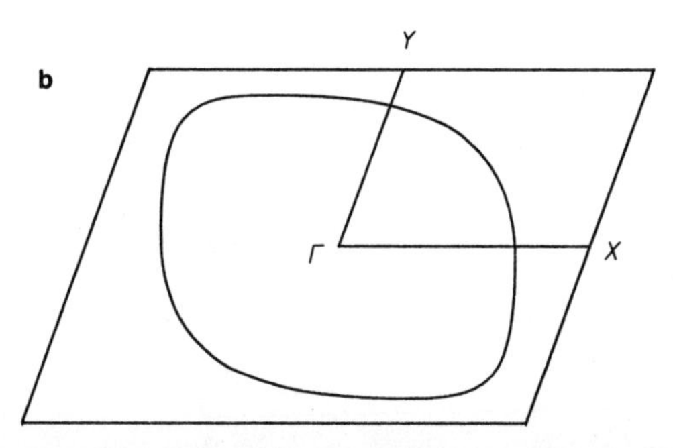

Figure 2. (a) Dispersion relations of the two highest occupied bands and (b) Fermi surface calculated for β-$(BEDT\text{-}TTF)_2I_3$ on the basis of its crystal structure determined at 9K. The dashed line in (a) refers to the Fermi level.

order of 0.1 and 0.5, respectively. Among the several factors affecting the magnitude of T_c in Eq. 1, the most important one is λ. In general, T_c increases with increasing λ. For a lattice with atoms of mass M, λ is expressed as[11]

$$\lambda = \frac{n(e_f) < I^2 >}{M < \omega^2 >} \propto \frac{I}{M < \omega^2 >} \qquad (2)$$

where $n(e_f)$ is the electronic density of states at the Fermi level e_f, $<I^2>$ is the square of the electron-phonon matrix elements averaged over the Fermi surface, and the $<\omega^2>$ is the square of the phonon frequency averaged over the phonon band. As already discussed, the band electronic structures of β-(BEDT-TTF)$_2$X ($X^- = I_3^-$, AuI_2^-, IBr_2^-) are essentially identical and their $n(e_f)$ values are very similar (See Table 1). Thus the $n(e_f)<I^2>$ term should

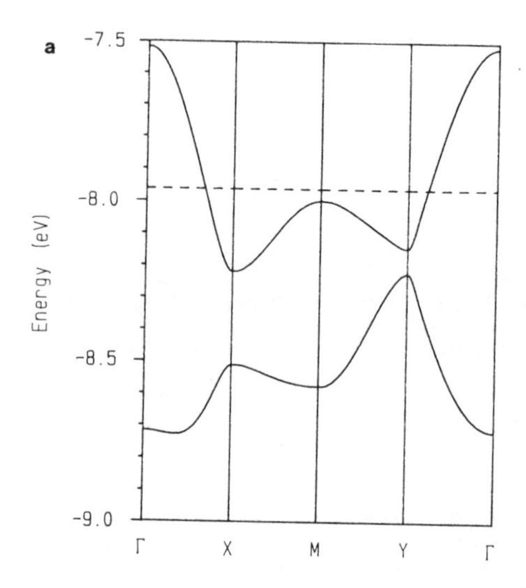

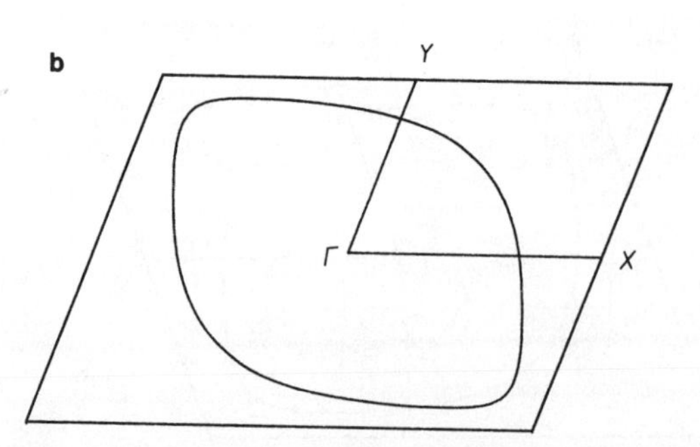

Figure 3. (a) Dispersion relations of the two highest occupied bands and (b) Fermi surfaces calculated for β*-(BEDT-TTF)$_2$I$_3$ on the basis of its crystal structure determined at 4.5K under 1.5 kbar.

Table 1. T_c, $n(e_f)$ and λ values of β-(BEDT-TTF)$_2$X

Salt	T_c (K)	$n(e_f)$ (e^-/eV)	λ
β-(BEDT-TTF)$_2$IBr$_2$	2.8	3.69[a]	0.43
β-(BEDT-TTF)$_2$AuI$_2$	5.0	3.48[b]	0.52
β^*-(BEDT-TTF)$_2$I$_3$	8	3.53[c]	0.62
β-(BEDT-TTF)$_2$I$_3$	1.4	3.66[a]	0.37

[a] Based upon the crystal structure determined at 9 K under ambient presure.
[b] Based upon the crystal structure determined at 20 K under ambient presure.
[c] Based upon the crystal structure determined at 4.5 K under 1.5 kbar.

practically be constant[2,11] for the β-(BEDT-TTF)$_2$X salts, so that λ depends largely upon the denominator $M\langle\omega^2\rangle$, which has the dimension of a force constant. This means that a large λ value results when the lattice has soft phonons, which arise from vibrations with shallow potential wells.[2]

On the basis of the T_c values of β-(BEDT-TTF)$_2$X ($X^- = I_3^-$, AuI_2^-, IBr_2^-), along with the parameters $\mu^* = 0.1$, $\langle\omega\rangle/\omega_o = 0.3$,[12] and $\theta_D = 200$K,[13] we calculate the λ values of these salts by employing Eq. 1. As listed in Table 1, the T_c increases with λ, and the T_c jump in β-(BEDT-TTF)$_2$I$_3$ under pressure correlates with a sudden increase of λ (i.e., a sudden increase in the lattice softness) under pressure.[2] Phonons important for superconductivity are expected to have frequencies of the order of $\sim 8k_BT_c$.[14] Therefore, organic salt superconductors, phonons crucial for their superconductivity are translational and/or librational modes, which are strongly influenced by the C-H···donor and C-H···anion contact interactions.[2] Thus the lattice of an organic salt becomes softer for such phonon modes, when the C-H···donor and C-H···anion interactions possess shallow potential wells and also when the C-H···donor and C-H···anion contact distances become larger.

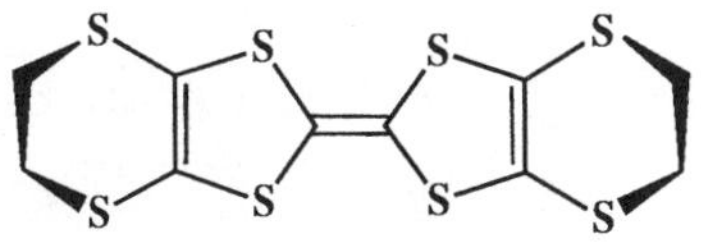

S S S S / S S S S

5a

S S S S / S S S S

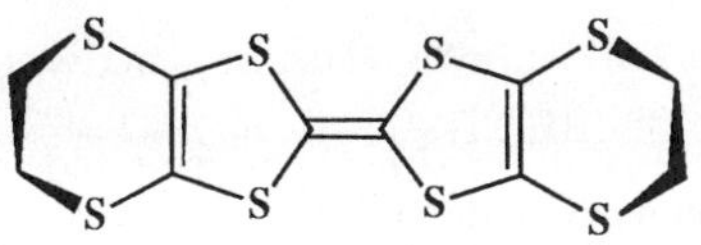

5b

As shown in the perspective diagram **6** each anion X^- of β-(BEDT-TTF)$_2$X (X^- = I_3^-, AuI_2^-, IBr_2^-) is enclosed in a hydrogen pocket made up of 12 BEDT-TTF molecules, and each terminal halogen atom of X^- is surrounded by six ethylene groups.[2] This packing of BEDT-TTF molecules around X^- gives rise to short intermolecular C-H···donor and C-H···anion contacts shorter in length than expected from the van der Waals (VDW) radii sum. When viewed along the central C = C bond of BEDT-TTF, the two ethylene groups of BEDT-TTF are ether eclipsed (E) or staggered (S) as illustrated in **5a** and **5b**, respectively.[2]

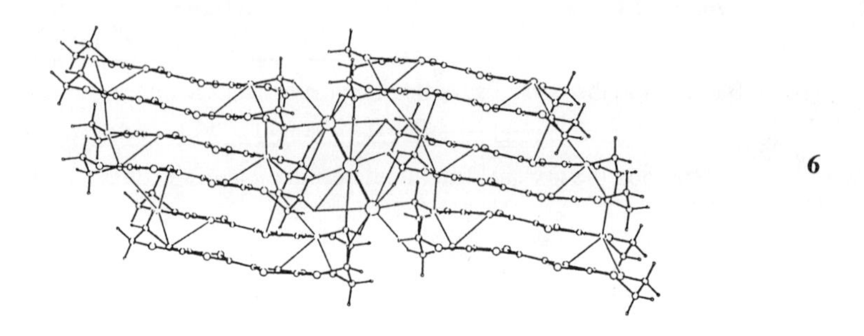

6

For X^- = AuI_2^- and IBr_2^-, the donor molecules of β-(BEDT-TTF)$_2$X have an eclipsed arrangement of ethylene groups. In β-(BEDT-TTF)$_2$I$_3$ both eclipsed and staggered donor molecules are present in nearly equal proportions. The most important structural change β-(BEDT-TTF)$_2$I$_3$ undergoes under pressure greater than 0.5 kbar is that all BEDT-TTF molecules adopt a staggered arrangement of ethylene groups, thereby altering the intermolecular contact distances associated with the C-H bonds.[2] Table 2 lists the shortest C-H···H and C-H···anion (with the terminal halogen atom) contact distances found for β-(BEDT-TTF)$_2$X (X^- = I_3^-, AuI_2^-, IBr_2^-) in their crystal structures below 20K.[2] These C-H···H and C-H···anion distances increases in the order β-(BEDT-TTF)$_2$IBr$_2$<β-(BEDT-TTF)$_2$AuI$_2$ < β*-(BEDT-TTF)$_2$I$_3$. Thus the lattice softness of these salts increase in the same order, which is consistent with the increase in λ in the same order. Under ambient pressure, β-(BEDT-TTF)$_2$I$_3$ has unfavorably short C-H···H and C-H···I contacts. The structural modulation of β-(BEDT-TTF)$_2$I$_3$, which occurs below 200K,[15] reflects the presence of structural strain that leads to the λ value smaller than that of β-(BEDT-TTF)$_2$IBr$_2$. Adoption of a staggered ethylene group arrangement by BEDT-TTF, under $P >$ 0.5 kbar, makes the lattice of β*-(BEDT-TTF)$_2$I$_3$ particularly soft (for the translational and/or the librational modes of vibration) because it makes the C-H···H and C-H···anion contacts longer.[2] A further increase in pressure beyond 0.5 kbar in β*-(BEDT-TTF)$_2$I$_3$ makes the lattice stiffer and lowers the T_c, just as in the case of β-(BEDT-TTF)$_2$X (X^- = AuI_2^-, IBr_2^-).

ISOTOPE EFFECT ON T_c

To discuss how an isotope substitution affects the magnitude of T_c, it is convenient to simplify Eq. 1 as

Table 2. Ethylene group arrangements and the shortest C-H···H and C-H···anion (with the terminal halogen) contact distances in β-$(BEDT\text{-}TTF)_2X$

Salts	Ethylene group arrangement	Shortest intermolecular distance (Å)	
		C-H···H	C-H···anion
β-$(BEDT\text{-}TTF)_2IBr_2$	E	2.169	2.887
β-$(BEDT\text{-}TTF)_2AuI_2$	E	2.224	2.966
β*-$(BEDT\text{-}TTF)_2I_3$	S	2.261	3.014
β-$(BEDT\text{-}TTF)_2I_3$	E & S	2.152	2.842

$$T_c \propto \theta_D \exp\left(-\frac{1}{\lambda}\right) \tag{3}$$

When the hydrogen atoms of BEDT-TTF are replaced with deuterium atoms, the resulting BEDT-TTF-d_8 has a higher mass than BEDT-TTF. According to the preexpotential term θ_D of Eq. 3, the isotope substitution H → D should lower the T_c, because the lowering of the vibrational frequencies associated with the isotope substitution will lower the Debye temperature θ_D. The T_c of an organic salt superconductor should decrease by the H → D substitution if the electron-phonon coupling constant λ is not affected by the isotope substitution. However, this assumption is not necessarily valid. The C-D bond has a lower stretching frequency than does the C-H bond so that, during the stretching vibration, the D atom displacement from the equilibrium position is smaller than expected for the H atom displacement. Thus, the C-D bond is effectively shorter than the C-H bond as far as the C-H(D)···donor and C-H(D)···anion contact interactions are concerned. Therefore, the H → D substitution may provide a softer lattice and hence lead to a larger λ value thereby raising the T_c according to the exponential term exp(-1/λ) of Eq. 3. That is, Eq. 3 shows that an isotope substitution such as H → D should lower the T_c according to the pre-expenential term θ_D but raise the T_c according to the exponential term exp(-1/λ). When the latter effect dominates, the isotope substitution H → D raises the T_c as observed for κ-$(BEDT\text{-}TTF)_2Cu(NCS)_2$ (i.e., 10.4 → 10.8 K).[16]

DONOR···DONOR AND DONOR···ANION INTERACTION

As discussed earlier, the lattice softness as measured by the shortest C-H···H and C-H···anion contact distances allows us to rationalize the pressure- and anion-dependence of the T_c in β-$(BEDT\text{-}TTF)_2X$ ($X^- = I_3^-$, AuI_2^-, IBr_2^-). For general organic donor salts with monovalent anions, donor···donor contacts shorter than the VDW distances (r_{vdw}) consist of

Table 3. Interaction energies ΔE_{vdw} and ΔE_{opt} of the chalcogen···chalcogen contacts calculated at their r_{vdw} and r_{opt} values

Contact	r_{vdw} (Å)	ΔE_{vdw} (kcal/mol)	r_{opt} (Å)	ΔE_{opt} (kcal/mol)
O···O	3.00	1.85	∞	0.00
S···S	3.60	-0.10	4.05	-0.41
Se···Se	3.80	-0.19	---	---
Te···Te	4.20	-1.02	---	---

Table 4. Interaction energies ΔE_{vdw} and ΔE_{opt} of the C-H···donor contacts calculated at their r_{vdw} and r_{opt} values

Contact	r_{vdw} (Å)	ΔE_{vdw} (kcal/mol)	r_{opt} (Å)	ΔE_{opt} (kcal/mol)
C-H···O	2.70	-2.35	2.61	-2.43
C-H···S	3.00	-0.24	3.43	-0.45
C-H···Se	3.10	-0.43	3.44	-0.67
C-H···Te	3.30	-0.31	3.72	-0.54
C-H···C(sp^2)	2.85	-0.55	3.21	-0.69
C-H···H	2.40	-0.11	2.69	-0.15

Table 5. Interaction energies ΔE_{vdw} and ΔE_{opt} of the C-H···anion contacts C-H···Y-I-Y^- calculated at their r_{vdw} and r_{opt} values

Contact	r_{vdw} (Å)	ΔE_{vdw} (kcal/mol)	r_{opt} (Å)	ΔE_{opt} (kcal/mol)
C-H···I	3.15	-0.98	3.50	-1.13
C-H···Br	3.00	-1.19	3.14	-1.27
C-H···Cl	2.90	-1.48	2.85	-1.58

chalcogen···chalcogen, C-H···chalcogen, C-H···C and C-H···H contacts, while donor···anion contacts shorter than r_{vdw} include only the C-H···anion contacts in most cases. Recently, the nature of such short intermolecular interactions has been examined[5,17-19] by performing ab initio MP2 calculations[20] on model molecular systems, essential results of which are summarized in this section.

Chalcogen···chalcogen (X···X) contact interaction energies, obtained by ab initio MP2 calculations[18] on $H_2X \cdots XH_2$ (X = O, S, Se, Te) with the 6-311G(3df, 3pd) basis sets at the VDW and optimum X···X distances (r_{vdw} and r_{opt}, respectively), are listed in Table 3. The O···O contact interaction is repulsive at all O···O distances, but other X···X (X = S, Se, Te) interactions are attractive at r_{vdw}. In the vicinity of r_{vdw}, the magnitudes of the attractive X···X interactions increase in the order: S···S, Se···Se < Te···Te. We note that $r_{opt} > r_{vdw}$ for the S···S interaction. The same trend is expected for the Se···Se and Te···Te interactions.

To evaluate the energies of the C-H···X (X = O, S, Se, Te), C-H···C(sp^2) and C-H···H contact interactions, ab initio MP2 calculations[5b,19] with the 6-311G(d, p) basis set are carried out for $H_3C\text{-}H \cdots XH_2$ (X = O, S, Se, Te) with a C_s-symmetry arrangement and the methane C-H bond aligned along the p_π-orbital or XH_2, $H_3C\text{-}H \cdots CH_2 = CH_2$ with a C_s-arrangement and the methane C-H bond aligned along one p_π-orbital of $CH_2{=}CH_2$, and $H_3C\text{-}H \cdots H\text{-}CH_3$ with a D_{3h}-symmetry arrangement. The interaction energies of these contacts calculated at their r_{opt} and r_{vdw} distances are summarized in Table 4. At r_{vdw} all these C-H···donor interactions are attractive, and their magnitudes increase in the following order: C-H···H < C-H···X (X=S, Se, Te), C-H···C(sp^2) << C-H···O.

Ab initio MP2 calculations with the 3-21+G(d, p) basis set are performed[5b,17] on $H_3C\text{-}H \cdots Y\text{-}I\text{-}Y^-$ (Y = I, Br, Cl) to estimate the interaction energies of the C-H···anion contacts C-H···Y (Y = I, Br, Cl). The interaction energies calculated at r_{opt} and r_{vdw} are summarized in Table 5. At r_{vdw} these C-H···anion interactions are attractive and their magnitudes increase in the order: C-H···I < C-H···Br < C-H···Cl.

From Tables 3-5 the magnitudes of the attractive donor···donor and donor···anion contact interactions at their r_{vdw} are found to increase in the following order:

$$\begin{aligned} &\text{C-H}\cdots\text{H, S}\cdots\text{S, Se}\cdots\text{Se} < \text{C-H}\cdots\text{X (X = S, Se, Te, C }(sp^2)) < \\ &< \text{Te}\cdots\text{Te, C-H}\cdots\text{Y-I-Y}^- \text{ (Y = I, Br, Cl)} < \text{C-H}\cdots\text{O} \end{aligned} \tag{4}$$

Except for the C-H···O and Te···Te interactions, the C-H···anion interactions C-H···Y-I-Y^- are more attractive than all the C-H···donor interactions. The C-H···O interaction is more attractive than the C-H···Y-I-Y^- interactions, while the Te···Te interaction is comparable to the C-H···anion interactions C-H···Y-I-Y^-. The C-H···donor interactions C-H···X (X = S, Se, Te, C(sp^2)) are more attractive than the C-H···H, S···S, and Se···Se interactions. In the following two sections we show that the thermal phase transitions of some organic salt and the unique donor-layer packing of the BEDO-TTF salts are easily explained on the basis of the intermolecular contact interaction energies discused above.

Under thermal tempering, α-$(BEDT\text{-}TTF)_2I_3$ undergoes a phase transition to α_t-$(BEDT\text{-}TTF)_2I_3$,[6a,21-22] which exhibits a superconducting transition near 8K. Several experimental studies strongly suggest that α_t-$(BEDT\text{-}TTF)_2I_3$ is structurally very similar to β*-$(BEDT\text{-}TTF)_2I_3$. Conversion of α-$(BEDT\text{-}TTF)_2I_3$ to a β-like structure by thermal

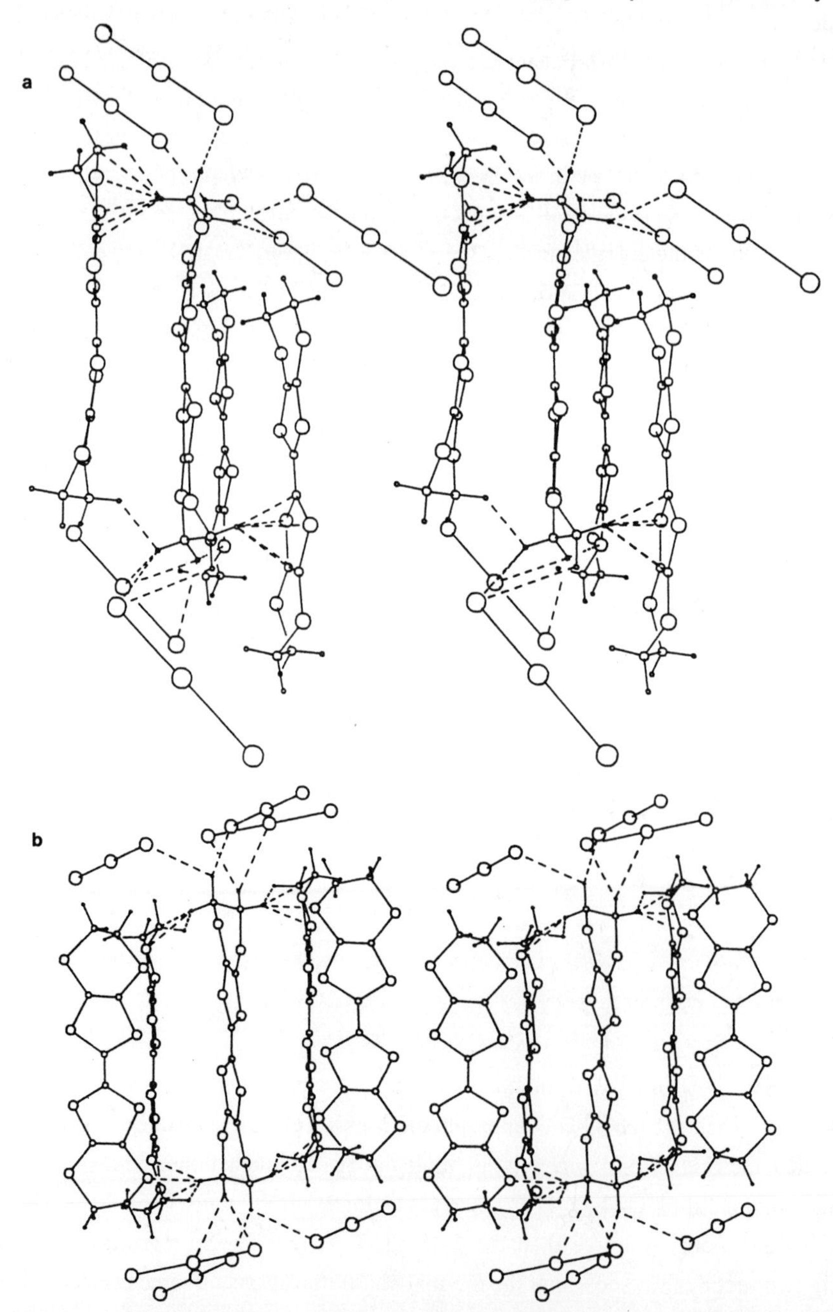

Figure 4. Stereodiagrams of how the donor molecules make C-H···anion and C-H···donor contacts in (a) β*-$(BEDT\text{-}TTF)_2I_3$ and (b) α-$(BEDT\text{-}TTF)_2I_3$.

tempering means that α-$(BEDT\text{-}TTF)_2I_3$ is thermodynamically less stable than β*-$(BEDT\text{-}TTF)_2I_3$. Figures 4a and 4b show the stereodiagrams of how the C-H bonds of one BEDT-TTF interact with the surrounding donor molecules and I_3^- anions in β*-$(BEDT\text{-}TTF)_2I_3$ and α-$(BEDT\text{-}TTF)_2I_3$, respectively.[6a] Each ethylene group uses three C-H Bonds to make C-H···anion contacts in β*-$(BEDT\text{-}TTF)_2I_3$ but only two C-H bonds in α-$(BEDT\text{-}TTF)_2I_3$. Therefore, every BEDT-TTF makes more C-H···anion contacts in β*-$(BEDT\text{-}TTF)_2I_3$ than α-$(BEDT\text{-}TTF)_2I_3$. Since the C-H···anion interaction C-H···I is more attractive than the C-H···donor interactions such as C-H···S, C-H···C(sp^2) and C-H···H, β*-$(BEDT\text{-}TTF)_2I_3$ is expected to be thermodynamically more stable than α-$(BEDT\text{-}TTF)_2I_3$, in agreement with experiment.

δ-$(BEDT\text{-}TTF)_2AuBr_2$ is converted to α'-$(BEDT\text{-}TTF)_2AuBr_2$ at 420K thereby indicating that the latter is thermodynamically more stable.[6b] Figures 5 and 6 show the C-H···anion contacts of δ-$(BEDT\text{-}TTF)_2AuBr_2$ and α'-$(BEDT\text{-}TTF)_2AuBr_2$, respectively. Clearly, the donor molecules have more C-H···anion contacts in α'-$(BEDT\text{-}TTF)_2AuBr_2$ than in δ-$(BEDT\text{-}TTF)_2AuBr_2$.[6b] Since the C-H···anion interaction C-H···Br is more attractive than the C-H···donor interactions such as C-H···S, C-H···(sp^2) and C-H···H, α'-$(BEDT\text{-}TTF)_2AuBr_2$ is expected to be thermodynamically more stable than δ-$(BEDT\text{-}TTF)_2AuBr_2$. This is consistent with the phase transition δ-$(BEDT\text{-}TTF)_2AuBr_2$ → α'-$(BEDT\text{-}TTF)_2AuBr_2$ under thermal tempering.

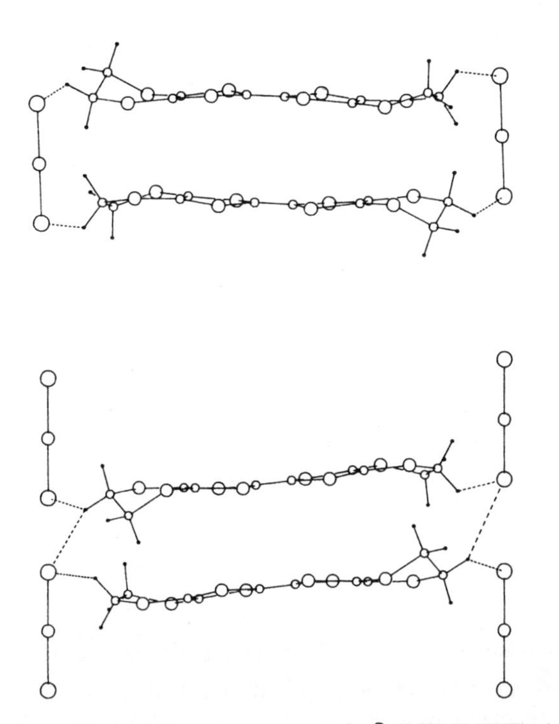

Figure 5. Short C-H···anion contacts in δ-$(BEDT\text{-}TTF)_2AuBr_2$.

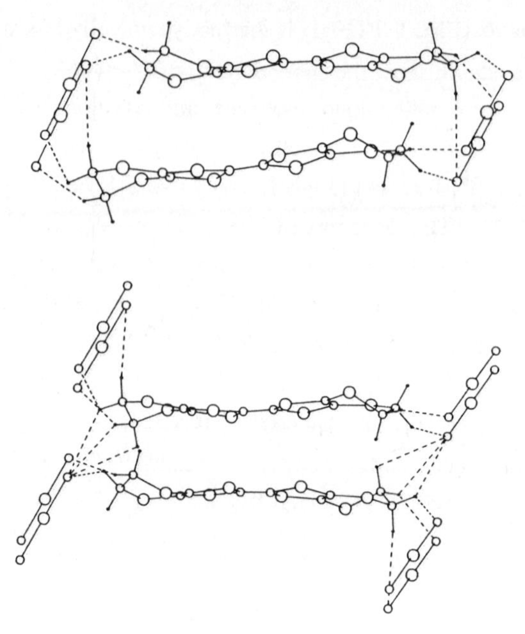

Figure 6. Short C-H···anion contacts in α'-(BEDT-TTF)$_2$AuBr$_2$.

UNIQUE DONOR-LAYER PACKING OF BEDO-TTF SALTS

In a donor stack of (BEDO-TTF)$_2$AuBr$_2$ the donor molecules BEDO-TTF (**2**) slip along the direction perpendicular to the donor molecule central C=C bond, as depicted in **7**.[5, 23] This slipping allows the BEDO-TTF molecules of (BEDO-TTF)$_2$AuBr$_2$ make short C-H···O contacts within each donor stack as well as between adjacent donor stacks as shown in Figures 7. This pattern of donor-molecule packing is common to all known BEDO-TTF salts with different anions including the superconducting phase[24] β_m-(BEDO-TTF)$_2$Cu$_2$(NCS)$_3$. This reflects the fact that the C-H···donor interaction C-H···O is strongly

7

attractive, even more attractive than the C-H···anion interactions such as C-H···Y-I-Y$^-$ (Y = I, Br, Cl) as discussed in the earlier section. For many BEDO-TTF salts it is difficult to resolve the structures of their anion layers due to their disorder. It appears that the particular donor

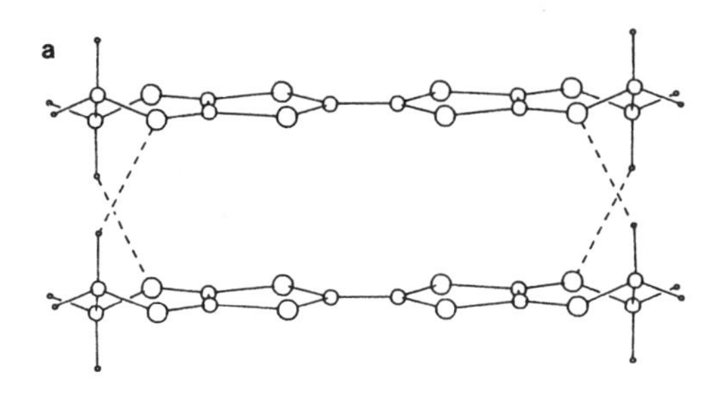

Figure 7. Short C-H···O contacts in the donor layers of (BEDO-TTF)$_2$X (a) within a donor stack and (b) between adjacent donor stacks.

layer packing of the BEDO-TTF salts does not provide good C-H···anion contacts to stabilize their anion layer structures.

ELECTRONIC STRUCTURES OF κ-PHASE

Ambient pressure κ-phase superconductors include κ-(BEDT-TTF)$_2$X (X^- = $Cu[N(CN)_2]Br^-$,[25] $Cu(NCS)_2^-$,[26] I_3^- [27]), κ-(MDT-TTF)$_2$AuI$_2$ [28] and κ-(DMET)$_2$AuBr$_2$.[29] The band dispersion relations of κ-(BEDT-TTF)$_2$Cu[N(CN)$_2$]Br and κ-(BEDT-TTF)$_2$Cu(NCS)$_2$ calculated on the basis of their ~120K structures[30] are shown in Figures 8a and 8b, respectively. In both salts, the lowest-unoccupied (LU) bands derived from the anion chains overlap with the HO, half-filled, bands derived from the donor-molecule layers. This is in contrast to the case of β-(BEDT-TTF)$_2$I$_3$ in which the LU band derived from the I_3^- anions does not overlap with the HO band of the donor molecules as shown in Figure 9.[31] The LU bands of β-(BEDT-TTF)$_2$Cu[N(CN)$_2$]Br and β-(BEDT-TTF)$_2$Cu(NCS)$_2$ have primarily the π* orbital character of the N≡C-N-C≡N$^-$ and N≡C-S$^-$ units, respectively.

The Fermi surfaces associated with the HO bands of κ-(BEDT-TTF)$_2$Cu[N(CN)$_2$]Br and κ-(BEDT-TTF)$_2$Cu(NCS)$_2$ are shown in Figures 10a and 10b, respectively. Essentially, these Fermi surfaces have a shape of overlapping circles. In κ-$(BEDT\text{-}TTF)_2Cu(NCS)_2$, the degeneracy of the intended crossing points of overlapping circles is lifted due to the presence of two nonequivalent donor molecules. Therefore its Fermi surface is split into one-dimensional wavy lines and two-dimensional ellipses. More explicitly, the Fermi surface of κ-$(BEDT\text{-}TTF)_2Cu(NCS)_2$ contains the hole surface (Figure 11a) and the electron surface (Figure 11b), which are respectively associated with the lower and upper subbands of the HO band in Figure 8b. This explains why the nature of the

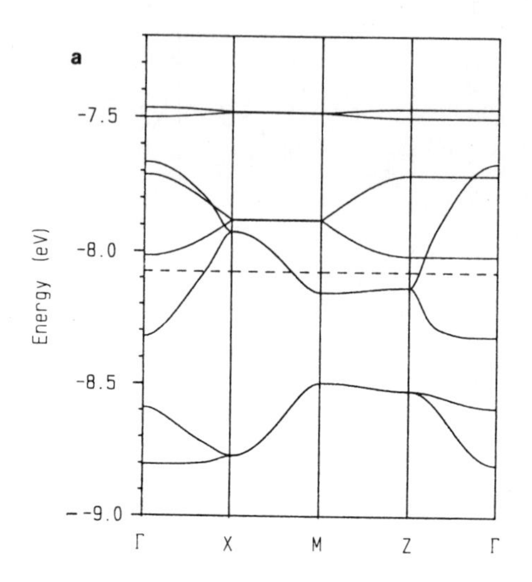

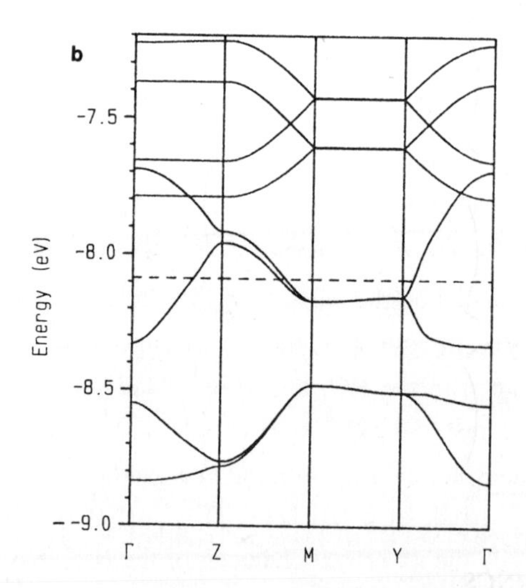

Figure 8. Dispersion relations of the occupied and unoccupied bands around the Fermi level in (a) κ-$(BEDT\text{-}TTF)_2Cu[N(CN)_2]Br$ and (b) κ-$(BEDT\text{-}TTF)_2Cu(NCS)_2$.

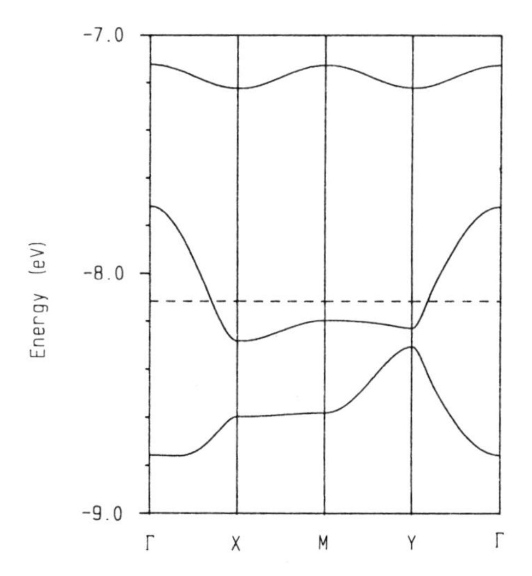

Figure 9. Dispersion relations of the occupied and unoccupied bands around the Fermi level in β-$(BEDT\text{-}TTF)_2I_3$.

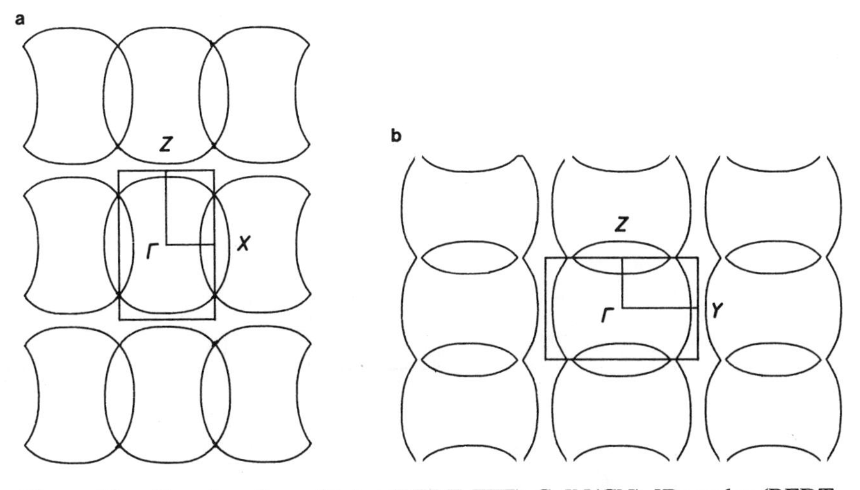

Figure 10. Fermi surface of (a) κ-$(BEDT\text{-}TTF)_2Cu[N(CN)_2]Br$ and κ-$(BEDT\text{-}TTF)_2Cu(NCS)_2$.

carriers, as determined by the thermoelectric power measurements, is hole-like along the $\Gamma \rightarrow$ Z direction (i.e., the crystallographic c-direction) but electron-like along the $\Gamma \rightarrow$ Y direction (i.e., the crystallographic b-direction.)[32] In the Fermi surface of κ-(BEDT-TTF)$_2$Cu[N(CN)$_2$]Br, the degeneracy of the crossing points of the overlapping circles is not lifted. Even in this case, the Fermi surface can be separated into the hole surface (Figure 12a) and the electron surface (Figure 12b), which are respectively associated with the lower and upper subbands of the HO band in Figure 8a. Consequently, the nature of the carriers in κ-(BEDT-TTF)$_2$Cu[N(CN)$_2$]Br, as determined by thermoelectric power measurements, should be hole-like along the $\Gamma \rightarrow$ X direction (i.e., the crystallographic a-direction) but electron-like along the $\Gamma \rightarrow$ Z direction (i.e., the crystallographic c-direction).[33] All known κ-phase superconductors should exhibit similar thermoelectric characteristics since their

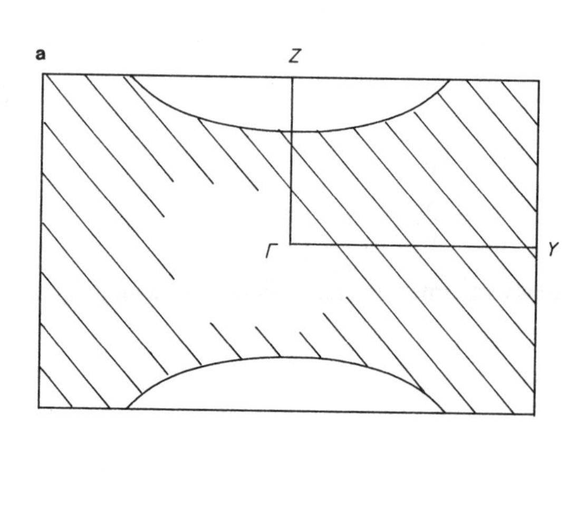

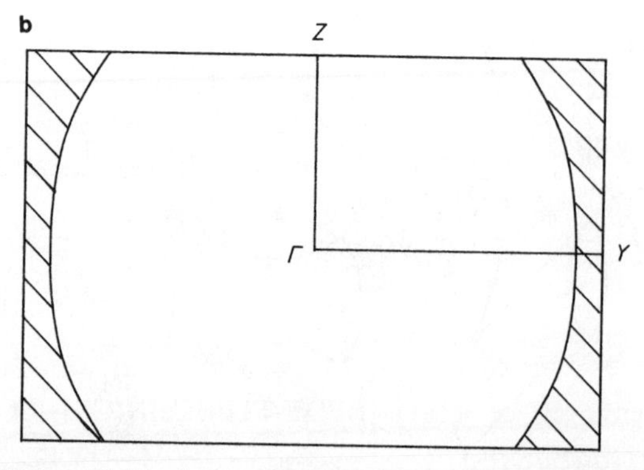

Figure 11. (a) Hole and (b) electron Fermi surfaces of κ-(BEDT-TTF)$_2$Cu(NCS)$_2$, where the occupied wave vectors are indicated by shading.

Table 6. T_c and $n(e_f)$ values of κ-phase superconducting salts

Salt	T_c (K)	$n(e_f)$ (e^-/eV)
κ-$(DMET)_2AuBr_2$	1.8	8.92[a]
κ-$(BEDT\text{-}TTF)_2I_3$	3.6	7.09[a]
κ-$(MDT\text{-}TTF)_2AuI_2$	4.5	7.98[a]
κ-$(BEDT\text{-}TTF)_2Cu(NCS)_2$	10.4	7.39[b]
κ-$(BEDT\text{-}TTF)_2Cu[N(CN)_2]Br$	11.6	7.27[b]

[a] Based upon the crystal structure determined at room temperature.
[b] Based upon the crystal structure determined at ~120 K.

Fermi surface, having the shape of overlapping circles[3], can be separated into hole and electron surfaces. As reported elsewhere,[3] the electronic structures of κ-phase salts are quite similar.

The T_c and calculated $n(e_f)$ values of the superconducting κ-phase salts are listed in Table 6. As in the case of the β-phase superconductors, there exists no apparent correlation between the T_c and $n(e_f)$ values.[3] In the κ-phase superconductors as well, the magnitude of the T_c appears to be largely governed by factors associated with their crystal structures.

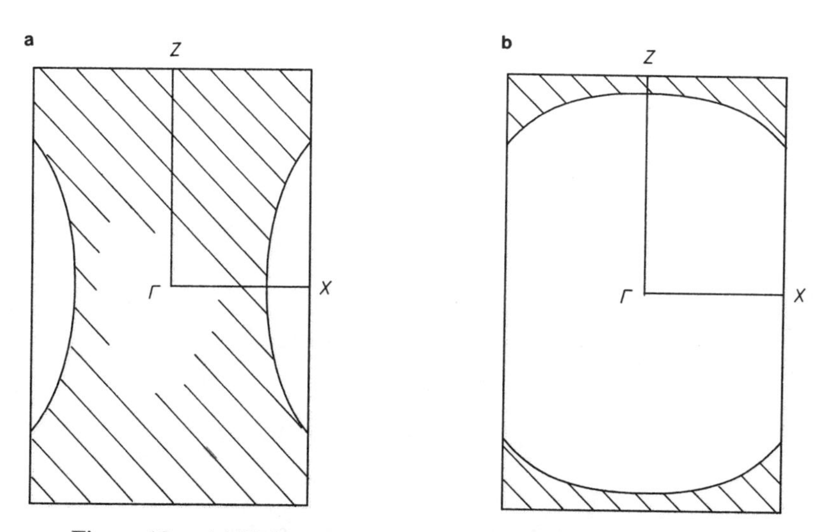

Figure 12. (a) Hole and (b) electron Fermi surfaces of κ-$(BEDT\text{-}TTF)_2Cu[N(CN)_2]Br$.

The κ-phase superconductors are made from different types of donor molecules and different types of anions. Therefore, it is difficult to single out a geometrical parameter that reflects their lattice softness. Nevertheless, the T_c's of the three κ-phases κ-$(DMET)_2AuBr_2$, κ-$(BEDT\text{-}TTF)_2I_3$ and κ-$(MDT\text{-}TTF)_2AuI_2$ may be rationalized on the basis of their C-H···donor and C-H···anion interactions: Since iodine is more polarizable than bromine the C-H···anion contact C-H···I should lead to a softer lattice than does the C-H···anion contact C-H···Br. The latter is consistent with the observation that κ-$(BEDT\text{-}TTF)_2I_3$ and κ-$(MDT\text{-}TTF)_2AuI_2$ have a higher T_c than does κ-$(DMET)_2AuBr_2$. κ-$(BEDT\text{-}TTF)_2I_3$ has short C-H···donor contacts such as C-H···S and C-H···C(sp^2) while such C-H···donor contacts are absent in κ-$(MDT\text{-}TTF)_2AuI_2$. Thus κ-$(MDT\text{-}TTF)_2AuI_2$ would have a softer lattice and hence a higher T_c than does κ-$(BEDT\text{-}TTF)_2I_3$, as observed.

The $Cu[N(CN)_2]Br^-$ anions of κ-$(BEDT\text{-}TTF)_2Cu[N(CN)_2]Br$ and the $Cu(NCS)_2^-$ anions of κ-$(BEDT\text{-}TTF)_2Cu(NCS)_2$ are ribbon-like chains. In κ-$(BEDT\text{-}TTF)_2Cu[N(CN)_2]Br$ one ethylene group of each donor primarily has the C-H···anion contacts (C-H···C(sp) and C-H···N while the other ethylene group primarily has the C-H···anion contact C-H···Br (See Figure 13). In κ-$(BEDT\text{-}TTF)_2Cu(NCS)_2$ one of the two nonequivalent donors makes primarily the C-H···anion contacts C-H···C(sp) and C-H···N

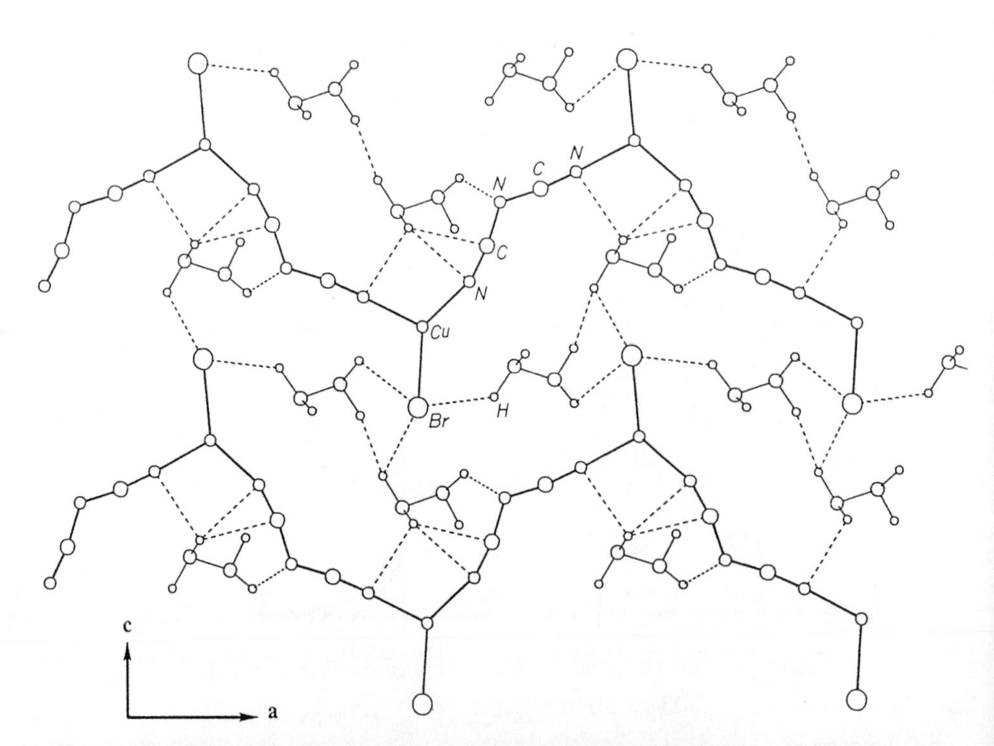

Figure 13. Schematic diagram of how the ethylene groups make short C-H···anion and C-H···H contacts in κ-$(BEDT\text{-}TTF)_2Cu[N(CN)_2]Br$.

while the other donor primarily has the C-H···anion contact C-H···S (See Figure 14). All these C-H···anion contact interactions cannot be softer than the C-H···anion interaction C-H···I present in κ-$(BEDT-TTF)_2I_3$ and κ-$(MDT-TTF)_2AuI_2$, but κ-$(BEDT-TTF)_2X$ (X^- = $Cu[N(CN)_2]Br^-$,$Cu(NCS)_2^-$) has a much higher T_c. Therefore the ribbon-like chain anions must provide a phonon spectrum conducive for high-T_c superconductivity.

Table 7 summarizes some important structural properties of the κ-$(BEDT-TTF)_2X$ (X^- = $Cu[N(CN)]Br^-$, $Cu(NCS)_2^-$) salts obtained from their crystal structures at ~120K. The number of all donor···donor and donor···anion contacts shorter than r_{vdw} is greater in κ-$(BEDT-TTF)_2Cu(NCS)_2$ than in κ-$(BEDT-TTF)_2Cu[N(CN)_2]Br$. This suggests that κ-$(BEDT-TTF)_2Cu[N(CN)_2]Br$ has a softer lattice than does κ-$(BEDT-TTF)_2Cu(NCS)_2$, in agreement with the observation that κ-$(BEDT-TTF)_2Cu[N(CN)_2]Br$ has a higher T_c. It is desirable to devise an index by which to measure a lattice softness. For an intermolecular contact with the actual and VDW distances r and r_{vdw}, respectively, we define the strain index (SI) as follows:

$$SI = \frac{r_{vdw} - r}{r_{vdw}} \times 100 \tag{5}$$

Among the numerous intermolecular contacts of organic donor salts, the most important contacts associated with the lattice strain and hence the lattice hardness are the shortest ones in various categories of the donor···donor and the donor···anion interactions. For such contacts, $r < r_{vdw}$ and the strain indices are positive. Since a contact with a greater amount of strain should have a larger strain index, a softer crystal lattice would have more

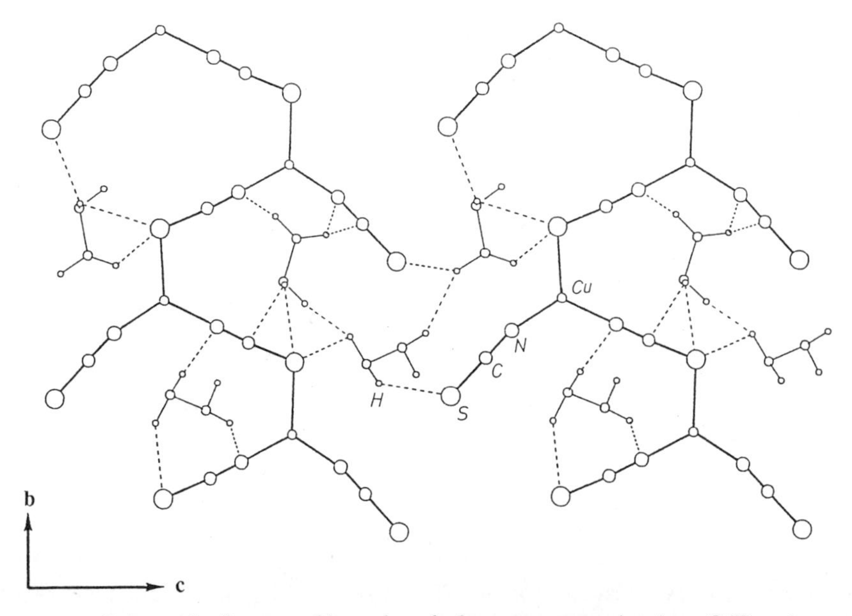

Figure 14. Schematic diagram of how the ethylene groups make short C-H···anion and C-H···H contacts in κ-$(BEDT-TTF)_2Cu(NCS)_2$.

Table 7. Structural properties of κ-$(BEDT\text{-}TTF)_2X$

Property	$X^- = Cu[N(CN)_2]Br^-$	$X^- = Cu(NCS)_2^-$
Anion shape	Ribbon-like chain	Ribbon-like chain
Intradimer spacing	3.30 Å	3.32 Å
Ethylene group arrangement	Eclipsed	Staggered
Donor leaning direction	Parallel to the anion chain	Perpendicular to the anion chain
Number of intermolecular contacts shorter than r_{vdw}	Fewer	More

Table 8. Strain indices (SI) calculated for the shortest donor···donor and donor···anion contacts of κ-$(BEDT\text{-}TTF)_2X$

Class	Contacts	r_{vdw}	$X^- = Cu[N(CN)_2]Br^-$		$X^- = Cu(NCS)_2^-$	
		(Å)	r(Å)	SI	r(Å)	SI
Donor···donor	S···S	3.60	3.45	4.2	3.51	2.5
	C-H···H	2.40	2.15	10	2.18	9.2
	C-H···S	3.00	2.77	7.7	2.64	12
Donor···anion	C-H···Br	3.00	2.98	0.7	-----	-----
	C-H···S	3.00	-----	-----	2.71	9.7
	C-H···C	2.85	2.74	3.9	2.71	4.8
	C-H···N	2.75	2.69	2.2	2.56	7.4

intermolecular contacts with smaller strain indices. Table 8 summarizes the strain indices calculated for κ-$(BEDT\text{-}TTF)_2X$ (X^- = $Cu[N(CN)_2]Br^-$, $Cu(NCS)_2^-$) on the basis of their ~120K crystal structures. Except for the donor···donor contacts S···S and C-H···H, all the shortest contacts have smaller strain indices in κ-$(BEDT\text{-}TTF)_2Cu[N(CN)_2]Br$ thereby suggesting that the crystal lattice of κ-$(BEDT\text{-}TTF)_2Cu[N(CN)_2]Br$ is softer than that of κ-$(BEDT\text{-}TTF)_2Cu(NCS)_2$. This is consistent with the finding that κ-$(BEDT\text{-}TTF)_2Cu[N(CN)_2]Br$ has a higher T_c.[25]

CONCLUDING REMARKS

The electronic properties of the β-phase superconducting salts are very similar, and so are those of the κ-phase superconductors. Therefore, the anion- or pressure-dependence of the T_c's of these salts is not controlled by their electronic structures but by the softness of their lattices toward the translational and/or librational modes of vibrations (with frequencies of the order of $\sim 8k_BT_c$). Such vibrations are strongly governed by the intermolecular C-H···donor and C-H···anion contact interactions. In general, a softer lattice has a larger electron-phonon coupling constant λ and thus a higher superconducting transition temperature T_c.

As a semi-quantitative measure of the lattice softness, we defined the strain indices of short intermolecular contacts and examined the softness of the κ-phase salts κ-(BEDT-TTF)$_2$X (X^-= Cu[N(CN)$_2$]Br^- and Cu(NCS)$_2^-$ in terms of their softness indices. In general, a softer lattice is expected to possess more short contacts with smaller strain indices.

An isotope substitution such as H → D in a donor molecule induces two opposing effects on the T_c of its salt, i.e., a T_c-lowering due to a decrease in the Debye temperature and a T_c-raising due to an increase in the lattice softness. When the latter dominates, the T_c increases by an isotope substitution such as H → D.

The C-H···donor and C-H···anion contact interactions are also crucial for the packing of the organic salts and hence for their electronic structures. At the VDW distances, the donor···donor contacts X···X (X = S, Se, Te), C-H···X (X = O, S, Se, Te, C (sp^2)) and C-H···H as well as the donor···anion contacts C-H···Y-I-Y^- (Y = I, Br, Cl) are all attractive. The magnitude of these interactions at the VDW distances increase in the order: C-H···H, S···S,Se···Se < C-H···X (X = S, Se, Te, C(sp^2)) < Te···Te, C-H···Y-I-Y^- (Y = I, Br, Cl) < C-H···O. The thermal transformations α- to α_t-(BEDT-TTF)$_2$I$_3$ and δ- to α'-(BEDT-TTF)$_2$AuBr$_2$ are both found to be a phase transition which increases the number of the C-H···anion contacts per molecule in the lattice thereby leading to a thermodynamically more stable crystal packing. The BEDO-TTF salts have a unique packing in which the donor molecules make short C-H···O contacts within each donor stack as well as between adjacent donor stacks. This packing is a consequence of the fact that the C-H···donor interaction C-H···O is strong compared with the C-H···anion interactions present in the salts.

ACKNOWLEDGMENT

Work at North Carolina State University and at Argonne National Laboratory is supported by the Office of Basic Energy Sciences, Division of Materials Sciences, U. S. Department of Energy, under Grant DE-FG05-86ER45259 and Contract W-31-109-ENG-38, respectively. We express our appreciation for computing time on the ER-Cray Computer, made available by DOE, and that on VAX-8700 at Argonne National Laboratory provided by the Chemistry Division. J. J. N. thanks the NATO and Ministerio de Educacion y Ciencia (Spain) for Fellowships, which made it possible to visit North Carolina State University.

REFERENCES

1. Williams, J. M.; Wang, H. H.; Emge, T. J.; Geiser, U.; Beno, M. A.; Carlson, K. D.; Thorn, R. J.; Schultz, A. J.; Whangbo, M. -H. Prog. Inorg. Chem. **1987**, 35, 51.
2. (a) Whangbo, M. -H.; Williams, J. M.; Schultz, A. J.; Emge, T. J.; Beno, M. A. J. Am. Chem. Soc. **1987**, 109, 90.
   (b) Whangbo, M. -H.; Williams, J. M.; Schultz, A. J.; Beno, M. A. Organic and Inorganic Low-Dimensional Crystalline Materials, Delhaes, P.; Drillon, M., Eds.: Plenum; New York, 1989; p 333.
3. (a) Jung, D.; Evain, M.; Novoa, J. J.; Whangbo, M. -H.; Beno, M. A.; Kini, A. M.; Schultz, A. J.; Williams, J. M.; Nigrey, P. J. Inorg. Chem. **1989**, 28, 4516.
   (b) Whangbo, M. -H.; Jung, D.; Wang, H. H.; Beno, M. A.; Williams, J. M.; Kikuchi, K. Mol. Cryst. Liq. Cryst. **1990**, 181, 1.
4. (a) Williams, J. M.; Wang, H. H.; Kini, A. M.; Carlson, K. D.; Beno, M. A.; Geiser, U., Whangbo, M. -H.; Jung, D.; Evain, M.; Novoa, J. J. Mol Cryst. Liq. Cryst. **1990**, 181, 59.
   (b) Williams, J. M.; Kini, A. M.; Geiser, U.; Wang, H. H.; Carlson, K. D; Kwok, W. K.; Vandervoort, K. G.; Thompson, J. E.; Stupka, D. L.; Jung, D.; Whangbo, M. -H. This volume.
5. (a) Whangbo, M. -H.; Jung, D.; Ren, J.; Evain, M.; Novoa, J. J.; Mota, F.; Alvarez, S.; Williams, J. M.; Beno, M. A.; Kini, A. M.; Wang, H. H.; Ferraro, J. R. The Physics and Chemistry of Organic Superconductors; Saito, G.; Kagoshima, S., Eds.: Springer-Verlag, in press.
   (b) Novoa, J. J.; Whangbo, M. -H.; Willaims, J. M. Mol. Cryst. Liq. Cryst. **1990**, 181, 25.
6. (a) Wang, H. H.; Ferraro, J. R.; Carlson, K. D.; Montgomery, L. K.; Geiser, U; Williams, J. M.; Whitworth, J. R.; Schlueter, J. A.; Hill, S.; Whangbo, M. -H.; Evain, M.; Novoa, J. J. Inorg. Chem. **1989**, 28, 2267.
   (b) Montgomery, L. K.; Wang, H. H.; Schlueter, J. A.; Geiser, U.; Carlson, K. D.; Williams, J. M.; Rubinstein, R. L.; Brennan, T. D.; Stupka, T. D.; Whitworth, J. R.; Jung, D.; Whangbo, M. -H. Mol. Cryst. Liq. Cryst. **1990**, 181, 197.
7. (a) Schirber, J. E.; Azevedo, L. J.; Kwak, J. F.; Venturini, E. L.; Leung, P. C. W.; Beno, M. A.; Wang, H. H.; Williams, J. M. Phys. Rev. B **1986**, 33, 1987.
   (b) Schirber, J. E.; Azevedo, L. J.; Kwak, J. E.; Venturini, E. L.; Beno, M. A.; Wang, H. H.; Williams, J. M. Solid State Commun. **1986**, 59, 525.
8. (a) Schultz, A. J.; Wang, H. H.; Williams, J. M.; Filhol, A. J. Am. Chem. Soc. **1986**, 108, 7853.
   (b) Schultz, A. J.; Beno, M. A.; Wang, H. H.; Williams, J. M. Phys. Rev. B **1986**, 33, 7823.

9. Emge, T. J.; Wang, H. H.; Geiser, U.; Beno, M. A.; Webb, K. S.; Williams, J. M. J. Am. Chem. Soc. **1986**, 108, 3849.
10. (a) Bardeen, J.; Cooper, L. N.; Schrieffer, J. R. Phys. Rev. **1987**, 108, 1175.
(b) Solymar, L.; Walsh, D. Lectures on the Electrical Properties of Materials; Oxford University Press: Oxford, 1988; 4th Ed., Ch. 14.
11. McMillan, W. L. Phys. Rev. **1968**, 167, 331.
12. Nowack, A.; Weger, M.; Schweitzer, D.; Keller, H. J. Solid State Commun. **1986**, 60, 199.
13. (a) Stewart, G. R.; O'Rourke, J.; Crabtree, G. W.; Carlson, K. D.; Wang, H. H.; Williams, J. M. Phys. Rev. B **1986**, 33, 3046.
(b) Stewart, J. R.; Williams, J. M.; Wang, H. H.; Hall, L. N.; Perozo, M. T.; Carlson, K. D. Phys. Rev. B **1986**, 34, 5609.
14. (a) Mitrovic, B.; Leavens, C. B.; Carbotte, J. P. Phys. Rev. B **1980**, 21, 5048.
(b) Hawley, M. E.; Gray, K. E.; Terris, B. D.; Wang, H. H.; Carlson, K. D.; Williams, J. M. Phys. Rev. Lett. **1986**, 57, 629.
15. (a) Leung, P. C. W.; Emge, T. J.; Beno, M. A.; Wang, H. H.; Williams, J. M.; Petricek, V.; Coppens, P. J. Am. Chem. Soc. **1984**, 106, 7644.
(b) Leung, P. C. W.; Emge, T. J.; Beno, M. A.; Wang, H. H.; Williams, J. M.; Petricek, V.; Coppens, P. J. Am. Chem. Soc. **1985**, 107, 6184.
16. (a) Ferraro, J. R.; Wang, H. H.; Geiser, U.; Kini, A. M.; Beno, M. A.; Williams, J. M.; Hill, S.; Whangbo, M. -H.; Evain, M. Solid State Commn. **1988**, 68, 917.
(b) Sugano, T.; Hayashi, H.; Kinoshita, M.; Nishikida, K. Phys. Rev. B **1989**, 37, 11387.
(c) Tokumoto, M.; Anzai, H.; Takahashi, K.; Hayakawa, Y.; Nagamori, H.; Nagasaka, K. Synth. Met. **1988**, 27, A171.
(d) Oshima, K.; Urayama, H.; Yamochi, H.; Saito, G. J. Phys. Soc. Jpn. **1988**, 57, 730.
17. Novoa, J. J.; Mota, F.; Whangbo, M. -H.; Williams, J. M. Inorg. Chem., in press.
18. Novoa, J. J.; Whangbo, M. -H.; Williams, J. M. This Volume.
19. Novoa, J. J.; Whangbo, M. -H. Williams, J. M. In preparation.
20. Details of our ab initio MP2 calculations are described in Ref. 5b.
21. Baram, G. O.; Buravov, L. I.; Degtyarev, L.S.; Kozlov, M. E.; Laukhin, V. N.; Laukhina, E. E.; Onishchenko, V. G.; Pokhodnya, K. I.; Sheinkman, M. K.; Shibaeva, R. P.; Yagubskii, E. B. JETP Lett. **1986**, 44, 376.
22. (a) Schweitzer, D.; Brunner, H.; Gogu, E.; Haeberten, U.; Hennig, I.; Klutz, I.; Swietlik, R.; Keller, H. J. Z. Phys. B **1987**, 67, 489.
(b) Schweitzer, D.; Keller, H. J. Organic and Inorganic Low-Dimensional Crystalline Materials; Delhaes, P.; Drillon, M., Eds.: Plenum; New York, 1987; p 219.
(c) Swietlik, R.; Schweitzer, D.; Keller, H. J. Ibid; p 325.
23. Beno, M. A.; Wang, H. H.; Carlson, K. D.; Kini, A. M.; Frankenbach, G. M.; Ferraro, J. R.; Larson, N.; McCabe, G. D.; Thompson, J. E.; Purnama, C.; Vashon,

M.; Williams, J. M.; Jung, D.; Whangbo, M. -H. Mol. Cryst. Liq. Cryst. **1990**, 181, 145.

24. Beno, M. A.; Wang, H. H.; Kini, A. M.; Carlson, K. D.; Geiser, U.; Kwok, W. K.; Thompson, J. E.; Williams, J. M.; Ren, J.; Whangbo, M. -H. Inorg. Chem. **1990**, 29, 1599.
25. Kini, A. M.; Geiser, U.; Wang, H. H.; Carlson, K. D.; Williams, J. M.; Kwok, W. K.; Vandervoort, K. G.; Thompson, J. E.; Stupka, D. L.; Jung, D.; Whangbo, M. -H. Inorg. Chem. **1990**, 29, 2555.
26. (a) Urayama, H.; Yamochi, H.; Saito, G.; Nozawa, K.; Sugano, T.; Kinoshita, M.; Sato, S.; Oshima, K.; Kawamoto, A.; Tanaka, J. Chem. Lett. **1988**, 55.
    (b) Urayama, H.; Yamochi, H.; Saito, G.; Sato, S.; Kawamoto, A.; Tanaka, A.; Mori, T.; Maruyama, Y.; Inokuchi, H. Chem. Lett. **1988**, 463.
    (c) Gärtner, S.; Gogu, E.; Heinen, I.; Keller, H. J.; Klutz, T.; Schweitzer, D. Solid State Commun. **1988**, 65, 1531.
    (d) Carlson, K. D.; Geiser, U.; Kini, A. M.; Wang, H. H.; Montgomery, L. K.; Kwok, W. K.; Beno, M. A.; Williams, J. M.; Cariss, C. S.; Crabtree, G. W.; Whangbo, M. -H.; Evain, M. Inorg. Chem. **1988**, 27, 965, 2904.
27. Kobayashi, A.; Kato, R.; Kobayashi, H.; Moriyama, S.; Nishio, Y.; Kajita, K.; Sasaki, W. Chem. Lett. **1987**, 459.
28. (a) Papavassiliou, G. C.; Mousdis, G. A.; Zambounis, J. S.; Terzis, A.; Hountas, S.; Hilti, B.; Mayer, C. W.; Pfeiffer, J. Synth. Met. **1988**, 27, B379.
    (b) Kini, A. M.; Beno, M. A.; Son, D.; Wang, H. H.; Carlson, K. D.; Porter, L. C.; Welp, U.; Vogt, B. A.; Williams, J. M.; Jung, D.; Evain, M.; Whangbo, M. -H.; Overmyer, D. L.; Schirber, J. E. Solid State Commun. **1988**, 69, 503.
29. (a) Kikuchi, K.; Honda, Y.; Ishikawa, Y.; Saito, K.; Ikemoto, I.; Murata, K.; Anzai, H.; Ishiguro, T.; Kobayashi, K. Solid State Commun. **1988**, 66, 405.
    (b) Kikuchi, K.; Ishikawa, Y.; Saito, K.; Ikemoto, I. Synth. Met. **1988**, 27, B391.
30. Geiser, U. et al. In preparation.
31. First-principles self-consistent-field band electornic structure calculations give essentially an identical result. See: Kasowski, R. V.; Whangbo, M. -H. Inorg. Chem. **1990**, 29, 260.
32. Urayama, H.; Yamochi, H.; Saito, G.; Sugano, T.; Kinoshita, M.; Inabe, T.; Mori, T.; Maruyama, Y.; Inokuchi, H. Chem. Lett. **1988**, 1057.
33. This prediction is consistent with the preliminary results of the thermoelectric power measurements carried out by Chaikin, P. M. et al.
34. All band electronic structures reported in the present work are based upon extended Hückel tight-binding calculations. For details, see:
    (a) Whangbo, M. -H.; Hoffmann, R. J. Am. Chem. Soc. **1978**, 100, 6093.
    (b) Whangbo, M. -H.; Williams, J. M.; Leung, P. C. W.; Beno, M. A.; Emge, T. J.; Wang, H. H.; Carlson, K. D.; Crabtree, G. W. J. Am. Chem. Soc. **1985**, 107, 5815.

WHAT CAN WE LEARN FROM HIGH T_c CERAMIC SUPERCONDUCTORS FOR THE DESIGN OF ORGANIC SUPERCONDUCTORS?

János J. Ladik

Chair for Theoretical Chemistry at the Friedrich-Alexander-University Erlangen-Nürnberg, Egerland-str. 3, D-8520 Erlangen, FRG

ABSTRACT

From the breakthrough in the field of inorganic superconductors (ceramic materials) we can see that the system has to be

1. two-dimensional
2. it should have a broader partially filled band
3. some filled narrow bands.

The latter two conditions fulfil the requirements of a generalized polarization model which includes as special case the excitonic model of superconductivity.

At the end of the lecture some examples to design possible new organic superconductors will be presented.

INTRODUCTION

High T_c superconductivity has been predicted by different theoreticians since 1950. London has raised the question in the introduction of his book "Superfluids" [1] already in 1950 "whether a superfluid-like state might occur in certain macromolecules which play an important role in biochemical reactions". Little [2] has postulated in 1964 high T_c (room-temperature) superconductivity in a long chain with mobile electrons if it has side chains with localized electrons (whose polarization can be described in first order with the help of virtual excitons; "excitonic theory of high T_c superconductivity") which play the same role as the phonons in the usual theory of superconductivity. Applying a BCS-type theory [3] for this case, due to the mass ratio $(M_{nucleon}/m_e)$ = 1840 he has estimated T_c to be about 2000 K.

There were two important counterarguments against this theory. First of all, as was pointed out by several authors (see, for instance, Ref. 4), there cannot be long-range order in quasi-1-dimensional systems, because of thermodynamic time-dependent fluctuations would destroy it. Little, in a subsequent paper [5], came to the conclusion that, in 1-dimensional (1D) systems with nonzero two other dimensions (quasi-1D

Organic Superconductivity, Edited by V.Z. Kresin and W.A. Little, Plenum Press, New York, 1990

systems) and due to the effects of environment, still a greatly enhanced conductivity can occur in some local règions of the chain.

Actually as Landau and Lifschitz write in their book [6] on "Statistical Mechanics" that the probability of fluctuations in a quasi-1D system is the higher, the larger is the ratio of the large dimension as compared to the smaller two other dimensions. According to the theorem of Wagner and Mermin the fragment fluctuations hinder the possibility of a phase transition. As far as the author knows these considerations were not applied in detail for the case of superconductivity [7]. On the other hand it seems so, that this argument is valid because one can notice in stacked systems (see below) that T_c increases with the number of rings on the same "floor" of the stack. In this way $T_c = 10K$ was reached [8] in the (BEDT-TTF) $Cu(NCS)_2$ system and 11.6K in the $(ET)_n X_m$ system which contains bridging molecules between the stacks making it more 2-dimensional [9].

A further argument of Little was that the interaction of a chain with its environment and with other chains may make the system 3-dimensional or at least quasi-2D [10]. Of course the validity of such a model is depending on the particular case: how strong is the direct coupling and coupling through tunneling between a chain and the solvent molecules surrounding it, or the coupling between the solvent and the other chains (if the system consists of not only one type of chains) as well as the coupling between the different chains. In the author's opinion to decide for a given system the role of fluctuations one has to perform detailed quantum chemical calculations on all these interactions. There is a similar situation to the superconductivity in a 3D material which consists of small grains which are strongly coupled by Josephson tunneling. A close cooperation between organic and quantum chemists as well as physicists would be needed to try to design such an organic molecular crystal built up from chains and having an organic solvent (with possibly larger solvent molecules) which would give enough coupling between the constituents to provide a quasi-2D analogy of the ceramic materials.

The other difficulty with Little's original theory was, as the author, Biczó and Zawadowski have pointed out already in 1965 [11], that due to the spatial separation of the mobile electrons from the localized ones, as detailed quantum chemical calculations have shown it, one does not obtain in this way a larger attraction term in absolute value than the screened Coulomb term. To overcome this difficulty, it was pointed out in the same paper that one can more probably expect high T_c superconductivity in a system of stacked planar molecules. In this case the π-electrons are the mobile ones and the σ-electrons (forming covalent bonds) or lone pair electrons are the localized ones which therefore can be polarized and in this way provide attraction between a pair of mobile π-electrons. This model does not separate in space the mobile and polarizable localized electrons, but distinguishes between the two electronic systems only in a quantum mechanical sense. It should be pointed out that most of the organic superconductors, though they do not show a very high T_c, are actually stacked systems [12].

Subsequently, in a series of papers using semiempirical and ab initio band structures [13-15], it was shown that in a nucleotide base stack (cytosine stack) the above sketched σ-excitonic theory of high T_c superconductivity does not exclude the existence of a superconductive state (for instance, in the ab initio case [14] at the lower limit of the conduction band the attractive term is larger by 2 eV than the repulsive Coulomb term). However, due to fluctuations the superconducting regions in the stacks of the nucleotide base pairs are most probably so small, that one can hardly design an experiment to discover them (due to the sugar-phosphate backbone DNA is a polyelectrolyte).

Already in 1964 Ginzburg has supposed high T_c superconductivity on the surface of a crystal [16]. Subsequently in 1970 [17] he has published a 3D model with alternating macroscopic conductor and insulator layers. He has assumed that the mobile electrons of the conductor layers tunnel over to the insulators and polarize the localized electrons of these layers. In this way the polarization of the electronic distribution of the electrons in the insulator provide again an attractive interaction between two mobile electrons of the conductor layers. One should observe also that by assuming the tunneling of the electrons of the conductor into the insulator the two electronic systems (the mobile and the localized electrons) are again not separated in space. The Ginzburg model was investigated in detail by Allender, Bray, and Bardeen [18] assuming only one conductor and one insulator layer. They have come to the conclusion that high T_c superconductivity in such a system is rather probable.

CERAMIC HIGH T_c SUPERCONDUCTORS

The experimental discovery of superconductivity in $La_{2-x}Sr_xCuO_4$ with T_c at 30-40K by Bednorz and Müller in 1986 [19] and in $YBa_2Cu_3O_{7-x}$ ("1 2 3"compound) at T_c 90-100K by the Chu's group in 1987 [20] despite all the previous theoretical predictions has come as a great surprise to the physicists' community. After this also Bi and Tl containing compounds were found with T_c up to 107 and to 125K [21], respectively. All these complicated ceramic systems possessing 4 or 5 different types of atoms as X-ray analyses show it, have one common feature: they contain Cu-O planes in a tetragonal arrangement of the two kinds of atoms. In addition to their crystal structures, a very large nubmer of other experiments have been performed on these materials.

In the same way in the last 2 ½ years also a large number of theoretical models have been suggested. Without going into details a few of them should be mentioned. The resonating valence bond theory of Anderson et al. [22], the spin bag model of Schrieffer et al. [23], the plasmon model of Kresin [24]. Furtheron there are still models which assume the leading role of phonons or vibrational bipolarons, respectively (see for instance Yamaguchi et al. [25]), Tachibana's model with vibronic interactions [26] and different forms of magnetic (spin polaron) models [27,28]. Finally, there are theories for high T_c superconductivity based on the excitonic mechanism [29] or more generally on the narrow band with polarizable electrons-broad band with mobile electrons model [30,31].

Analyzing the most important experiments, Little in a recent paper [32] came to the conclusion that superconductivity in the ceramic materials is due to singlet, s-state, Cooper electron- or hole pairs which are weakly coupled and therefore a BCS-type [3] theory can be applied for them. As a source of coupling he postulates an excitonic mechanism which is based on the polarization of the narrow d band electrons which gives rise to an attraction between the positive holes in the broad band due to oxygen 2p-orbitals. One should point out, however, that, at least in the "1 2 3 compound", whether there exists an excitonic spectrum is controversal (until 3 eV no such spectrum was found [33], though recent experiments [34] show in this system a strong absorption in the 4-5 ev region). The authors interpret this part of the spectra as a charge transfer spectrum due to Cu 3d $\rightarrow$ O 2p transition. Since, however, charge transfer spectra are usually weak, it is much more probable that this part of the spectrum is due to Cu 3d $\rightarrow$ Cu 4s transitions (see the band structure of the Cu-O plane below). If there are no real excitons in this system, this would not change, however, much the basic considerations, because the description of the polarization of localized electrons via virtual excitons corresponds only to the first-order diagrams of the polarization propagator. If there are no excitons, one can still describe polarization of the narrow 3d bands [36] using second order terms [35].

Further one has the possibility that the polarization of O^{2}-ions (α = 3.9 10^{-24} cm^3) of the other (non Cu-O planes) play also a role in the effective attraction between the positive holes as well there it is possible that the Y atom in the "1 2 3 compound" does not give away all the 3 electrons and therefore can be polarized also. It can very well happen that besides the usual phonon mechanism also the polarization of the ions outside the Cu-O plane plays a role. However, the author agrees with W.A. Little that the main source of attraction between holes forming Cooper pairs in the ceramic materials is the polarization of the electrons in the narrow Cu 3d bands.

There are several band structure calculations in the literature for the ceramic materials using the local density functional formalism or the linearized APW method. For a review of these calculations see the article of Pickett [37]. Guo, Langlois, and Goddard [38] performed a generalized valence bond (GVB) calculation for $YBa_2Cu_3O_{7-x}$. In our opinion it is rather questionable, whether usual solid state physical methods (as local density functional formalism or APW), which are based on the electron gas idea, can be applied for systems containing four or five different kinds of atoms, including light and heavy ones (small and large electronic densities and large density gradients). The calculation of Guo, Langlois, and Goddard seems to be more realistic but their method (which contains already a great amount of correlation) is very tedious, because one needs a separate calculation for each band.

CLUSTER AND BAND STRUCTURE CALCULATIONS

For these reasons we have started a quantum chemical approach to determine the electronic structure of the ceramic materials. Ab initio SCF LCAO cluster and crystal orbital (CO)

[39] calculations with a non-local exchange have been performed.

For these calculations a double-zeta basis set [40] has been applied. After only Cu and O containing clusters the calculations were extended also to La-Sr-Cu-O clusters. We have found in this way [41] that there is a charge transfer from the La-O plane to the Cu-O plane. The net charge on an La-O unit is (if one uses 1500 point charges to simulate the Madelung potential) +0.91, while that of a CuO unit is -1.82. Using these values a CO calculation has been performed [42] for the 2D Cu-O plane taking into account the Madelung potential both of the Cu-O plane and those of the neighboring La-O planes. The resulting band structure is shown in Fig. 1.

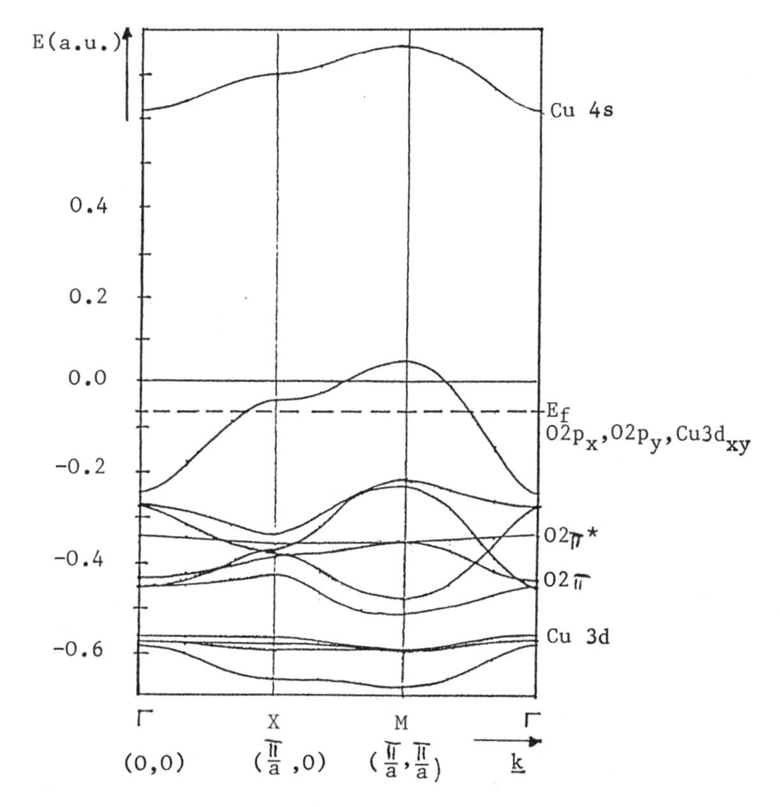

Fig. 1. Band structure of a CuO_2 plane including Madelung potentials in the Cu-O plane and in the neighboring La-O planes. Net charge in an CuO_2 unit -1.82, in an La-O unit +0.91. [q_{Cu} = +1.855 (in plane), q_O = -1.837 (in plane), q_{La}= +2.77 (out of plane), q_O = -1.86 (out of plane)].

From the Figure one can see that there are several very narrow 3d bands and the Fermi level goes through a broad essentially O $2p_x$ or p_y band. In this way the conditions for the "2 bands" polarization (excitonic) mechanism are fulfilled.

To take into account the major part of correlation on the band structure, the 2D Hartree-Fock (HF) crystal orbital calculations have to be repeated by applying also polarization functions (a set of 3d functions on O and a set of 4p functions on Cu).

Having these more accurate HF band structures and wave functions one can apply a generalized electronic polaron model to calculate correlated quasiparticle band structures [43] of the Cu-O plane at least on the MP/2 MBPT level. The generalization of our 2D quasiparticle band structure program to 3d functions has been done (quasiparticle band structure calculations on 2D graphite or boron-nitride layers have been already performed [44]).

According to the experiences obtained on 1D systems [43] correlation effects are to be expected to narrow both the distance between bands and also the bands themselves. Therefore, we can expect that the 3d bands of the Cu-O layers will be still narrower than in the HF case and therefore their electrons become still more polarizable. On the other hand, the 2D DODS quasiparticle band structures (the corresponding method has been formulated [45] and its programming is also in progress) and the corresponding wave functions will give an answer, whether besides the "2 bands" (excitonic) also other (spin) effects play an important role in the occurrence of high T_c superconductivity in ceramic materials (for this purpose one has to solve the BCS equation and determine T_c assuming first separately the validity of the "two-band model" and of other mechanisms and afterwards the role of both).

For instance after having the correlated functions one can perform also an intermediate exciton calculation whether are there any exciton bands corresponding to Cu 3d$\rightarrow$Cu 4s transitions [46].

Finally one should point out that if one desires to design on the basis of these experiences an organic superconductor, one can try with the help of molecular epitaxy to construct alternating 2D graphite- boron nitride layers. If there is a charge transfer between the two different layers, the mobile π-electrons (holes) in the boron nitride could form Cooper pairs due to the attraction caused by the polarization of the localized σ-electrons in these systems. One would expect that such a charge transfer is more probable, if BN and graphite layers are not alternating, but 2 BN layers are followed by a graphite layer (it is also more easy to build such 2:1 layers than alternating layers [46]). One can try also to dope the system by K which forms an ordered layer between the boron nitride and graphite layers [46] and therefore would not supress superconductivity by disorder. The suggested experiments are in progress at the Institutes of Inorganic Chemistry and Experimental Physics of the University Erlangen-Nürnberg.

ACKNOWLEDGMENTS

The author is very much indebted to Professor W.A. Little for his invitation which made his participation on this Conference possible. He should like to express his gratitude to Professors M.T. Adrian, K. Brodersen, T.C. Collins, G. Saemann-

Ischenko, R.F. Wood and to Drs. M.A. Abdel-Raouf, W. Förner, H.-U. Hummel and C.-M. Liegener as well as to Mr. P. Saalfrank for the many very fruitful discussions and advice. He would like to express his special thanks also to Mr. P. Saalfrank for putting some of his results to his disposal before publication. The author would like to thank very much the "Deutsche Forschungsgemeinschaft" (Project No. La 371/24-1), the National Foundation for Cancer Research and the "Fonds der Chemischen Industrie" for financial support.

REFERENCES

[1] F. London, "Superfluids" (Wiley New York, 1950) Vol. 1.
[2] W.A. Little, Phys. Rev. 134A:1416 (1964).
[3] J. Bardeen, L.N. Cooper, and J.R. Schrieffer, Phys. Rev. 108:1175 (1957).
[4] R.A. Ferrell, Phys. Rev. Lett. 13:330 (1964).
[5] W.A. Little, Phys. Rev. 156:396 (1967).
[6] L.D. Landau and E.M. Lifschitz, "Lehrbuch derr Theoretischen Physik", Vol. 5, Academic Press, Berlin (1971), p. 522.
[7] J. Hajdu (personal communication).
[8] See for instance: S. Ravy, J.P. Puget, C. Lenoir and P. Batael, Solid State Comm. 37:73.
[9] J.E. Schirberg, H.H. Wang, and J.W. Williams, Lecture at the Int. Conf. on "Organic Superconductors, South Lake Tahoe, CA (1990).
[10] W.A. Little, in "Novel Superconductivity" (S.A. Wolf and V.Z. Kresin, eds.) Plenum Press, New York-London (1987) p. 341.
[11] J. Ladik; G. Biczó, and A. Zawadowski, Phys. Lett. 18:257 (1965).
[12] See the papers in Chapter II (Organic Superconductors) in the book cited at ref. [10].
[13] J. Ladik, G. Biczó, and J. Rédly, Phys. Rev. 188:710 (1969).
[14] J. Ladik and A. Bierman, Phys. Lett. 29A:636 (1969).
[15] J. Ladik, R.D. Singh, and S. Suhai, Phys. Lett. 81A:488 (1981).
[16] V.L. Ginzburg, Sov. Phys. Usp. Dec. 1964.
[17] V.L. Ginzburg, Sov. Phys. Usp. 13:335 (1970) JETP Lett. 14:393 (1970).
[18] D. Allender, J. Bray, and J. Bardeen, Phys. Rev. B7:1020 (1973); 8:4433 (1973).
[19] J.G. Bednorz and K.A. Müller, Z. Phys. B64:189 (1986).
[20] C.W. Chu, P.H. Hor, R.L. Meng, L. Gao, Z.J. Huang, and Y.O. Wang, Phys. Rev. Lett. 58:405 (1987); P.H. Hor, R.L. Meng, Y.O. Wang, L. Gao, Z.J. Huang, J. Bechtold, K. Forster and C.W. Chu, Phys. Rev. Lett. 58:1891 (1987).
[21] R. Dogani, Science, May 16:24 (1988).
[22] P.W. Anderson, G. Baskaran, Z. Zou, and T. Hsu, Phys. Rev. Lett. 58:2790 (1987).
[23] J.R. Schrieffer, X.-G. Wen, and S.-G. Zhang, Phys. Rev. Lett. 60:944 (1988).
[24] V.Z. Kresin, Phys. Rev. B35:8716 (1987).
[25] K. Yamaguchi, Y. Takahara, T. Fueno, and K. Nasu, Jpn. J. Appl. Phys. 26:2037 (1987).
[26] A. Tachibana, Phys. Rev. A35:18 (1987).
[27] R.F. Wood, Phys. Rev. B165:97 (1990).
[28] G. Chen and W.A. Goddard III, Science 239:899 (1988);

H. Kamimura, S. Matsumo, and R. Saito, Solid State Comm. 67:363 (1988); V.L. Emery and G. Reiter, Phys. Rev. B38: 4547 (1988).
[29] W.A. Little in "Novel Superconductivity" (S.A. Wolf and V.Z. Kresin, eds.) Plenum Press, New York-London (1987) p. 341.
[30] A. Zawadowski, Phys. Rev. B, to appear; M. Jarrell, H.R. Krishmannuty, and D.L. Cox, Phys. Rev. B38:4584 (1988).
[31] J.J. Ladik and T.C. Collins, in "Novel Superconductivity", (S.A. Wolf and V.Z. Kresin, eds.) Plenum Press, New York-London (1987), p. 181; T.C. Collins, A.B. Kunz, and J.J. Ladik, ibid, p. 378.
[32] W.A. Little, Science 242:1390 (1980).
[33] I. Bozovic, D. Kirilov, A. Kapitulnik, K. Char, M.R. Hahn, M.R. Bedslev, T.H. Geballe, Y.H. Kim, and A.J. Heeger, Phys. Rev. Lett. 59:2219 (1987).
[34] H.P. Geserich, G. Scheiber, J. Geerk, H.C. Li, G. Linker, W. Assmuss, and W. Weber, Europhysics Letters 6:1277 (1988).
[35] A.L. Fetter and J.D. Walecka, "Quantum Theory of Many-Particle Systems" (McGraw-Hill, New york, 1971) p. 558.
[36] T.C. Collins, J. Ladik, and P. Saalfrank, Phys. Rev. B (submitted).
[37] W.E. Pickett, Rev. Mod. Phys. 61:433 (1988).
[38] Y. Guo, J.-M. Langlois, and W. Goddard, Science 239:896 (1988).
[39] G. Del Re, J. Ladik, and G. Biczó, Phys. Rev. 155:997 (1967); J.-M. André, L. Gouverneur, and G. Leroy, Int. J. Quant. Chem. 1:427, 451 (1967); P. Saalfrank, J. Ladik, and R.F. Wood, Phys. Rev. B (accepted).
[40] (a) For O ((9s5p)/[4s2p]): T.H. Dunning and J.P. Hay, in "Modern Theoretical Chemistry", (H.F. Schaeffer III, ed.) Plenum Press, New York-London (1977), Vol. 3, p. 1;
(b) For Cu ((14s8p5d)/[8s4p2d]) optimized for ^2D state: S. Huzinaga, in "Physical Science Data 16, Gaussian Basis Sets for Molecular Calculations", (S. Huzinaga, ed.) Elsevier, New York (1985) p. 192.
[41] P. Saalfrank, M.A. Abdel-Raouf, J. Ladik, and R.F. Wood, Phys. Rev. B (accepted).
[42] P. Saalfrank, J. Ladik, and R.F. Wood, Phys. Rev. B (to be published).
[43] S. Suhai, Phys. Rev. B27:3506 (1983).
[44] M. Vracko, C.-M. Liegener, and J. Ladik, Chem. Phys. Lett. 126:255 (1988); Int. J. Quant. Chem. 37:241 (1990).
[45] J. Ladik and C.-M. Liegener, J. Mol. Struct. (THEOCHEM) 206:39 (1990).
[46] H.-U. Hummel (personal communication).
[47] See for instance: J.J. Ladik, "Quantum Theory of Polymers as Solids", Plenum Press, New York-London (1988) p. 271.

QUASI-PARTICLES IN POLYMERIC CONDUCTORS

Jorge Medrano* and Douglas Dudis**

* University of Dayton, Chemistry Department
300 College Park
Dayton, OH 45469

**Wright Research and Development Center
Materials Laboratory
Wright-Patterson AFB, OH 45433-6533

ABSTRACT

The studies reported here are part of an effort to understand the electronic structure and statistical properties of quasi-particles in polymeric materials. These are solitons, antisolitons, polarons, and bipolarons, and in one way or another each of them is known to play a role in the conduction of electric current in those materials. We have used pristine and doped polyacetylene as our example to study the geometrical and electronic distortions associated with quasi-particles in oligomers and infinite polymers. The computational approach has been based on the semi-empirical self-consistent LCAO-MO theory at the AM1 level of approximation for finite systems and the corresponding LCAO-CO theory in the tight binding approximation for infinite systems.

INTRODUCTION

Although conducting polymers in general and polyacetylene in particular have received much attention in recent years, there are comparatively few first-principles theoretical calculations aimed at understanding the electronic structure of quasi-particles. Most of the calculations performed have used the Su, Schrieffer, and Heeger (SSH) hamiltonian [1]. This work was indeed a very important stepping stone, mainly because it showed that the soliton model of polyacetylene could be used to explain a number of properties of this material; optical, electrical, and magnetic. No other model can claim such an accomplishment. This success prompted other work and different extensions and modifications of the SSH hamiltonian, too numerous to quote. Most notably however, H. Fukutome has very recently performed a very complete study of a Pariser-Parr-Pople (PPP) unrestricted Hartree-Fock (UHF) model of polyacetylene [2]. The SSH approach however, is still too crude for the kind of accuracy that is needed since it is basically a Huckel-type approximation with sigma bond compressibility and electron-electron Coulomb interactions are completely neglected. Moreover, the SSH model does not have an analytical solution and needs a number of empirical parameters specific to polyacetylene (or the particular polymer under study) in order to be able to make some quantitative predictions.

On the other hand, a few self-consistent LCAO-MO calculations have been performed on finite oligomers of polyacetylene, both semi-empirical [3] and ab initio [4]. The excellent agreement that we report below between one of our semi-empirical calculations and the ab initio result reported in ref. 4, convinced us that the SCF-LCAO-MO semi-empirical theory is completely adequate for the purpose of a systematic study of the electronic structure associated with quasi-particles. This methodology is still first-principles based and although there are indeed parameters incorporated into the theory, these are atomic in nature. Therefore, the same theory with the same parameters

Organic Superconductivity, Edited by V.Z. Kresin and
W.A. Little, Plenum Press, New York, 1990

can be used to study any molecule or infinite polymer without the need for information about the specific system.

We can mention here the only calculation of the infinite polymer of which we are aware; that of Stafstrom and Bredas [5], who used the Valence Electron Hamiltonian (VEH) method to study the band structure at various doping levels.

Many fundamental aspects of the electronic and structural properties of the conducting polymers (in particular the highly conducting ones) are not well established, and at this time theoretical understanding lags behind the rapid experimental developments. In what follows we expect to contribute information that will ultimately prove useful for a thorough understanding of the mechanisms of conductivity in polymers.

METHOD

The theoretical method used for the calculations was the semi-empirical SCF-MO theory at the AM1 level of approximation [6] as implemented in the MOPAC 5.0 program (QCPE #455). The AM1 hamiltonian has been shown to correct for some deficiencies present in the older MNDO hamiltonian. The geometries of all the systems studies were **fully optimized** with due allowance for the symmetry constraints in each case. Convergence of the calculation was difficult in some of the examples, especially for doublet and triplet states. In each case the optimization was very carefully performed by running a first step with the keyword PRECISE (thus increasing the criteria for terminating all optimizations, electronic and geometric by a factor of 100). In subsequent runs, the gradient was decreased even further by using the SIGMA method due to Komornicki and McIver [7] while at the same time making the criterion for self-consistent convergence even stricter (using the keyword SCFCRT = 10^{-10} or SCFCRT = 10^{-11}). In this way the gradient norm was usually reduced to less than 1 (in kcal/A or kcal/degree) even when more than seventy geometrical variables were simultaneously optimized. For infinite systems we used the "cluster approximation" developed by Stewart [8] (the name is misleading because Born-von Karman cylic boundary conditions are actually used; it is thus an approximation to a full crystal orbital (CO) calculation). This approach is known to be an excellent approximation to the full CO result when the unit cell used is large enough to ensure that atoms at one end of the repeat unit have a negligible density matrix element with atoms at the other end. Since the repeat units we used are almost 27Å long, this condition is very rigorously met even in the examples studied, where there is an extended π system. However, in at least one example (see below) the full CO calculation was performed as a check (using the MOSOL program, QCPE #495), thus verifying that the results were the same to four significant digits for the optimized geometrical variables, heats of formation and degrees of bonding. Obtaining the band structure however, requires the full solid state calculation involving sampling of the Brillouin zone using a regular mesh. The advantage of the approximation is of course that it takes 1/5 to 1/10 of the CPU time required by the full CO calculation.

We mimicked the effect of a dopant in the repeat unit by using the "sparkles" provided in the MOPAC program. These are uncharged species which immediately ionizes the polymer. They can be regarded as unpolarizable ions of diameter 1.4Å. We used an acceptor dopant in the polaron calculations; the effect of the sparkle being that a net positive charge is left in the π system.

The degrees of bonding were calculated from the density matrix using the definitions given by Medrano et al [9], for molecular and infinite polymers as well. These are known to be an accurate measure of the integrated electron density associated with a bond, and have the important advantage that the same basic formalism (based on the first-order reduced density operator) is used both for molecules and infinite polymers. In the cases we studied, the degrees of bonding give a much better description of the bond alternation (or lack of it) than the bond lengths, since the former are more directly associated with the electronic structure.

For almost all the examples, we carried out calculations both at the open-shell restricted Hartree-Fock (ROHF) and unrestricted Hartree-Fock (UHF) levels of the theory. The differences between both descriptions is discussed below.

DESCRIPTION OF THE SYSTEMS STUDIED

As already stated, it was our purpose to study the electronic structure of quasi-particles such as solitons, polarons, etc. A static neutral soliton (S) can be considered the same as a radical. The effect of the soliton on the geometrical structure is to flip the bond alternation from one to the

Table 1. Charge and Spin Relationships of Quasi-Particles

	Charge	Spin
Neutral Soliton (S)	0	1/2
Charged Soliton (S^+) or (S^-)	±1	0
Soliton - Antisoliton Pair (SS)	0	0, ±1
Polaron (P^+) or (P^-)	±1	1/2
Bipolaron (bP^{++}) or (bP^{--})	±2	0

other of the two degenerate phases in the material. A soliton can be positively charged (carbenium ion) or negatively charged (carbanion). The three situations are depicted in Fig. 1. These schemes in Fig.1 should not be taken too strictly however, since they do not show the essentially delocalized nature of the quasi-particles, but are intended to allow easy electron counting. Now, solitons are known to be topological particles. Therefore, in infinite systems they are always created in pairs, called soliton-antisoliton pairs (SS^*) (This can also be seen as arising from stability conditions.) Any of the two members of the pair will cancel the effect of the other member next to it, so to left and right of the pair the bond alternation will be the same (cf. Fig. 2). What happens in between, and how that depends on the distance between soliton and antisoliton, can only be determined by calculation and will be discussed below. One such pair can also be regarded as a (singlet or triplet) biradical.

One electron can be removed from or added to a region of the material where a soliton-antisoliton pair has been created, giving rise to a positive or negative polaron respectively (P^+ or P^-). (They can also be regarded as a radical cation or radical anion respectively in the static case.) Removal or addition of two electrons results in a positive or negative bipolaron ($bP^{\ddagger}$ or $bP^{=}$) or a dication or dianion respectively. Figure 3 attempts to give a graphical representation of polarons and bipolarons, while at the same time emphasizing the delocalized nature of the particles. Again, both at right and left of a polaron or bipolaron, the bond alternation is the same; i.e. we have either phase A or phase B at both sides of the quasi-particle in a polyacetylene chain (cf. Fig 3). Once again the detailed geometrical and electronic structure can only be determined by calculation.

These particles have very unusual relations of charge-spin, in most cases at variance with elementary particles such as protons and electrons. Table 1 gives a summary of the charge and spin associated with solitons, soliton-antisoliton pairs, polarons, and bipolarons.

The actual systems we used as case studies are: (i) The polyacetylene oligomer chain $C_{21}H_{23}$ for studying the isolated soliton and the repeat unit $C_{22}H_{22}$ for studying the (SS^*) pair in an infinite system. We used the same repeat unit to study the polaron P created from the (SS^*) by including an acceptor "sparkle" in the calculation. These systems are depicted in Figure 4.

RESULTS AND DISCUSSION

1. Solitons

We fully optimized the geometry of the $C_{21}H_{23}$ oligomer with a neutral soliton at the central carbon atom, under C_{2v} symmetry as described above. The ground state of the system is of course a doublet, and we performed the calculation both at the UHF and ROHF level of theory. The UHF result gives no bond length alternation and no degree of bonding alternation, except close to the ends of the chain, due to the terminal effects. Close to the soliton site, all bond

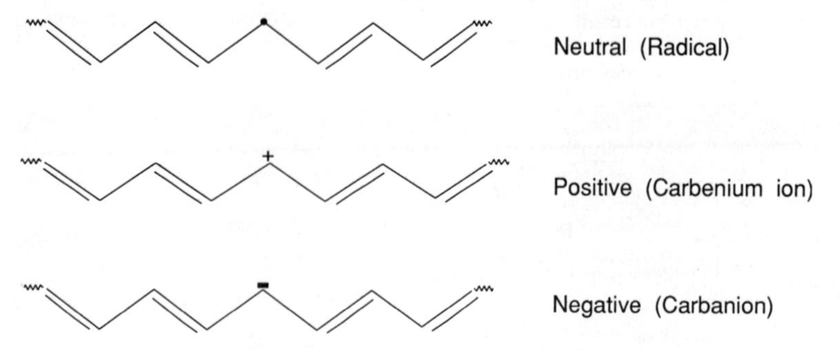

Figure 1. Solitons in polyacetylene.

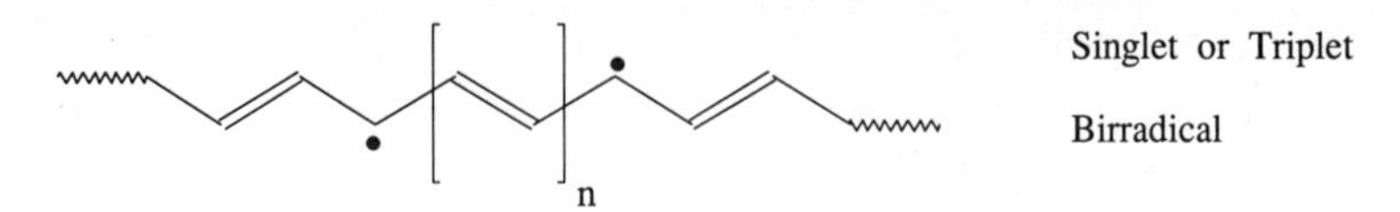

Figure 2. Soliton-antisoliton pair.

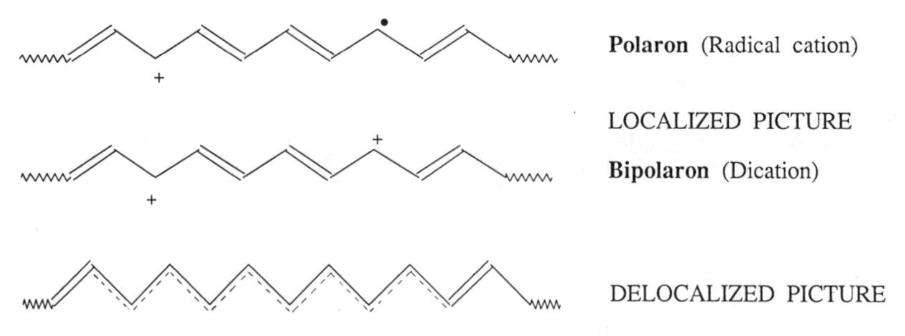

Figure 3. Polaron and bipolaron.

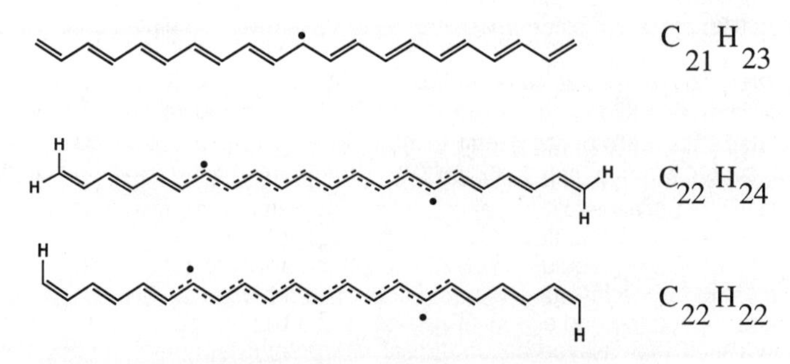

Figure 4. Systems used for the studies.

lengths are 1.40Å and all degrees of bonding are 1.33. Now, the UHF wavefunction is not an eigenfunction of the operator that represents the spin squared, S^2. The theoretical value for this case is $S^2 = 0.75$. The UHF result is $S^2 = 3.66$, i.e. there is a very large spin contamination. This is probably due to low lying excited states of higher spin multiplicity such as quartet, sextet, etc. These excited states probably have no bond length alternation. In any case this level of spin contamination indicates that the UHF wavefunction is not a sufficiently accurate representation of the doublet. It appears the UHF description exaggerates the opening of the shells due to the magnetic interactions that originates in the unpaired spin. On the other hand, because of the way it is obtained, in the ROHF wavefunction, only the singly occupied orbital contributes to the spin density wave away from the soliton, which is unrealistic. So neither type of wavefunction can give an accurate description of all the properties of interest, when the number of α and β spins is not the same. It will probably be necessary to resort to some method for incorporating correlation to improve over both the UHF and ROHF descriptions. Some preliminary calculations along that direction have already been carried away in our laboratory at the MP2 level, and they confirm, as expected, the above discussion. As regards the bond lengths and degrees of bonding alternation, the ROHF calculation gives quite accurate results. At the same time the agreement between those fragment-based approximate ab initio calculations [10] and ours, lends additional credibility to the semi-empirical approach in these situations.

Therefore, we used the ROHF/AM1 calculations hereafter, which is of course the same as closed-shell RHF/AM1 in the cases where there is the same number of α and β spins.

Figure 5 shows the difference in adjacent bond lengths Δr, plotted against carbon site distance from the soliton, for both the neutral and the positively charged quasi-particles. Only the right half of the molecule is included in the plot, since the other half can be derived by symmetry. $\Delta r = 0$ implies no bond alternation, and this happens precisely at the soliton center.

The plot in Figure 5 also shows a horizontal line drawn at the value of $\Delta r_\infty = 0.095$ Å corresponding to infinite polyacetylene, as obtained through a full CO calculation with the MOSOL program. This value is in very good agreement with the experimental result of $\Delta r_\infty = 0.104$Å. The end effect is clearly visible in Figure 5 in that the values of Δr for the last carbon atoms exceed the infinite polymer value.

Both curves for the neutral and positive solitons, can be approximated by a function of the form $\Delta r = \Delta r_\infty \tanh(n/l)$ as proposed by Su et al, in ref. 1. Taking for Δr_∞ the value quoted above, we obtain the best fit for the neutral soliton when $l=2$. This is a measure of the extent of the quasi-particle, and is in very good agreement with the result obtained by Boudreaux et al. [3], who found $l=3$ through a calculation on $C_{41}H_{43}$ using the MNDO hamiltonian and $\Delta r_\infty = 0.106$Å (in ref. 1, the authors came up with $l=7$).

The values of Δr for the positive soliton are in excellent agreement with those obtained by Villar et al. [4] who carried out an ab initio calculation on the same oligomer $C_{21}H_{23}$ as we did, using a double-zeta 6-31G basic set. All the bond lengths we obtained by geometry optimization agree within 1% with the ab initio ones. The hyperbolic tangent that best interpolates among the calculated points, is calculated with $l=7$ if $\Delta r_\infty = 0.112$ is used (as in ref. 5) or with $l=6$ if $\Delta r_\infty = 0.095$ is used, as we did for the neutral soliton. In any case this is somewhat larger than the value $l=5$ obtained in ref. 3. The difference could be due to the fact that we used the AM1 hamiltonian and Boudreaux et al. used the MNDO one. The agreement is still very good however.

We also calculated the degrees of bonding B_{AB} between consecutive carbon atoms, and ΔB_{AB} analogously to Δr for both the neutral and positive soliton. Except for the last three points where the discrepancy is slightly larger, we again have an agreement within 1% with the ab initio values in ref. 4 for the positive soliton. This, and the agreement for the bond lengths mentioned above, gave us an assurance that the semi-empirical method we used is sufficiently reliable for this purpose. The result for ΔB_{AB} versus carbon site distance from the center, are shown in Figure 6. Once again the value for the infinite polymer is shown as a horizontal line, as calculated with the MOSOL program. Again, the end effects are visible in the last few points. It is an interesting and completely new result that a hyperbolic tangent interpolates very well among the calculated points, with $l=3$ for the neutral soliton, and $l=6$ for the positive one.This means that the relation between these curves and the solution of the solitary wave equation in ϕ^4 field theory [11] goes beyond the geometrical distribution caused by the defect, and reflects in the associated electron-density distribution.

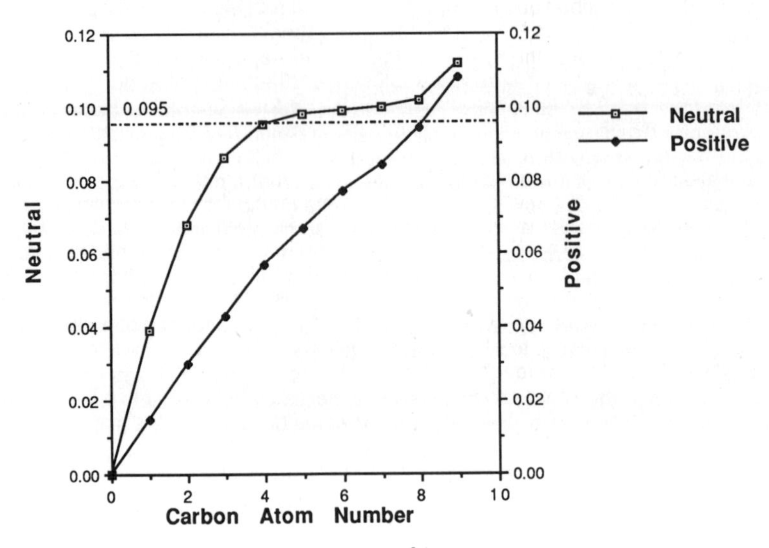

Figure 5. Bond length alternation in Å in neutral and positive solitons in C21H23 oligomer.

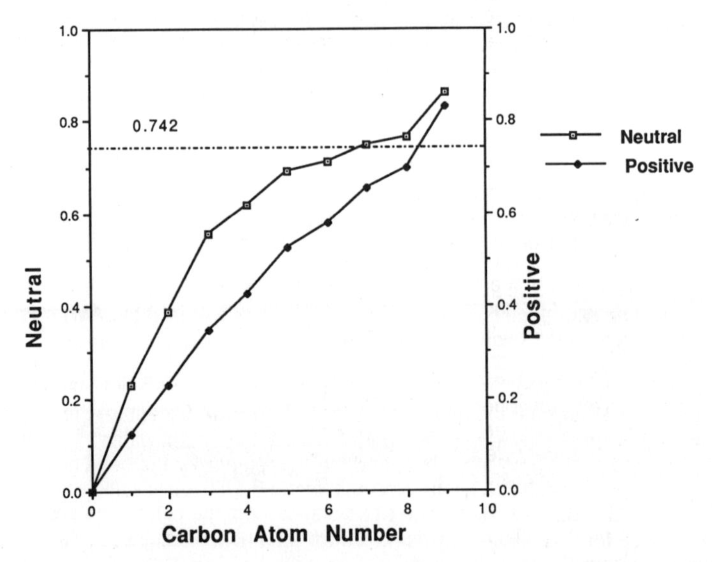

Figure 6. Degree of bonding alternation in neutral and positive solitons in C21H23 oligomer.

2. Soliton-Antisoliton Pairs

We calculated the singlet state of neutral soliton-antisoliton pairs in infinite systems using the cluster approximation described above, and $C_{22}H_{22}$ as the repeat unit. Besides, we repeated the calculation for separation of R=1,5,9 and 11 bonds between the soliton and antisoliton. Table 2 shows the heats of formation of the system for the different separations and the heat of formation of the $C_{22}H_{22}$ repeat unit when it is used to represent pristine polyacetylene. This comes out to be 141.65 kcal/mol or 12.87 kcal/mol per CH unit. The full CO calculation with the MOSOL program gives 12.93 kcal/mol for the same quantity; showing that the cluster approximation is providing the correct energies to within hundredths of a kcal/mol. As Table 2 shows, the calculated heats of formation are independent of the soliton-antisoliton separation, within the uncertainty of the method of a few hundredths of a kcal/mol. This result is at variance with a previous calculation by Bredas et al. [12], who used a Huckel-type method with sigma bond compressibility and cyclic polyene molecules with about 110 carbon atoms to represent the system. These authors find an attractive interaction between soliton and antisoliton in the electronic ground state of the system, i.e. the energy of the system when the soliton and antisoliton are in close proximity is about 0.9eV lower than that corresponding to a large separation. This is tantamount to stating that when they are in close proximity we have the perfectly dimerized state, and when they are separated by a distance larger than approximately twice the width of a soliton, then we have two transitions: for instance, $A \rightarrow B \rightarrow A$. (This meaning that the system is in phase A left and right of the pair and in phase B in between.) The energy would then rise rapidly when R goes to infinity, to its asymptotic value of twice the energy of one isolated soliton. The two limits $R \rightarrow 0$ and $R \rightarrow \infty$ cannot be argued of course, but the total energy should depend on the exact shape of the soliton pair at intermediate distances. Interestingly, we obtained an unexpected result regarding this point. The two bond lengths adjacent to the soliton and to the antisoliton are constrained to be equal, but in spite of that, the bond length alternation is not changed by either particle. Their alternation to the right of each soliton for example, recovers to what it was to the left two bonds past the quasi-particle. This picture is even clearer when the degree of bonding alternation is examined. It decreases somewhat together with the bond length alternation, without ever vanishing, and recovers together with the latter, too. In fact it decreases from its infinite polymer value of 0.74 to 0.65 at the center of each soliton, i.e. less than 0.1! This implies that the two bonds at each side of either soliton have the same length but very different degrees of bonding: one is a double bond, B_{AB}=1.75 and the other is single, B_{AB}=1.10. Figure 7 shows precisely that, for the case when both quasi-particles are nine bonds apart, but the same is true irrespective of the separation. In case there were an influence of a different environment left and right of each particle, we checked carefully the situation when they are eleven bonds apart. Then each of the particles has its left and right neighbor at exactly the same distance in a $C_{22}H_{22}$ repeat unit. The same picture is obtained also in this case. This is of course consistent with our finding that the energy is the same irrespective of the separation. It is also consistent with the small energy of creation of the soliton-antisoliton pair. According to the results in Table 2, this is only 5.55 kcal/mol (or 0.24 eV). Exactly at what distance, both parts of the pair start behaving as independent particles we of course do not know. Trying larger separations would entail using larger repeat units, making the system intractable at this level of the theory. The question immediately arises regarding the nature of the interaction between both particles. In other words, how does each of them know about the presence of the other when their separation is as large as eleven bonds? (It should be kept in mind that their width is two bonds or three bonds to each side of the center.) The answer to this question is not clear to us at the present time.

The description based on the more simplistic picture (based on a SSH-type of calculation) described above has already been published in a book on the subject [13] by S. Kivelson. According to their study however, the details of the shape of the pair are at the very least debatable matter. Moreover, we can conclude that due consideration of electron-electron interactions is absolutely necessary for an accurate description of the details of the electronic structure of these systems.

3. Polarons

Although the polaron calculations are still being run, the two values for the energy that we already have, one for the situation with the two parts of the system nine bonds apart and the other five bonds apart, show that there is indeed an attractive interaction in this case. The values are E(R=9)=271.10 kcal/mol and E(R=5)=266.30 kcal/mol. Their difference amounts to 4.8 kcal/mol, or 0.21 eV. In this case, we find a better agreement with Bredas et al. calculation [12] of an attractive interaction of 0.35 eV.

Table 2. Energies of the Soliton-Antisoliton Pair for Different Separations

		ΔH_f (kcal/mol)	$\Delta\Delta H_f$
Normal $C_{22}H_{22}$		141.65	
n	R		
0	1	147.20	5.55
2	5	147.19	5.54
4	9	147.26	5.61
5	11	147.28	5.63

In both our calculations, we placed the dopant above the plane of the polymer, 3.5Å on top of the hydrogen bonded to the carbon atom where the center of the left soliton is. This of course destroys the C_{2h} symmetry and gives rise to a large dipole moment, of about 18 debyes.

Energetically, it takes about 121 kcal/mol (or 5.2 eV) to introduce the dopant (i.e. ionize the polymer) and create the polaron, as compared with the energy of the system with a neutral soliton-antisoliton pair. A more detailed study of the electronic structure is presently underway.

LINES OF FUTURE RESEARCH

This research can be extended to include negative polarons and positive and negative bipolarons in polyacetylene. Several issues regarding this subject need clarification.

Also, other interesting highly conducting polymers can be studied with these techniques, including (SN)x which is known to be superconducting at very low temperature.

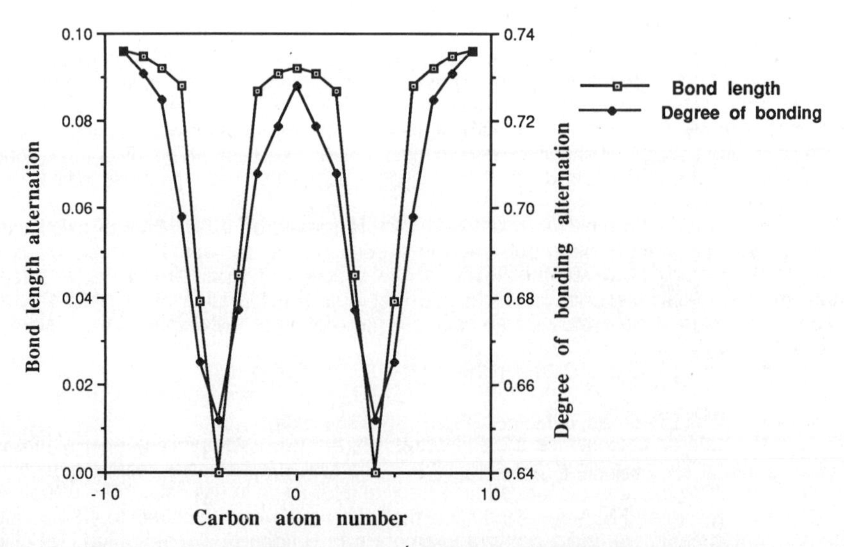

Figure 7. Bond length and bond order alternation in polyacetylene (C22H22 repeat unit) with a neutral soliton-antisoliton pair.

Then, the next step would be the detailed study of the band structure associated with quasi-particles. The few publications on this subject are not in agreement with one another. For example, Fukutome's calculation [2] (see above) gives a structure of levels in the band gap very different from what is routinely accepted from the SSH-type of work. This is a challenging calculation in any case, because Hartree-Fock based methods, whether ab initio or semi-empirical are know to exaggerate notoriously the band gap and not give very accurate densities of states.

There are also some very important theoretical issues to explore. For instance, which is the relation (if any) between conductivity and hyperpolarizabilities. This subject can be investigated with the same tools described here.

The statistical mechanical aspects of quasi-particles, are also essentially unexplored, for instance the coupling of two fermious into a boson, as in the formation of a soliton-antisoliton pair. Can this be considered a pairon? Does it obey a strict Bose Einstein statistic? Does it bear any relation with Cooper-type pairs? These are only a few examples of unanswered questions.

Finally, it would certainly be worth exploring in more detail an interesting, very recent suggestion by Mazunder and Ramasesha [14] about a bipolaronic mechanism for superconductivity in organics to see if the same idea can be applied to polymers of other structures and compositions.

Some of these issues will be addressed in our laboratory, and the results published elsewhere.

CONCLUSIONS

We have been able in this investigation to use some new tools and ideas and have found some previously unreported aspects of the electronic structure of quasi-particles.

Ours is one of the first calculations of infinite polymers when quasi-particles are present, and the first to our knowledge in which the dopant is explicitly taken into account. This is also the first calculation (at this level of theory) of the interaction between the two parts of the system in a soliton-antisoliton pair or in a polaron.

The difference between the ROHF and UHF descriptions had not been pointed out before. The question still remains, if correlation has to be included, which is the minimum level of theory necessary for an accurate description of all the properties of the system. We intent to address this issue.

We also believe that the use of the "degree of bonding wave" (DOBW) is the most descriptive tool available to understand, in the most pictorial way, the characteristics of the electron density distribution. We have shown as well that the DOBW also can be described by a hyperbolic tangent functional dependence with the distance from the center of the soliton.

Finally, we have found some unexpected features of the electron structure associated with a soliton-antisoliton pair, that we believe are important for the detailed understanding of this, and probably other quasi-particles.

REFERENCES

1. W. P. Su, J. R. Schrieffer, and A. J. Heeger, Phys. Rev. Lett. 42 ,1698 (1979).

2. H. Fukutome, J. Molec. Struct. (Theochem), 188, 337 (1989).

3. D. S. Boudreaux, R. R. Chance, J. L. Bredas, and R. Silbey, Phys. Rev. B 28, 6927 (1983).

4. H. O. Villar, M. Dupuis, and E. Clementi, Phys. Rev. B 37, 2520 (1988).

5. S. Stafstrom and J. L. Bredas, J. Molec. Struct. 188, 393 (1989).

6. M. J. S. Dewar, E. G. Zoelrich, E. F. Healy, and J. J. P. Stewart, J. Am. Chem. Soc. 107, 3902 (1985).

7. (a) A. Komornicki and J. W. McIver, Chem. Phys. Lett. 10, 303 (1971); (b) A. Komornicki and J. W. McIver, J. Am. Chem. Soc. 94, 2625 (1981).

8. J. J. P. Stewart, New Polymeric Mat. 1, 53 (1987).

9. (a) M. A. Natiello, H. F. Reale, and J. A. Medrano, J. Comp. Chem, 6, 108 (1985); (b) R. C. Bochicchio, H. F. Reale, and J. A. Medrano, Phys. Rev. B 40, 7186 (1989).

10. G. P. Das, Chem. Phys. Lett. 147, 591 (1988).

11. A. S. Davidov in "Solitons in Molecular Systems," D. Reidel Publishing Co., 1985.

12. J. L. Bredas, R. R. Chance, and R. Silbey, Phys. Rev. B 26, 5846 (1982).

13. S. Kivelson, "Soliton Model of Polyacetylene," in "Solitons," page 325; V. M. Agranovich and A. A. Maradudin, editors; North Holland Publishing Co., 1986.

14. S. Mazumdar and S. Ramasesha, Synth. Metals 27, A105 (1988).

PAIR CORRELATION IN ORGANIC MOLECULES

Vladimir Kresin

Materials and Chemical Sciences Division
Lawrence Berkeley Laboratory
University of California, Berkeley, CA 94720

INTRODUCTION

This paper is concerned with pair correlation in π-electron systems of large aromatic molecules; this phenomenon is different from macroscopic organic superconductivity. The problem has been discussed by the author and co-workers in [1,2]. The present paper contains a description of the approach and contains several new results.

Polyatomic aromatic molecules contain a finite number of delocalized π-electrons. For example, the hexabenzcoronene molecule (Fig.1a) has 42 π-electrons, ovalene has 40, etc. π-electrons form a so-called finite Fermi system, similar to that found in atomic nuclei [3]. Pair correlation of π-electrons is analogous in many respects to the well-known phenomenon of pair correlation in nuclei (see below); this is related to the aforementioned similarity. On the other hand, π-electron systems have a number of similarities with conduction electrons in metals: Coulomb forces, the presence of light (electrons) and heavy (ions) particles. As a result, one can employ the adiabatic approximation and introduce electron-vibrational coupling (see, e.g., the reviews [4]) which is similar to the electron-phonon interaction in metals.

Like nucleons or conduction electrons, the π-electrons display pair correlation. In the next section we describe some peculiar properties of this correlation.

Organic Superconductivity, Edited by V.Z. Kresin and W.A. Little, Plenum Press, New York, 1990

Energy gap. As is known, pairing leads to the appearance of an energy gap. However, for nuclei the concept of an energy gap is more complicated than for metals (see, e.g., [3,5]). The important distinction is that the spectrum is discrete , because of the finite size. The same is valid for π-electron systems. The following question arises: what is the meaning of the energy gap concept in such system? It turns out that pair correlation leads to a situation when the spacing between the highest occupied energy level and the first non-occupied level (0-0′ transition) greatly exceeds the interval between the first and second unoccupied levels (0′-0′′), etc. (see Fig.2). This disproportion is treated in nuclear physics as a manifestation of pair correlation. In addition, the finite size of the system leads to a crucial difference in spectra for the cases of odd and even numbers of particles. Namely, the disproportion in level spacing (energy gap) has been observed only for nuclei with even numbers of protons (neutrons). The presence of an unpaired particle (odd number) leads to the disappearance of the energy gap, that is, to the possibility of absorbing γ-quanta of a smaller frequency.

Similar regularities have been observed for π-electron systems. For example, in the coronene molecule the 0-0′ spacing is 22500 cm^{-1}, whereas the 0′-0′′ distance is only 5500 cm^{-1}. The addition of one π-electron (making the total number of π-electrons odd) leads to a drastic decrease in the 0-0′ spacing.

Anomalous diamagnetism. Another effect which is directly related to pair correlation is anomalous diamagnetism. Consider a molecule in an external weak magnetic field; the field is perpendicular to the plane of the molecule. Based on the theory of finite Fermi systems [3], one can evaluate the density matrix, and we obtain for the orbital magnetic susceptibility [1,2]:

$$\chi = \chi_P + \chi_D,$$

where

$$\chi_P = \frac{e^2}{8m^2} \sum_{\lambda\lambda'} F(\lambda,\lambda',\Delta)\,|\hat{L}_{z\ \lambda\lambda'}|^2,$$

$$F(\lambda,\lambda',\Delta) = \frac{E_\lambda E_{\lambda'} - \xi_\lambda \xi_{\lambda'} - \Delta^2}{2E_\lambda E_{\lambda'}(E_\lambda + E_{\lambda'})}\ ; \qquad E_\lambda = (\xi_\lambda^2 + \Delta^2)^{1/2}.$$

and χ_D is the usual diamagnetic term (λ and λ' are quantum numbers, and ξ_λ is the one-particle energy in the absence of pair correlation, referred to the highest occupied level) . If Δ=0, we recover the expression for the usual Van Vleck paramagnetism.

In a normal metal, the terms χ_P and χ_D cancel each other, and the metal displays only a small residual Landau diamagnetism. In molecules such a cancellation does not take place, but the quantities are of the same order of magnitude. Nevertheless, experimentally one observes a depression of the paramagnetic term and, consequently, anomalous diamagnetism.

One should note that χ_P vanishes in molecules with axial symmetry. Indeed, the operator $\hat{L}_z$ is diagonal in this case, and $F(\lambda,\lambda',\Delta)=0$, regardless of the presence of pair correlation. Then the magnetic susceptibility of such molecules and consequently the π-electron current, are determined exclusively by the diamagnetic part of the Hamiltonian. This occurs,for example, in the benzene molecule whose symmetry is frequently assumed to correspond to the $D_{\infty K}$ group. However, anomalous diamagnetism has also been observed in molecules which do not possess axial symmetry (e.g., ovalene, Fig. 1b).

The origin of this anomalous diamagnetism is similar to the Meissner effect. The presence of pair correlations, and hence the gap parameter Δ, result in a depression of χ_P. One can show that $F(\lambda,\lambda',\Delta) << F(\lambda,\lambda',\Delta=0)$; this is due to the inequality $\delta E<<\Delta$, where δE is the level spacing in the absence of correlation. As a result, $\chi \cong \chi_D$, in agreement with the experimental data.

One should stress that each one of these features (the appearance of an energy gap, anomalous diamagnetism), taken separately, can be explained without invoking pair correlation. Indeed, the concept of an energy gap is natural for the insulating state. Of course, such a state does not contain free delocalized carriers. Note that the conventional quantum chemical treatment (Hartree-Fock theory) leads to the appearance of an energy gap dividing the bonding and antibonding states. As for anomalous diamagnetism, it can be explained on the basis of the metallic model with currents moving along the perimeter (see, e.g., [6]). However, such an approach, which is suitable for describing the delocalization of π-electrons, is unable to produce an energy gap.

The pair correlation picture allows us to describe both phenomena: the appearance of an energy gap and anomalous diamagnetism. This concept is important for understanding superconductors (energy gap, Meissner effect), nuclear matter (energy gap, anomalous decrease of moments of inertia; the latter effect is similar to anomalous diamagnetism), and for π-electron systems.

Charge transfer. Electron transfer in biological membranes is an example of a tunneling phenomenon. Consider a row of aromatic molecules with a fixed distance between them. Pair correlation inside each molecule leads to phase coherence in its π-electron system. This coherence results in Josephson tunneling in the membrane and in the possibility of simultaneous transfer of two π-electrons. This implies the existence of dissipationless transfer in this large system arising from the motion of solitons, in analogy with [7].

MECHANISMS OF PAIR CORRELATION

Pairing in π-electron systems is due to interactions similar to those in usual superconductors. The major mechanism of correlation is polarization of the σ-core. This polarization is caused by a number of interactions.

Let us discuss, first of all, the electron-vibrational interaction. This interaction can be introduced in a rigorous way on the basis of the adiabatic theory; see the paper by W.Lester and the present author [4]. In connection with the question of the contribution of the electron-vibrational interaction to π-electron pairing, one should note that in usual superconductors the value of the energy gap is smaller than the characteristic phonon energy $\hbar\tilde{\Omega}$ ($\tilde{\Omega} \cong \Omega_D$, Ω_D is the Debye frequency). The situation in molecules is different. Namely, the spacing between electronic levels and the energy gap (0-0′ transition) exceed the vibrational spacing. Nevertheless, according to the theory of superconductivity such a situation may also occur in bulk metals if we are dealing with very strong electron-lattice coupling. The effect of strong coupling on T_c has been studied numerically in [8], and then analytically (including the Coulomb term) by H.Gutfreund, W.A.Little and the present author in [9]. The relationship between 2Δ and T_c, and thus between 2Δ and the coupling strength g, has been studied by the author in [10]. According to [9,10], in the limit of large g, $2\Delta = \alpha g^{1/2}\tilde{\Omega}$; $\alpha \cong 2.5$. As a result, 2Δ may exceed the value of $\tilde{\Omega}$. Such a situation occurs in large aromatic molecules. Indeed, in the adiabatic theory electron-vibrational interaction is directly related to non-adiabaticity which is described by the parameter $\tilde{\Omega}/\delta E$. An increase in the size of a π-electron system leads to a decrease in δE, whereas the characteristic vibrational energy does not strongly depend on the size of the molecule. As a result, in large molecules $\tilde{\Omega}$ and δE become comparable, which corresponds to strong electron-vibrational coupling.

In addition to the electron-vibrational interaction, pairing can be also caused by an electronic mechanism, namely, by σ-π virtual excitations (see [2]). In addition, if the molecule contains several different π-electron groups (see, e.g., Fig. 1c), the Little mechanism [11], based on the presence of spatially separated groups of electrons, may also be very efficient.

SMALL METAL CLUSTERS

There exists another example of a finite Fermi system, where the pair correlation is similar to that in the π-electron system of aromatic molecules. Namely, this effect may occur in small metal clusters. If we study, for example, small clusters of Al, Pb, etc., we are dealing with materials which are superconducting in the bulk, i.e., macroscopic form. There arises the very interesting question about the critical cluster size at which the pair correlation state appears (see the discussion in [12]). Small metal clusters are similar to aromatic molecules; everything described in [1,2] and in this paper is applicable to them. It would be interesting to carry out

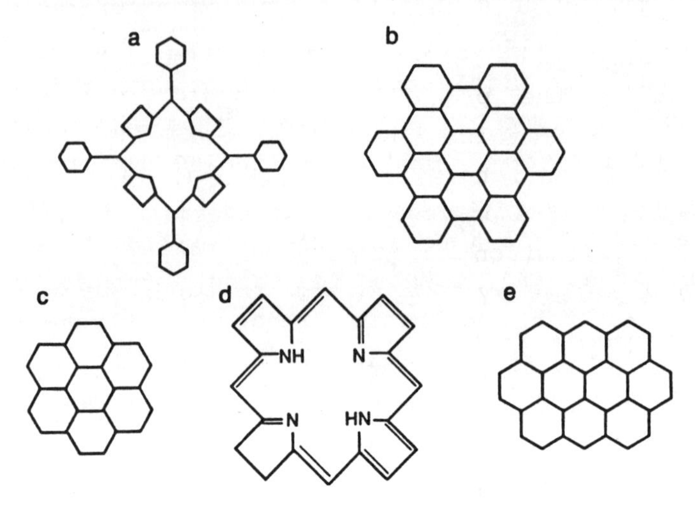

Figure 1. Molecules with delocalized electrons: a) tetraphen
a) tetraphenylporphin, b) hexbenzcorenene,
c) corenene, d) porphin, e) evalen.

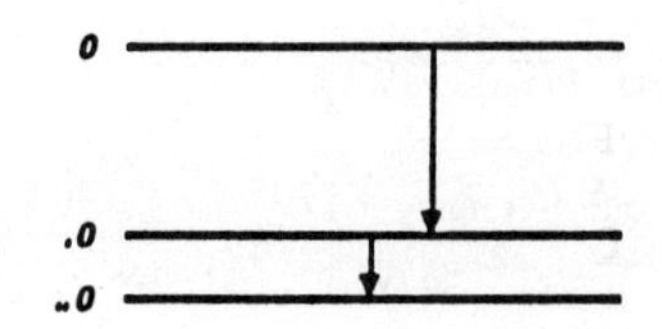

Figure 2. π-Electron energy levels.

an analysis of their energy spectra and magnetic properties as a function of cluster size and, consequently, of the number of delocalized electrons.

CONCLUSION

In summary, we have discussed the effect of pair correlation on the properties of π-electron systems in aromatic molecules. It is manifested in the simultaneous appearance of an energy gap for an even number of π-electrons and of anomalous diamagnetism. The pairing is caused by polarization of the σ-core. Charge transfer in biologically active systems is due to tunneling through a chain of Josephson barriers.

The author is grateful to W.A.Little and E.Teller for valuable discussions. The work was supported by the Director, Office of Energy Research, Office of Basic Energy Sciences, Chemical Sciences Division of the U.S.Department of Energy under Contract No. DE-AC03-76SF00098.

REFERENCES

1. V.Z.Kresin, Phys.Lett. A24, 749 (1967); Sov.Phys. - Doklady 12, 1147 (1968).
2. V.Z.Kresin, V.Litovchenko, and A.Panasenko, J.Chem.Phys. 63, 3613 (1975); J. de Phys. 39, C6-479 (1978).
3. A.Migdal, Theory of Finite Fermi Systems and Application to Atomic Nuclei (Wiley, New York, 1967).
4. V.Z.Kresin and W.A.Lester,Jr., in Mathematical Analysis of Physical Systems, ed. by R.Mickens (Van Nostrand Reinhold, New York, 1985), p.247; Int.J.Quan.Chem.Symp. 23, 17 (1989).
5. A.Bohr and B.Mottelson, Nuclear Structure, v.1 (Benjamin, New York, 1969)
6. L.Pauling, J.Chem.Phys. 4, 673 (1936); F.London, J.Phys.Rad. 8, 397 (1937); L.Salem, The Molecular Orbital Theory of Conjugated Systems (Benjamin, New York, 1966).
7. E.Ben-Jacob, K.Mullen, and M.Amman, preprint. I am grateful to D. Van Harlingen for bringing this paper to my attention.
8. P.Allen, R.Dynes, Phys.Rev.B 12, 905 (1975).
9. V.Z.Kresin, H.Gutfreund, and W.A.Little, Solid State Comm. 51, 339 (1984).
10. V.Z.Kresin, Solid State Comm.63, 725 (1987).
11. W.A.Little, Phys. Rev 134, A1416 (1964).
12. W.D.Knight,in Novel Superconductivity, ed. by S.Wolf and V.Z.Kresin (Plenum, New York, 1987),p.47.

SEARCH FOR NEW MATERIALS, STRUCTURE AND T_c, SYNTHESIS

PROSPECTS FOR NEW DISCOVERIES IN THE ORGANICS: SYNTHESIS PANEL

Dwaine Cowan (a), Ronald Elsenbaumer (b), Fred Wudl (c), James Collman (d) Gunzi Saito (e), and Peter Erk (d)

(a) Department of Chemistry, The Johns Hopkins University, Baltimore, MD 21218 (b) Allied Signal Inc., P.O. Box 1021R, Morristown, NJ 07962 (c) Institute for Polymers and Organic Solids, University of California at Santa Barbara, Santa Barbara, CA 93106 (d) Department of Chemistry, Stanford University, Stanford, CA 94305 (e) Department of Chemistry, Faculty of Science, Kyoto University, Kyoto 606, Japan.

I. INTRODUCTION

The objective for the panel discussion is to stimulate new ideas and ultimately encourage the synthesis and study of new organic metals and superconductors. In order to understand in detail the organic superconductors, minor and in some cases major structural modifications in the compounds need to be made and the new material studied to see how the physical properties, like the superconducting transition temperature, have been altered. However, the major breakthroughs in this field may well come from high risk ventures where totally new classes of compounds are devised, synthesized and studied. Wudl pointed out that it is best not to refer to the success/failure ratio in the preparation of new superconductors because we learn new chemistry and often physics whether the compound is a hit or a miss. In fact BEDT-TTF (see Figure 1), from which the highest T_c organic superconductors have been prepared, could have been considered a miss when Cava first prepared it in 1976. It was about six years later before the field had developed far enough to indicate that new salts from this compound should be studied. How many other compounds can be converted from a miss to a hit in 1990's?

Since the synthesis of a new organic π-donor, π-acceptor, or polymer can take from one to five years to develop it is important to optimize the

Organic Superconductivity, Edited by V.Z. Kresin and W.A. Little, Plenum Press, New York, 1990

number of hits. To aid in this endeavor it is possible to summarize some of what we have learned about organic metals with the following design considerations.

1. Stable open-shell (free radical) species or
2. Obtain band overlap; small HOMO-LUMO gap, large bandwidths
3. Molecules with delocalized π-molecular orbitals
4. Inhomogeneous charge and spin distribution
5. Segregated stacks or sheets of radical species
6. No periodic distortion which opens a gap in the density of states across the entire Fermi surface
7. Little or no disorder
8. Molecular components of appropriate or compatible size
9. Fractional charge transfer (Mixed-valence materials)
10. Strong interchain coupling or sheets to suppress metal to insulator (M–I) phase transitions
11. Cation and/or anion nominally divalent
12. Polarizable species to help reduce U

These considerations need to be constantly reexamined, refined and expanded as new materials are studied. Since this list has been discussed in detail elsewhere [1], only the most important considerations will be mentioned in reference to the design of superconducting compounds. Certainly any periodic distortion which opens a gap in the density of states across the entire Fermi surface will result in a metal-to-insulator phase transition (no. 6). If this takes place at a higher temperature than the superconducting transitions, we can either start over again with a new compound to try to suppress the M–I transition by applying pressure. If the materials are designed so that there are either relatively strong interchain coupling between the stacks or alternately sheets of open-shell species the M–I transition may be suppressed (no. 10). It is harder to move or reorder radical species in two dimensions than in one. All of the known organic metals and organic superconductors are mixed-valence or fractional charge transfer materials (No. 9). For example, HMTSF-TCNQ is an organic metal with a room temperature conducti·ity of about 2,000 $\Omega^{-1}cm^{-1}$ while isostructural salt HMTSF-$TCNQF_4$ is a semiconductor with a corresponding conductivity of about 2 x $10^{-4}\Omega^{-1}cm^{-1}$. The major difference is that HMTSF-TCNQ has a fractional charge transfer (Z = 0.74) while the $TCNQF_4$ salt does not (Z = 1). The degree of charge transfer has a dramatic effect upon the short-range coulombic repulsion between conduction electrons on the same molecular site (Ueff) and consequently a profound effect upon the electrical and magnetic properties of the material. With fractional charge transfer Ueff is reduced and it is possible to have two electrons or holes on a site giving rise to metallic transport properties. If it is not possible to place two electrons or holes on a site only a correlated type of motion is possible.

In summary, the compounds need to remain metallic as the temperature is lowered in order to observe the superconducting transition.

II. DONORS, ACCEPTORS AND THEIR SALTS

There has been a considerable amount of activity in the production of new donors and their salts in the last few years. Much of this work has been in four areas: the synthesis of (1) tetratellurafulvalenes (2) tetraoxafulvalenes (3) new salts of BEDT-TTF and new π-donors related to BEDT-TTF and (4) unsymmetrical donors.

In the last five years a smaller amount of work has focused on acceptors and their salts. This may be due in part to the realization that the charge density wave starts on the acceptor stack in TTF-TCNQ and that this induces a periodic distortion on the donor chain as the temperature is lowered. This Peierls distortion gives rise to the M–I transition observed in TTF-TCNQ and many related compounds. However, the recent work on N,N-dicyanoquinonediimine (DCNQI) and related acceptors indicates that this area should be probed further. For example, $Cu(2,5\text{-DM-DCNQI})_2$ has a room temperature conductivity of about 800 Ω^{-1} cm^{-1} and remains metallic as the temperature is lowered ($\sigma_{4K} = 5 \times 10^5\ \Omega^{-1}\ cm^{-1}$). In the future it may be possible to prepare new acceptors which are more polarizable (softer) but still good acceptors.

A. Tellurium π-donors (Tetratellurafulvalene) [2]

While there are about a dozen tellurium π-donors known at the present time, for various reasons (low solubility, high oxidation potentials due to the aromatic rings fused to the heterofulvalene moiety, and steric factors which preclude close stacking) only the recently prepared tetratellurafulvalene provided the wealth of information expected from this group of compounds. Several important trends are expected upon the replacement of sulfur or selenium with tellurium in the heterofulvalene salts. The larger more diffuse p and d orbitals centered on tellurium ought to give larger conduction bandwidths due to increased intrastack interaction and produce a material with reduced electron scattering and enhanced metallic electrical conductivity. In addition the tellurium p and d orbitals should increase the interchain interactions giving a more two-dimensional material. This increase in dimensionality should help suppress the various instabilities which often lead to insulating ground states in quasi-one-dimensional molecular metals. Also, the greater polarizability of the tellurium based heterocycle should help reduce the on-site coulombic repulsion (Ueff) and make doubly charged species more feasible. Finally, there could be a change in band filling because of the altered ionization potentials and polarizability of the new donors. Several interesting trends can be observed in Table 1 for the series, TTF-TCNQ, TSF-TCNQ and TTeF-TCNQ.

The acceptor bandwidths (W_A) remains about the same but the donor bandwidths (W_D) gradually increases as X changes from S to Se to Te. To a first approximation the electrical conductivity is proportional to one over the density of states at the Fermi level squared and this is proportional to the bandwidth squared.

$$\sigma \propto 1/[\,D(E_F)\,]^2 \propto [W_T]^2$$

The predicted conductivity ratio $\sigma(S):\sigma(Se):\sigma(Te)$ of 1:1.5:3.1 is in excellent agreement with the experimental ratio 1:1.2-2.3:3.2-6.3. The tight-binding electronic band structure calculations on TTeF-TCNQ using double-zeta Slater type orbitals indicate that the Fermi surface is warped

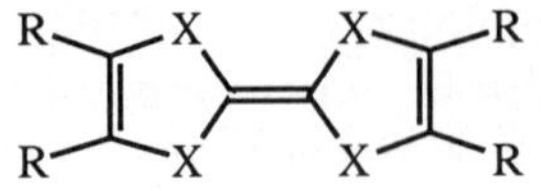

R X X R
R X X R

Tetrathiafulvalene (TTF): R= H, X= S
Tetraselenafulvalene (TSF): R= H, X= Se
Tetratellurafulvalene (TTeF): R= H, X= Te
Tetramethyltetrathiafulvalene (TMTTF):
R= CH_3, X= S
Tetramethyltetraselenafulvalene (TMTSF):
R= CH_3, X= Se

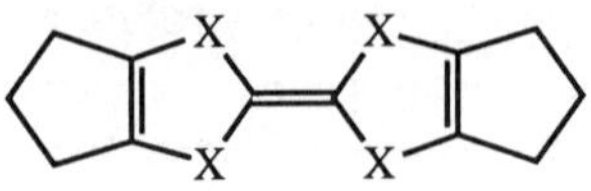

X X
X X

Hexamethylenetetrathiafulvalene (HMTTF):
X= S
Hexamethylenetetraselenafulvalene (HMTSF):
X= Se
Hexamethylenetetratellurafulvalene (HMTTeF):
X= Te

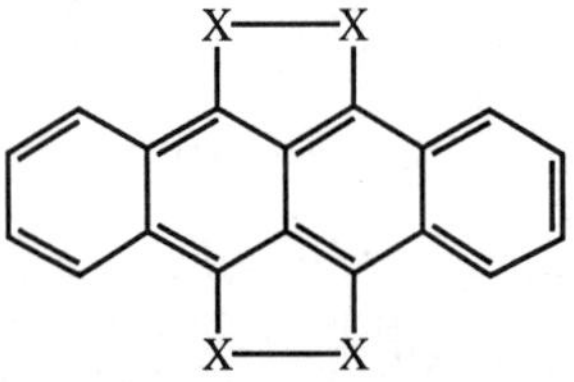

X—X
X—X

Tetrathiatetracene (TTF): X= S
Tetraselenatetracene (TST): X= Se
Tetratelluratetracene (TTeT): X= Te

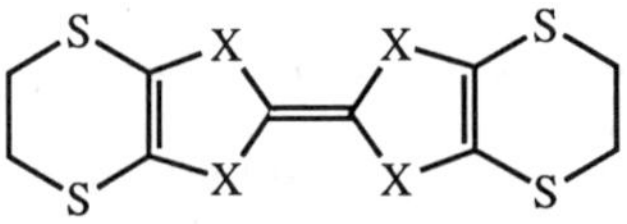

S X X S
S X X S

Bis(ethylenedithiolo)tetrathiafulvalene
(BEDT-TTF)

Figure 1. Organic metal and superconductor precursors.

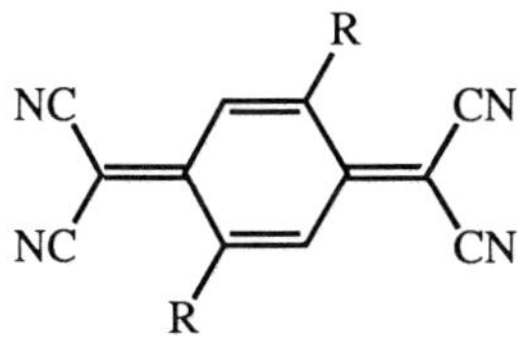

Tetracyanoquinodimethane (TCNQ):
R= H
Dimethyltetracyanoquinodimethane
(DMTCNQ): R= CH_3

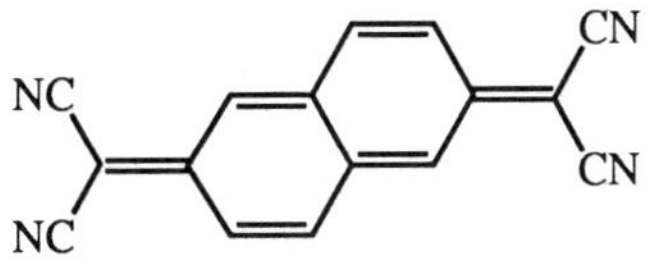

Tetracyanonaphthoquinodimethane
(TNAP)

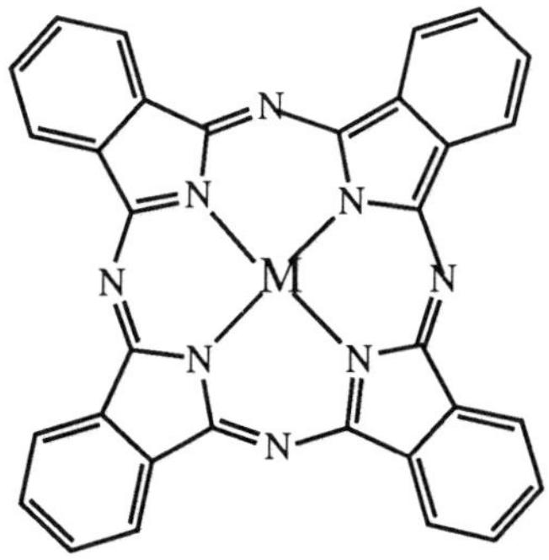

Metal Phthalocyanine [M(pc)]

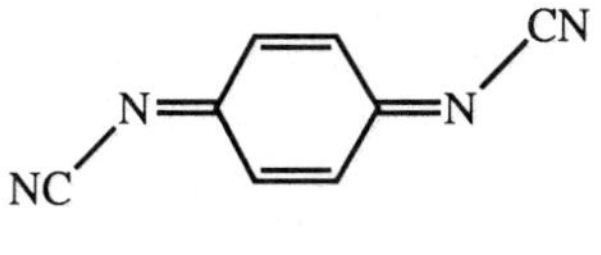

N,N-Dicyanoquinonediimine (DCNQI)

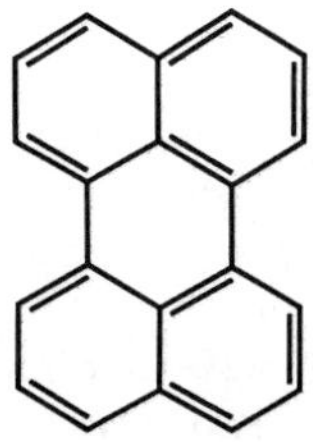

Perylene

Figure 1. (Continued.)

Table I. Physical and Theoretical Parameters of TXF-TCNQ

Parameter	X = S	X = Se	X = Te
σ_{300k} $(\Omega\ cm)^{-1}$	500 ± 100	800 ± 100	2,200 ± 300
Ratio σ	1	1.2 - 2.3	3.2 - 6.3
(a) Ratio $[W_T]^2$	1	1.5	3.1
(a) W_D (eV)	0.85	1.52	2.49
(a) W_A (eV)	1.37	1.31	1.36
$\sigma_{max}/\sigma_{300K}$	~14	~12	~9.5
T_{max} (k)	59 ± 1	40 ± 1	~2 (b)
Stacking axis (Å)	3.819	3.876	3.947
Z (e-)	0.59	0.63	0.71
X_{300} $(10^{-4}$ emu/mole)	4.0	3.0	-2.07

(a) double-zeta Slater type orbitals
(b) No M–I transition observed

relative to the more one-dimensional TTF-TCNQ and TSF-TCNQ. While the electrical conductivity of TTeF-TCNQ remains metallic to below 1.2K, no transition to a superconducting state could be found down to 0.5K. One possible reason that TTeF-TCNQ does not go superconducting is that the bands are too wide ($W_D \approx 2.5$eV). While this possibility is explored in the section on tetraoxafulvalenes we suspect that a more likely reason is that a two-dimensional material has yet to be constructed from the TTeF donor. Cowan believes what is needed is a β or κ phase TTeF salt similar to the BEDT-TTF salts. Other synthetic challenges in this area include the two tellurium analogues of the organic π-donors which have produced the largest number of organic superconductors – tetramethyltetratellurafulvalene (TMTTeF) and bis (ethylenedithiolo) tetratellurafulvalene (BEDT-TTeF).

B. Tetraoxafulvalenes

Much of the recent work on organic donors has been predicated on the notion that the substitution of Se and Te for S would increase the dimensionality (1D→ 2D→ 3D) in the salts formed from these donors and consequently minimize the instabilities so prominent in quasi-one-dimensional materials which frequently convert an organic metal into a

Peierls insulator. This hope was realized in the $(TMTSF)_2$ X series of salts. The transition to a semiconducting state was suppressed and many of the salts undergo a phase transition to a superconducting state.

While this reasoning is correct, it was pointed out many years ago for inorganic materials that the "poorest" metals often make the best superconductors, that is, have the highest transition temperatures (T_c). Certainly within the weak coupling limit ($U_oD(E_F) <<1$) of the Bardeen-Cooper-Schreiffer (BCS) mechanism of superconductivity it is possible to argue that T_c will be higher for materials which are not very good metals at room temperature. The BCS theory predicts for $U_oD(E_F) <<1$ that:

$$T_c = 1.14\,\theta \exp\left[-1/U_oD(E_F)\right]$$

where θ is the Debye temperature, $D(E_F)$ is the density of states at the Fermi level, and U_o is the electron lattice interaction. Note that since metals formed from the organic tellurium π-donors have wider bandwidths and lower density of states (same number of levels spread over a larger energy) than the corresponding selenium and tellurium compounds they might be expected to have lower superconducting transition temperatures. Also as the electron-phonon interaction term (U_o) increases, the room temperature scattering of the carriers ought to increase and again the materials should be poor metals. While the absolute value of the electrical conductivity is often not very accurately determined for the organic metals (due to the measurement of the rather small crystal dimensions), it is still instructive to examine the trends in T_c and σ_{RT} given in Table II.

Table II. Poorest Metals Make the Best Superconductors

Compound	$\sigma_{RT}(\Omega^{-1}cm^{-1})$	$T_c(K)$	P(K Bar)
$(TMSF)_2$ ClO_4	650	1.2	
$(TMTSF)_2$ PF_6	550	0.9	10
$(ET)_2$ ReO_4	200	1.5	5
κ-$(ET)_2$ $Cu[N(CN)_2]Br$	48*	11.6*	
β-$(ET)_2$ I_3	30	8	1.5
β-$(ET)_2$ IBr_2	20	2.8	
κ-$(ET)_2$ $Cu(SCN)_2$	14	10.4	

*Williams and co-workers, proceedings of this conference

It is interesting to note that the room temperature conductivities fall into two groups with values that are different by a factor of about 10 and that the salts in the lower conductivity group have Tc about 10 times higher than salts in the other group. This trend suggests that is ought to be worthwhile to explore both charge transfer and complex radical cation salts of tetraoxafulvalenes (TOF's). These certainly will have narrower bands and be poorer metals than the corresponding sulfur compounds and if the Peierls transition can be suppressed *via* 3D structural motifs like those found in the beta and kappa phases of BEDT-TTF (ET) salts, then higher superconducting transition temperatures may be realized. A number of potentially interesting but unknown TOF compounds suggested by Cowan as well as by others are shown in Figure 2. However, several important caveats need to be pointed out at this juncture. First, it is possible that these materials are not BCS superconductors. Second, even if the BCS mechanism is valid it seems unlikely that the weak coupling limit is appropriate. Tunnelling experiments, while not experimentally easy, give a direct way to evaluate the magnitude of the gap in the density of states caused by the condensation of carriers near the Fermi energy. This gap is given by the following equation:

$$2\Delta_0 / k_B T_c = 3.52$$

Many of the recent measurements indicate that this ratio is substantially larger than 3.52 for the organic superconductors. If the strong coupling limit is applicable then the reason why many of the organic metals do not go superconducting may have nothing to do with "wide" bandwidths but with the small number of k states available to scatter or couple into. McMillan's theory indicates that the electron coupling strength depends upon scattering integrals evaluated over the Fermi surface. The available number of states increases as the dimensionality of the material increases and as the Fermi surface extends into the second Brillouin zone (in the extended zone scheme).

C. New Salts of BEDT-TTF and New π-Donors Related to BEDT-TTF [3]

The electrochemical oxidation of BEDT-TTF produces ET salts with a diversity of stoichiometrics and crystal structures. Perhaps the most interesting of these are the kappa (κ) phase salts. The κ-phase of $(ET)_2Cu(NCS)_2$ is shown in Figure 3. Donor molecules in the κ-phase salts have an orthogonal packing motif where the donors and counter ions are arranged in sheets. While the largest interaction energy is found for the molecules within a dimer the interaction energy with an adjacent dimer is almost 1/2 as large. Since one electron is lost from every dimer the band formed from the HOMO ought to be 3/4 filled; but because of the structure the band is split into two bands and the upper one is 1/2 filled. Whangbo, Williams and co-workers calculate that the bandwidth for the upper band in κ-$(ET)_2Cu(NCS)_2$ is almost about 0.6 eV and the total bandwidth is about 1.0 eV. The Fermi surface associated with the half-filled band of κ-$(ET)_2Cu(NCS)_2$ is shown in Figure 4. The Fermi surface consists of overlapping "distorted circles" which interact to provide a noncrossing surface.

Since the highest T_c organic superconductors; κ-$(ET)_2Cu(NCS)_2$, 10.4K and κ-$(ET)_2Cu[N(CN)_2]Br$, 11.6K; are found among the κ-phase materials it is worthwhile to examine the question of why this is so. First, on the basis of the weak coupling limit of the BCS theory, a high T_c can indicate a large density of states at the Fermi level (narrow band), $T_c \propto \exp[-1 / D(E_F)]$. This was not found to be the case for a variety of κ–phase salts,

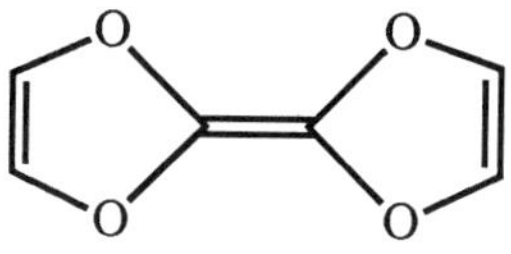

TOF

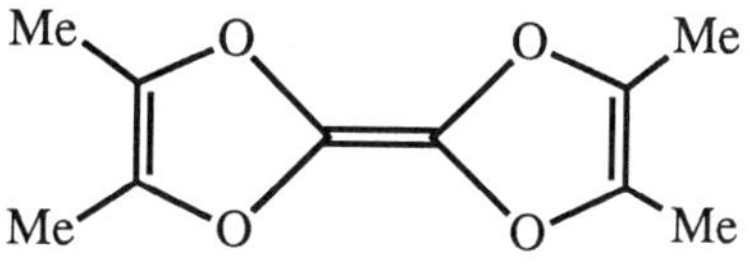

TMTOF

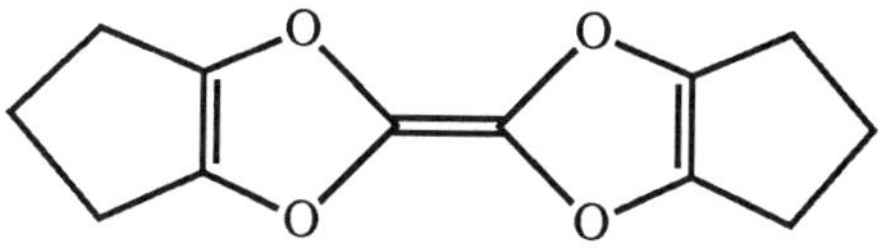

HMTOF

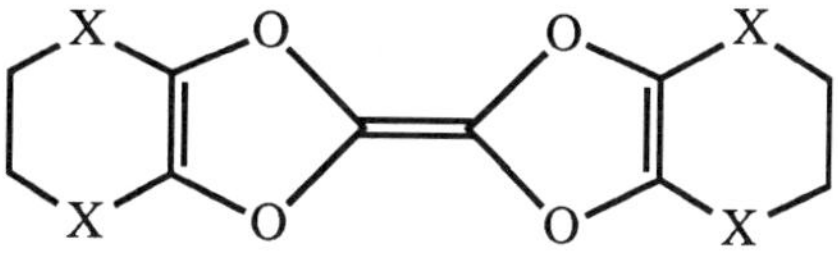

X = S,O

X = S BEDT-TOF (ET-TOF)

X = O BEDO-TOF

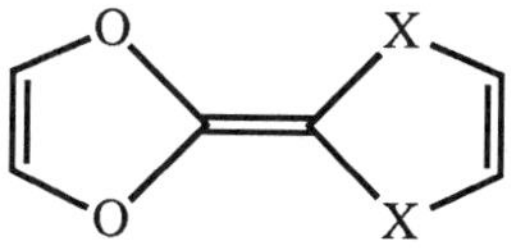

X = S, Se, Te

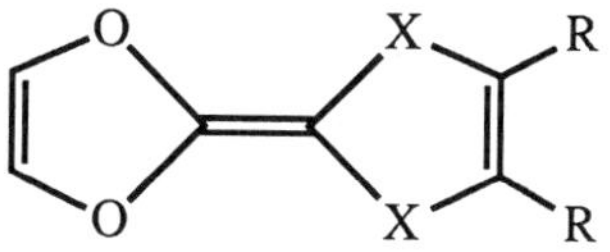

X = O, S, Se, Te

R = $-CH_3$; $-CH_2CH_2CH_2-$

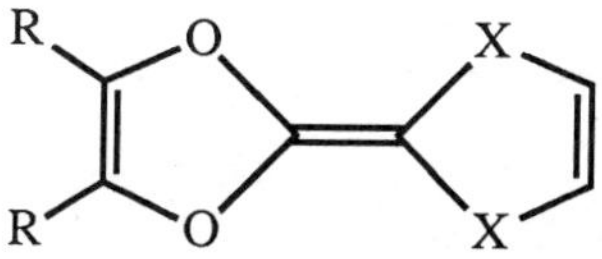

X = O, S, Se, Te

R = $-CH_3$; $-CH_2CH_2CH_2-$

Figure 2. Tetraoxafulvalenes.

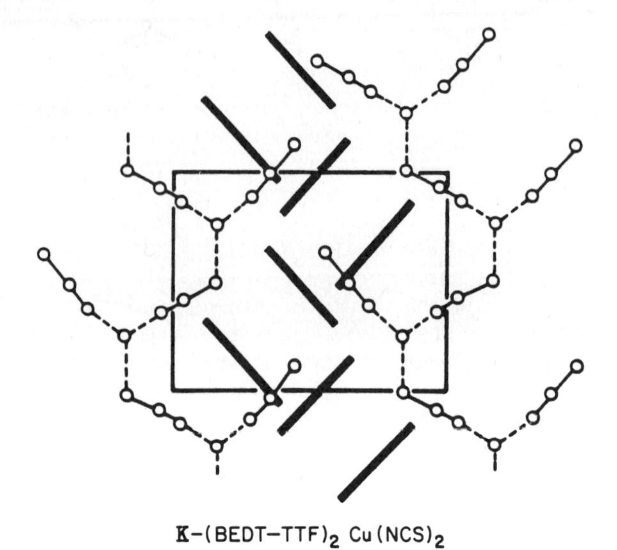

κ-(BEDT-TTF)$_2$ Cu(NCS)$_2$

Figure 3

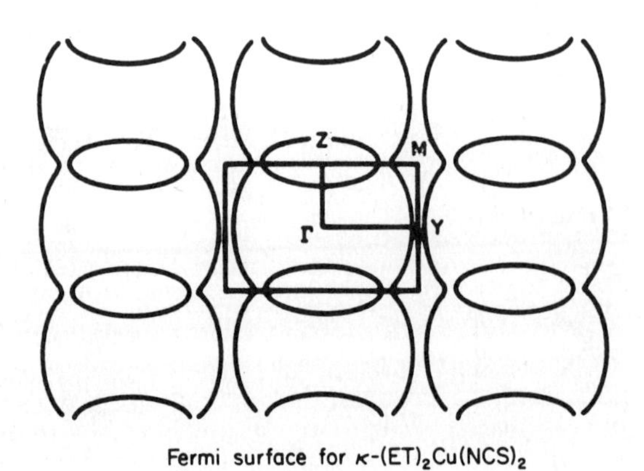

Fermi surface for κ-(ET)$_2$Cu(NCS)$_2$

Figure 4. (Redrawn after Whangbo.)

nor is it the case with the β-phase superconducting salts β-$(ET)_2$ X ($X^- = I_3^-$, AuI_2^-, IBr_2^-). For example, the density of states calculated for four k-phase compounds with T_c from nonsuperconducting to 10.4 were almost identical (7.1 to 8.0 e/eV). Whangbo, Williams and co-workers have proposed that the high Tc in the β-phase compounds, such as β*-$(ET)_2I_3$, 8K; and possibly the κ-phase material is due to the relative lattice softness which leads to a large electron-phonon coupling constant. This would presumably necessitate the use of the strong coupling limit of the BCS theory. While this is likely to be part of the explanation, one of us (DOC) would like to propose that the Fermi surface which extends into the second Brillouin zone, and the increased number of the states which are available to scatter (couple) into, could play an equally important role (see the discussion at the end of the TOF section). The calculated cylindrical Fermi surface corresponds to about 18% of the first Brillouin zone. This number is in good agreement with the value derived from the Shubnikov-deHass effect studies. The temperature dependence of the SdH oscillations gave an effective mass at the Fermi level of about 3.5 (m/m*). Williams and Saito have used either a correlation of anion length or effective unit cell volume, $V_{eff} = (V_{unit\ cell} - V_{anion})$ / (number of carriers (n) in unit cell) with T_c to help empirically predict new higher Tc superconductors. Saito's plot of V_{eff} vs T_c is given in Figure 5. It remains an important problem for the theorists to explain in detail why these correlations work as well as they do. The isotope effect on T_c of κ-$(ET)_2Cu(NCS)_2$ also poses a problem in that it is small and inverse (T_c for $D > C^{13} > H$). This is just the opposite of what would be expected from simple BCS theory.

D. ET Innovations

With the announcement at this conference by Williams and co-workers of the record setting superconductor; $ET_2Cu[N(CN)_2]Br$, T_c = 11.6K; we expect that a number of related salts will be prepared and studied. Further, we expect that continued examination of a variety of organic as well as inorganic counter ions which can form layers will expand our understanding of these materials and increase the superconducting transition temperature. Saito at this meeting, has asked the question of what function the thickness of the anion layers play. Do we want thin or thick anion layers? He also pointed out that if the κ-$(ET)_2Cu[N(CN)_2]Br$ could be converted into the β structure, one would predict a transition temperature of about 15K, because the donor molecules are more loosely packed in the β-phase than in the κ-phase. What factors influence the donor packing is a critical question that Whangbo and co-workers have been intensively examining. How important is the shape and cavity environment in which the anions reside in some phases? What are the relative importance of coulomb interaction, CH...anion contacts, CH...donor contacts, heteroatom...donor contacts, other factors? Does the relative "softness" or "hardness" of the CH...anion contact (interaction) play an important role in determining the superconducting transition temperature?

There are a number of interesting structural modifications which can be made on the ET framework. Wudl recently introduced the BEDO-TTF (or BO, see Figure 6) donor and one superconductor has already been reported by the Argonne group based on this donor, β_m-(BEDO-TTF)$_3Cu_2(NCS)_3$, T_c = 1.06K. One part of Wudl's rational for the synthesis of

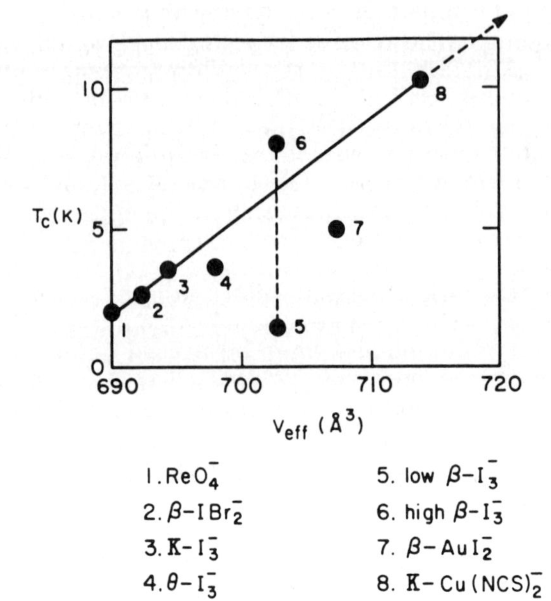

Figure 5. Donors related to BEDT-TTF which form superconducting salts.

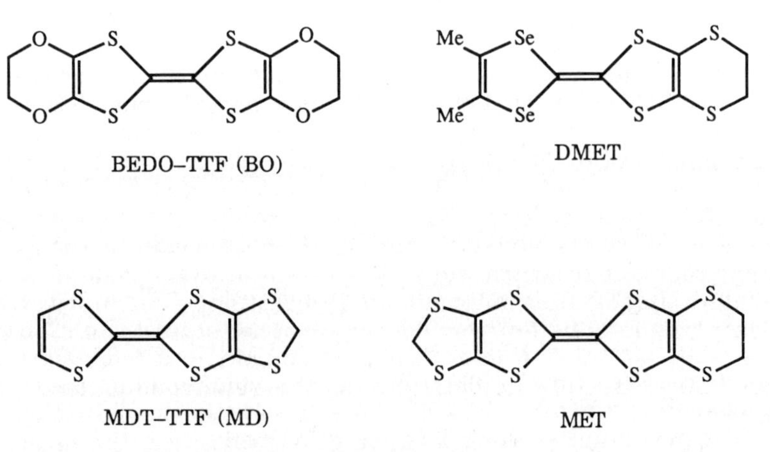

Figure 6

BO was an analogy with the high Tc inorganic superconductors. He believes that there could be hole states on oxygen in these donors just as there are in the ceramic superconductors [4].

Several noncentrosymmetric donors [5] consisting of half of a BEDT-TTF molecule and half TMTSF, TTF or BMDT-TTF have been prepared and shown to form superconducting salts or alloys. The recent synthetic work on the unsymmetrical donors has largely been carried out by Kikuchi (Tokyo); Kini (Argonne); Papavassiliou (Athens) and Lerstrup (Copenhagen). One of the major difficulties in the past has been to prepare the unsymmetrical donor in a fashion such that the two corresponding symmetrical donors are not also formed. If the symmetrical donors are also formed there is always a question regarding the purity of the unsymmetrical donor after isolation. The new synthetic methods are designed to give only the desired unsymmetrical product. The other question regarding the noncentrosymmetric donors is whether or not they would introduce disorder into the crystal structure and quench the superconductivity. Since superconducting salts are formed, the answer is no, but to understand why requires and examination of some of the crystal structures. For example, in the κ-$(DMET)_2AuBr_2$ (T_c = 1.9K) and κ-$(MD)_2AuI_2$ (T_c = 4.5K) the two unsymmetrical donor molecules of a dimer pair are related by inversion symmetry. In the β-$(DMET)_2I_3$ (T_c = 0.6K) the crystals have DMET donor stacks parallel to the b-axis. In the stack the unsymmetrical molecules alternate giving rise again to inversion symmetry in the crystal. The case which would almost demand some disorder is the alloy, $[(ET)_x(MET)_{1-x}]_2I_3$, T_c = 4.6K. If x was equal to 1 then T_c ought to be 8K (β^* structure). Possibly the small amount of MET (defects) lowers Tc from 8 to 4.6K.

The potential for the design and synthesis of new noncentrosymmetric donors is considerable.

E. New Acceptors for Multi-dimensional Organic Conductors [6]

During the last decade much synthetic effort has been directed towards the designing of new (fulvalene-like) donors. The synthesis of acceptor-based conductors has been thought generally to lead to systems with predominantly 1-D character. This is a consequence of the fact that the synthesis of molecules able to provide interstacking (e.g. chalcogen-chalcogen) interactions has been almost solely restricted to donor species.

In 1984 Aumüller and Hünig found a facile one step synthetic route to the then unknown class of N,N'-dicyanoquinonediimines (DCNQIs). The Lewis acid mediated condensation reaction of bis(trimethylsilyl)carbodiimide with carbonyl groups offers a general and facile means to prepare various N-cyanimines (see Figure 7). Substitution of the carbonyl group by the NCN-group raises the first reduction potential of quinoid type acceptors by 0.6 - 0.7 V, thus providing an acceptor-strength comparable to the TCNQs. This allows the synthesis of DCNQI radical anion salts by similar pathways as for TCNQ salts, e.g. the reaction with metal iodides and electrocrystallization. Whereas TCNQ and related polycyano acceptors react with metals to form semiconducting films on the metal surface, one can grow long single crystals of DCNQI radical anion salts by the immersion of copper or silver metal into solutions of DCNQIs in acetonitrile. This unexpected reactivity can only be explained with respect

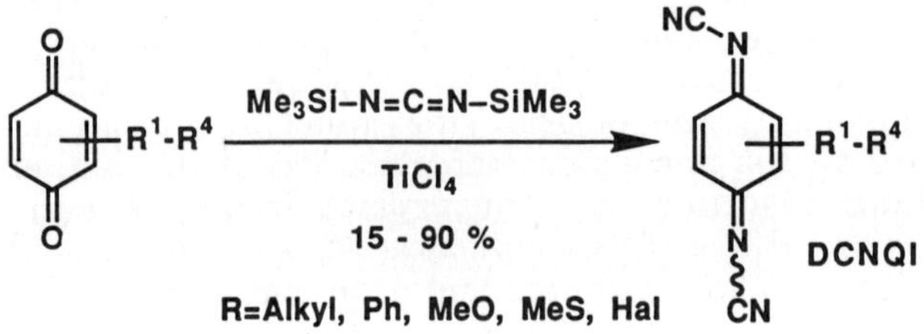

Figure 7. Synthesis of DCNQI.

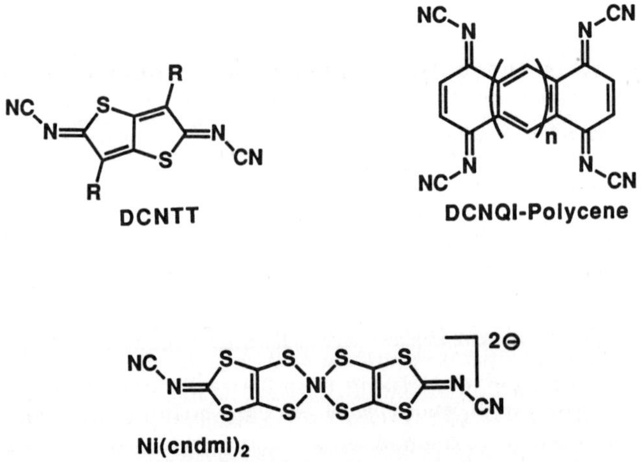

Figure 8. New acceptors.

to the ability of the DCNQI radical anions to act as strong ligands as demonstrated by their common chemical behavior and the structures of their radical anion salts in which they show coordination similar to the tetrakis(acetonitrile) complexes of copper and silver. The ability of the DCNQI radical anions to act as powerful ligands, determines their chemistry in solution and is the key for an understanding of the unusual physical properties of the DCNQI radical anion salts, e.g. their stable metallic character and extremely high conductivity up to $\sigma = 500{,}000$ $\Omega^{-1}cm^{-1}$.

We expect that the combination of coordinating ligands, and/or polarizable substituents, with acceptor type multi-step redox systems could produce new building blocks for superconducting materials. The application of these principles has lead to the synthesis of dicyanothienothiophenediimines (DCNTT: see Figure 8) which are currently under study [7]. Several other potential acceptors suggested by Erk are also shown in Figure 8. The introduction of NCN groups in the $Ni(dmit)_2$ system may allow the formation of radical anion salts with overlapping bands or closed Fermi surfaces.

The systematic investigation of metal-acceptor interactions and the participation of metal orbitals in the conduction band to form higher dimensional conductors will open up new frontiers and guide the design of materials with interesting properties.

III. POLYMERS, ORGANOMETALLICS AND INORGANIC MOLECULAR CONDUCTORS

A. Lessons from $(SN)_x$ [8]

Polythazyl is the only example of a covalently bonded polymer containing no heavy atoms which has been shown to be a superconductor. It is prepared by the solid state polymerization of S_2N_2. The $(SN)_x$ formed in this way exists as crystalline fibrils oriented in a common chain direction. Twinning is prominent and about 20% of the chains in each microfibril are in defect positions. The room temperature electrical conductivity in the chain direction is about $5 \times 10^3 \Omega^{-1}cm^{-1}$ with a $\sigma_{4K} / \sigma_{RT}$ of about 4. The IBM group found that there is a phase transition to the superconducting state at 0.26K. Addition of halogens, e.g., Br_2, forming $(SNBr_{0.4})_x$ removes electrons from the conduction band increasing the conductivity. This material is also superconducting below 0.3K. Since there are 3e/2 basis functions (1 on N and 2 on S) for the π-system the band formed from the π^* orbitals will be 1/2 filled. The bandwidths in the chain direction ($W_{\|}$) has been calculated to be on the order of 2 - 3 eV with a density of states at the Fermi level of about 0.18/eV spin molecule. The bandwidth in the direction perpendicular to the chain direction is about 0.5 eV. Consequently $(SN)_x$ can best be described as an anisotropic 3-D semimetal where both S···S and S···N interactions are important; and the transition to a superconducting state is reminiscent of a filamentary type II BCS superconductor.

B. Question from (CH)x [9]

Why is polyacetylene not a superconductor? It certainly is not as crystalline as $(SN)_x$ because of the way in which it is formed. However, the room temperature conductivity of doped $(CH)_x$ is from 500 to $10^3\Omega^{-1}cm^{-1}$ and

A. **ARE THE BANDS TOO WIDE?**
$D(E_F)$ TOO LOW?

Maybe, but it seems unlikely.

B. **HOWEVER, THE SOLITON, POLARON AND BIPOLARON BANDS FORMED UPON POLYMER DOPING ARE NARROWER, BUT FEW STATES--$D(E_F)$ TOO LOW?**

1. ***Help stamp out Solitons, Polarons and Bipolarons?***

2. ***Shoot the author and keep at least the Bipolarons?***

C. **ARE THE POLYMERS TOO 1-D?**

Too few k states to scatter into?

D. **IS THE INTERACTION ENERGY THAT FORMS THE COOPER PAIRS TOO WEAK?**

Always (This may be part of Murphy's law)!

E. **CRYSTALLINITY; AND DEFECTS DOPING UNIFORMITY**

May not be a problem with $\sigma > 10^4 \Omega^{-1} cm^{-1}$.

Figure 9. What are the critical considerations?

recently prepared material has a value of $2 \times 10^4 \Omega^{-1}cm^{-1}$. These values are as high as the values for $(SN)_x$. Wudl and Cowan at this time believe that the crystallinity, defects and doping uniformity are not the limiting factors. Elsenbaumer believes that they are. Elsenbaumer pointed out how thermal annealing of doped high conductivity (CH)x could pay rich dividends in terms of structural and electronic properties of the polymer. The band structure for $(CH)_x$ is derived from 1e/one basis function so that a filled band is formed from the π orbitals and an empty one from the π^* orbitals. The bandwidth ($W_{||}$) is on the order of 6.5 eV with a gap of about 1.4 eV. It is possible that the bandwidths are too wide [$D(E_F)$ too low] in these materials to support superconductivity. We suspect not, since many of the other conducting polymers have much narrower bands. Could it be a consequence of the doping to form either polarons, solitons, or bipolarons? The bands formed from, for example, the bipolaron states are narrow but there are relatively few states as well, so that the density of states at the Fermi level could be low. While this may be a problem at least one of us (DOC) thinks that it is the dimensionality of these polymers that is creating the problem. With $W_\perp$ of 0.1 eV there may be too few k states to scatter (couple into). (See earlier sections on BEDT-TTF and TOF). The final possibility is that the interaction energy that forms the Cooper pairs is too weak. This is hard to discuss since the mechanism for superconductivity in these materials is unknown.

If the dimensionality is the problem then the inclusion of heteroatoms and/or metal atoms in the polymer backbone could be critical. Several new systems for study which would fall into this category are described by Collman in the following section.

C. Inorganic and Organometallic Polymers

Among inorganic substances there are additional possibilities for conducting linear polymers. Some of these might become superconducting. Collman cited two examples from recent literature; neither has been developed nor studied as a molecular conductor.

A high molecular weight (3×10^5) linear polymer between germanium and hydroquinones [10] (Figure 10) has recently been reported. Prospects for doping this polymer or developing analogues with sulfur or selenium in place of oxygen should be explored.

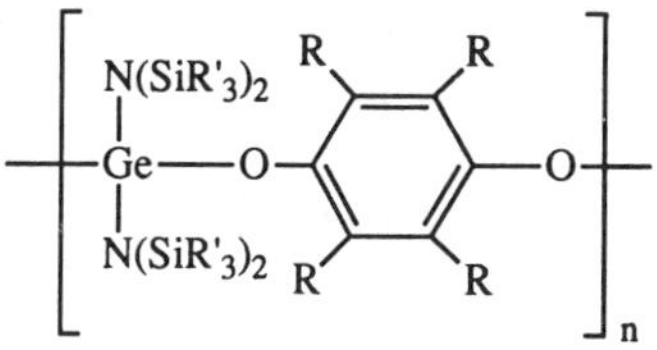

Figure 10

An unusual linear polymer containing P, N, and S has been described [10]. Collman suggested that chemical reduction of this polymer may result in a material which is electronically related to $(SN)_x$. The proposed scheme is shown in Figure 11. It is known that the chlorine atom on the polymer can be substituted with other functional groups [11].

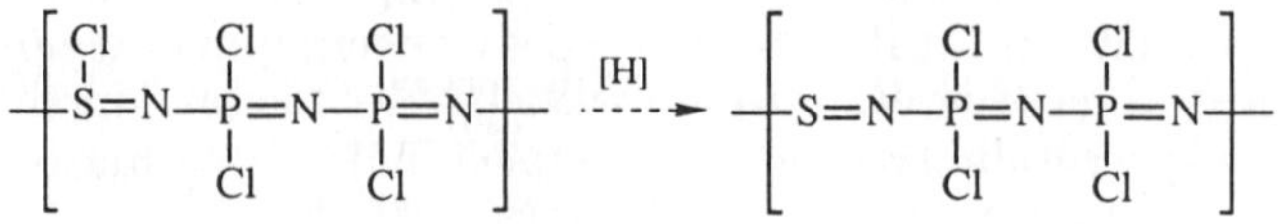

$$\left[-S(Cl)=N-P(Cl)_2=N-P(Cl)_2=N- \right] \xrightarrow{[H]} \left[-S=N-P(Cl)_2=N-P(Cl)_2=N- \right]$$

Figure 11

Secondly, he suggested that there are some interesting two dimensional systems that can be examined. These have less challenging problems with limited dimensionality and charge density waves. It is notable that the high T_c cuprates are quasi-two-dimensional. Graphite (doped) and TaS_2 already have been shown to exhibit superconductivity. Recent advances in graphite analogues such as (BN)x are promising.

Other layered compounds are known or feasible and these should be further explored. For example, black phosphorus is a layered structure semiconductor. Black phosphorus intercalated with iodine has been found to have metallic character. A new material, black phosphorus-germanium, has been prepared and shown to exhibit metallic behavior down to 1.5 K [12].

The majority of molecular superconducting materials are charge transfer salts. There are many unstudied charge transfer systems– especially among inorganic compounds. An important class which is virtually untapped concerns polymeric arrays of multiple metal-metal bonded dimers. Collman discussed multiply metal-metal bonded porphyrin dimers in his lecture [13]. Related systems have been proposed by Chisolm [14]. Two hypothetical polymers have been suggested (Figure 12A) and dimers containing each arrangement were prepared and examined in the context of electron transfer (Figure 12B).

In summary, there seem to be many unexplored prospects for new covalent conducting polymers; several of these involve inorganic materials.

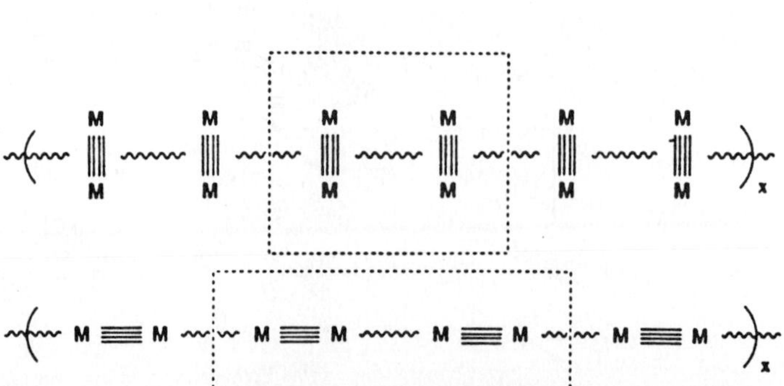

Figure 12A

Cowan pointed out that there are several proposed organometallic polymers that are being actively pursued for their electronic properties. A few of these are shown in Figure 13. Katz initially tried to use as-indacenide (2^-) ligand to form polymers but instead isolated quite interesting dimeric compounds [15]. This was a consequence of the fact that the metal cyclopentadienyl bonding can take place on the same side of the rings. Unlike the planar hydrocarbons, the helicenes cannot form dimers and he has been able to isolate low molecular weight polymers where M is $Co^+PF_6^-$[16]. Boekelheide has proposed using the $[2_n]$ cyclophanes since photoelectron and ESR studies of the radical cations and anions indicate that the π-systems of the aromatic ring strongly interact in these ions [17]. He has prepared several model bridged species to date. Lagowski has prepared the poly[(μ-η^6,η^6-naphthalene)chromium] *via* a new ligand exhange sequence of reactions. While this polymer is thermally sensitive and air-sensitive it is soluble in ND_3 presumably because of coordination of the solvent with the metal centers [18]. This work was discussed in detail in Lagowski's paper at this conference.

The macrocyclic metal complexes were to be discussed on the panel by Brian Hoffman. Because he had to cancel his trip just before the meeting this active and important area was not discussed in detail. Participants in this area have included Hoffman, Ibers, Marks, Hanack, as well as Collman and Little. It was pointed out that compounds like MPcI (where M = Ni, Cu, Co, H_2) have room temperature conductivity values in the range of 500-900Ω^{-1}cm^{-1} and that they remain metallic as the temperature is lowered reaching conductivity values near $10^4\Omega^{-1}$cm^{-1}, but do not undergo a phase transition to a superconducting state [19]. This area would seem to hold much promise if the interchain interactions can be increased.

Great opportunities in the design synthesis and study of new organic superconductors await the interested organic, organometallic and inorganic chemist.

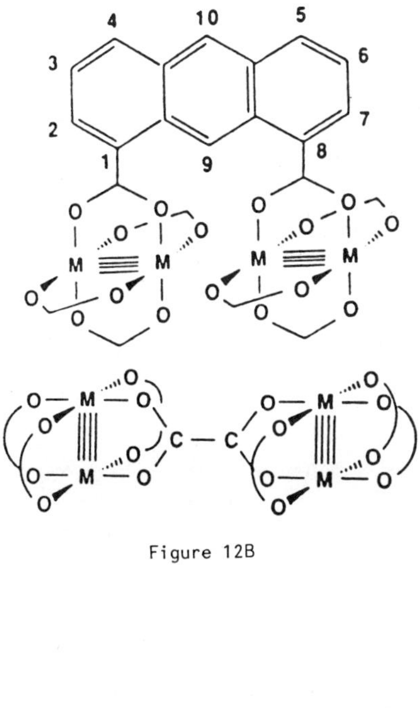

Figure 12B

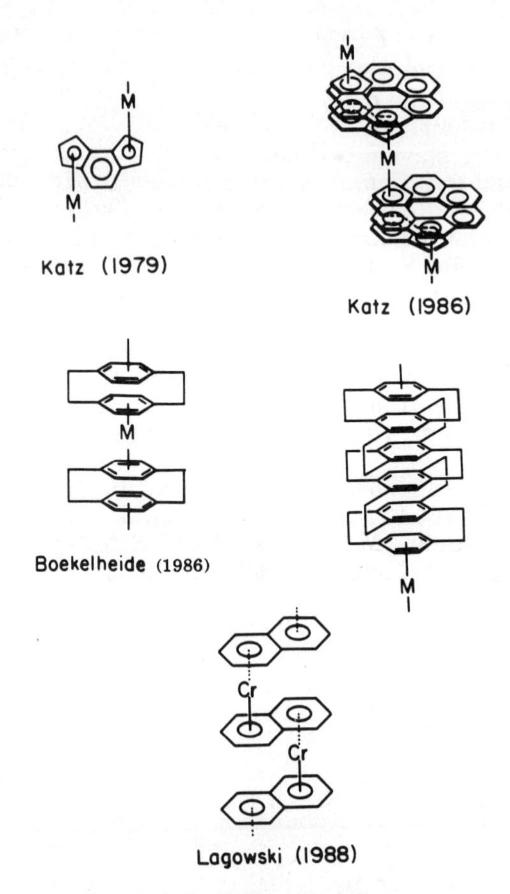

Figure 13. Organometallic polymers.

IV. ACKNOWLEDGEMENT

One of us (DOC) has greatly benefited from discussions of superconductivity with R.C. Dynes, Z. Tesanovic and W.A. Little. Work at the Johns Hopkins University was supported by NSF, Grant DMR-8921071; at University of California at Santa Barbara by NSF, Grant DMR-8820933; at Stanford by NSF, Grant CHE-8814949 and by DOE Grant DE-FG03-86ER45245.

V. REFERENCES

Inasmuch as there was less than two weeks between the time the panel took place and the deadline for the manuscript, a few references are given to aid the reader.

[1] Design considerations: Cowan, D.O. *In New Aspects of Organic Chemistry I. Proceedings of the Fourth International Kyoto Conference on New Aspects of Organic Chemistry* Yoshida, Z.; Shiba, T.; Oshiro, Y.; Eds.; VCH Publishers: New York, **1989**, pp 177-225.

[2] (a) Tellurium Donors: Cowan, D.O.; Mays, M.D.; Kistenmacher, T.J.; Poehler, T.O.; Beno, M.A.; Kini, A.M.; Williams, J.M.; Kwok, Y.-K.; Carlson, D.; Xiao, L; Novoa, J.J.; Whangbo, M.-H.; *Mol. Crys. Liq. Crys.* **1990**, *181*, 43; (b) McCullough, R.D.; Kok, G.; Lerstrup, K.A.; Cowan, D.O.; *J. Am. Chem. Soc.* **1987**, *109*, 4115, (and references cited).

[3] Kappa-phase: (a) Saito, G.; Uranyama, H.; Yamochi, H.; Oshima, K. *Synth. Metals* **1988**, *27*, A331; (b) Jung, D.; Evain, M.; Novoa, J.J.; Whangbo, M.-H.; Beno, M.A.; Kini, A.M.; Schultz, A.J.; Williams, J.M.; Nigrey, P.J. *Inorganic Chemistry* **1989**, *28*, 4516.

[4] DEDO-TTF: (a) Suzuki, T.; Yamochi, H.; Srdano, G.; Hinkelmann, K.; Wudl, F. *J. Am. Chem. Soc.* **1989**, *111*, 3108; (b) Beno, M.A.; Wang, H.H.; Kini, A.M.; Carlson, K.D.; Geiser, U.; Kwok, W.K.; Thompson, J.E.; Williams, J.M.; Ren, J.; Whangbo, M.-H. *Inorganic Chemistry* **1990**, *29*, 1599.

[5] Kijuchi, K.; Honda, Y.; Ishikawa, Y.; Saito, K.; Ikemoto, I.; Murata, K.; Anzai, H.; Ishiguro, T.; Kobayashi, K. *Solid State Commun.* **1988**, *66*, 405.

[6] (a) Hünig, S.; Aumüller, A.; Erk, P.; Meixner, H.; vonSchütz, J.U.; Gross, H-J.; Langohr, U.; Werner, H.-P.; Wolf, H.C.; Burschka, C.; Klebe, K.; Peters, K.; vonSchnering, H.-G. *Synth. Metals* **1988**, *27*, B181 and references cited therein; (b) Erk, P.; Hünig, S.; Meixner, H.; vonSchütz, J.H., Werner, H.-P. *Liebigs Ann. Chem.* **1988**, 157.

[7] Günther, E.; Hünig, S.; Peters, K.; Rieder, H.; von Schnering, H.G.; vonSchütz, U.; Söderholm, S.; Werner, H.-P.; Wolf, H.C. *Angew. Chem. Int. Ed. Engl.* **1990**, *29*, 204.

[8] Greene, R.L.; Street, G.B. in *Chemistry and Physics of One-Dimensional Metals* Keller, H.J.; Ed. Plenum Press, N.Y., **1977**.

[9] (a) Baughmann, R.H.; Brédas, J.L.; Chance, R.R.; Elsenbaumer, R.L.; Shacklette, L.W. *Chem. Rev.* **1982**, *82*, 209; (b) Patil, A.O.; Heeger, A.; Wudl, F. *Chem. Rev.* **1988**, *88*, 183.

[10] Kobayahhi, S.; Iwata, S.; Abe, M.; Shoda, S. *J. Am. Chem. Soc.* **1990**, *112*, 1625.

[11] Dodge, J.A.; Manners, I.; Allcock, H. *J. Am. Chem. Soc.* **1990**, *112*, 1268.

[12] Marugama, Y.; Inabe, T.; He, L.; Shirotani, I.; Willis, M.R. *Annual Review*, Institute for Molecular Science, **1989**, p 111.

[13] Collman, J.P.; McDevitt, J.T.; Kim, K.; Garner, J.M.; Zisk, M.B.; Leidner, C.R.; Yee, G.T.; Prodolliet, J.W.; Little, W.A., this symposium.

[14] Cayton, R.H.; Chisolm, M.H. *J. Am. Chem. Soc.* **1989**, *111*, 8921.

[15] Katz, T.J.; Slusarek, W. *J. Am. Chem. Soc.* **1979**, *101*, 4259.

[16] Sudhakar, A.; Katz, T.J.; Yang, B.-W. *J. Am. Chem. Soc.*, **1986**, *108*, 2790.

[17] Boekelheide, V. *Pure & Appl. Chem.* **1986**, *58*, 1.

[18] Bush, B.F.; Lagowski, J.J. *Organometallics* **1988**, *7*, 1945.

[19] Hoffman, B.M.; Ibers, J.A. *Acc. Chem. Res.* **1983**, *16*, 15.

METALLOPORPHYRIN LIQUID-CRYSTALLINE POLYMERS

James P. Collman[†], Matthew B. Zisk[†] and W. A. Little[‡]

Departments of [†]Chemistry and [‡]Physics
Stanford University
Stanford, California 94305

Introduction

The worked described in this paper involves refinements to a model for excitonic superconductivity. The evolution of this model is illustrated in figure 1. In 1964, Little proposed a model structure which might exhibit excitonic superconductivity at elevated temperatures.[1] This model was studied and further refined[2], based on the work of Krogmann[3], and incorporating the suggestion made by Collman[4] that a stacked array of square-planar transition metal complexes might also exhibit excitonic superconductivity. More recently, a model based on bridged group VIII metalloporphyrins has been explored in our laboratories.[5]

Our previous research involving bridged group VIII metalloporphyrins has demonstrated that these materials are conductors whose conductivity is supported by a bridged metal spine.[5] Much of the characterization of these materials was hampered by their intractability as insoluble polycrystalline powders. In an effort to produce materials that are better oriented and thus easier to study, we have undertaken the synthesis of porphyrinic materials which will self-assemble into well defined structures that still contain the essential features of the polycrystalline ones, as well as which may contain unprecedented features. This effort has led to the synthesis of a new porphyrin, octaundecylporphyrin (OUP) the structure of which is shown on the right-hand side of figure 1.

It had been our intention to insert group VIII transition metals into a known mesodiscotic liquid crystalline phthalocyanine. This material undergoes a phase transition from a so-called "herringbone" structure to a disordered columnar hexatic mesophase at between -60 and 60 °C, depending on the metal (see figure 2.) By small angle X-ray scattering, this hexatic phase has been shown to consist of macrocycles that stack one on top of another to form columns in one dimension, with the columns hexagonally closest packed in the other two dimensions.[6] We discovered after having gone through the lengthy synthesis of this ligand that it is all but impossible to insert second and third row transition metals into it. Therefore, we went about synthesizing analogous liquid crystalline porphyrins into which the larger metals should fit.

Organic Superconductivity, Edited by V.Z. Kresin and
W.A. Little, Plenum Press, New York, 1990

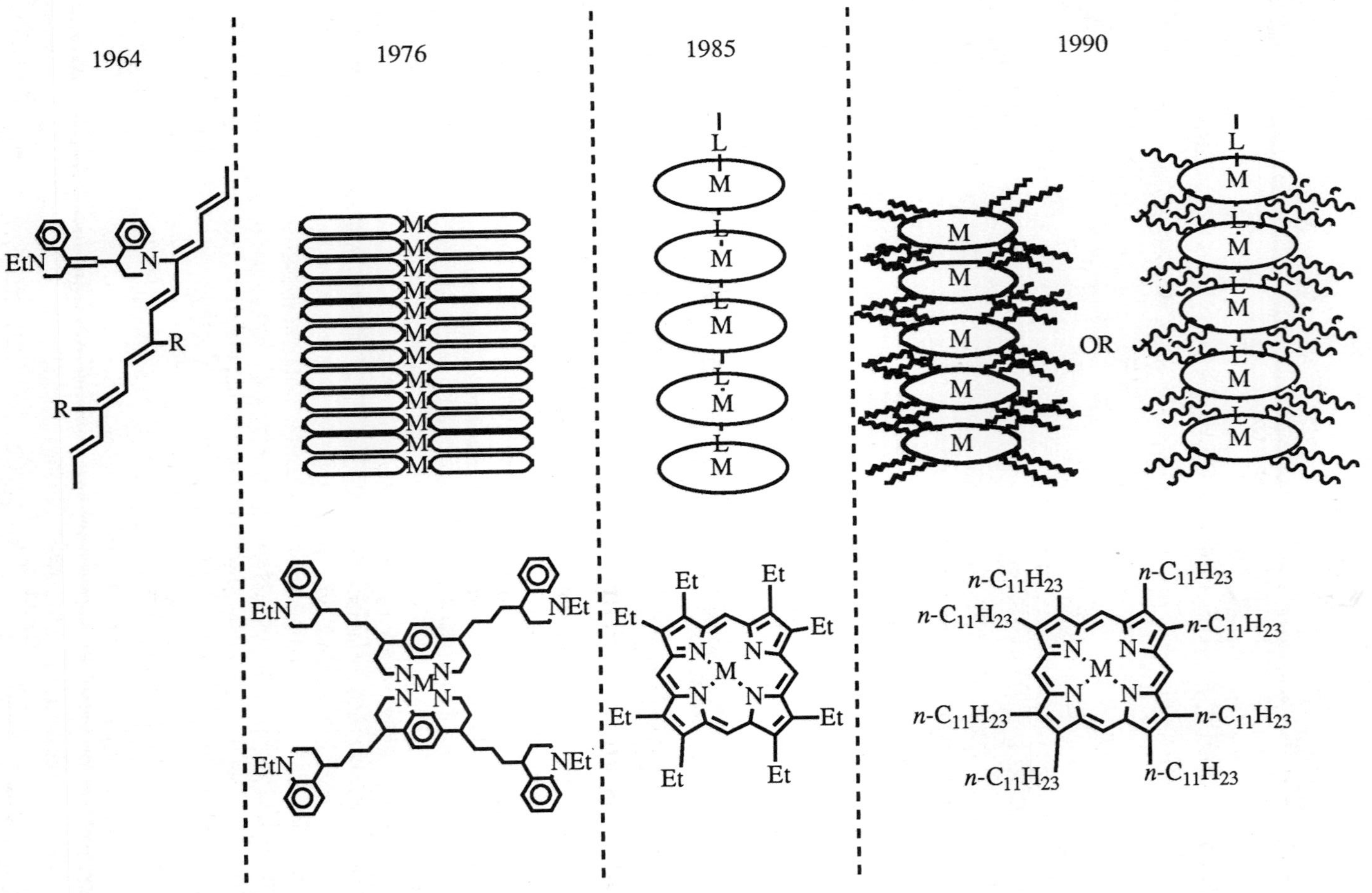

Figure 1. Evolution of model for excitonic superconductivity

RO
RO
OR
OR
N
N
HN
N
N
NH
N
N
RO
RO
OR
OR
R = CH2(CH2)nCH3, n = 6 - 10
Δ
"Herringbone"
Hexatic

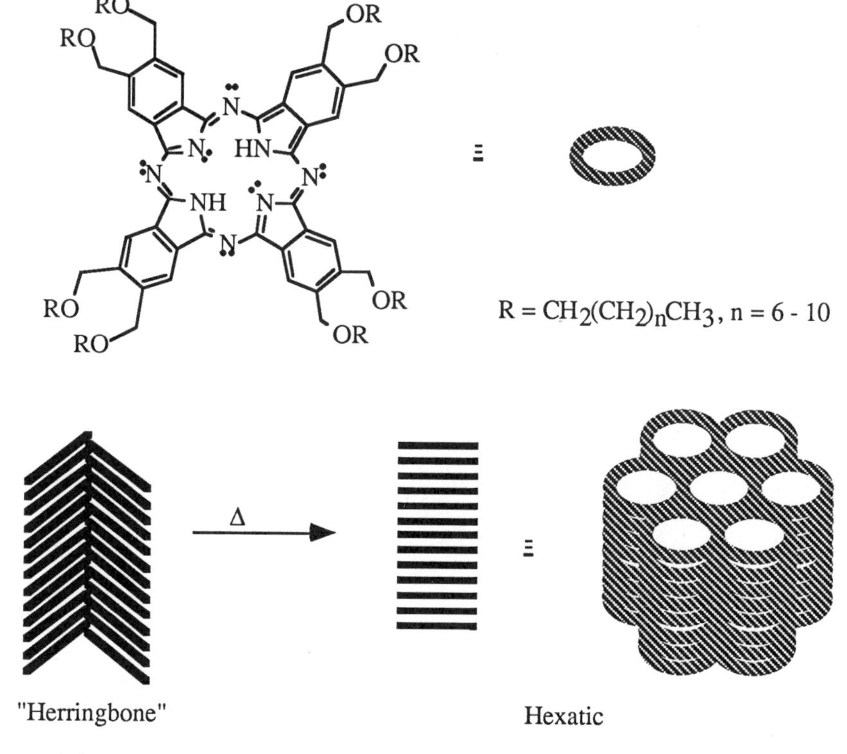

Figure 2. Known mesodiscotic liquid crystalline phthallocyanine

Synthesis and structure

Several examples of mesodiscotic liquid-crystalline porphyrins have appeared in the chemical literature in the past few years.[7] Of these, none has been structurally well characterized, and all have involved lengthy syntheses. Based on what is known about discotic liquid crystals in general, these materials have been presumed to have formed columnar hexatic phases (these assignments were made based on their optical textures in crossed polarized light; while optical textures are indicative of phase structures, they are not definitive.) In order to circumvent these lengthy syntheses, we have modified recently published[8] syntheses of octaethylporphine (OEP) so as to lengthen the β-pyrollic ethyl groups to undecyl groups (see Figure 3). This new porphyrin (OUP) is isolated as the zinc metallated material in order to facilitate its purification.

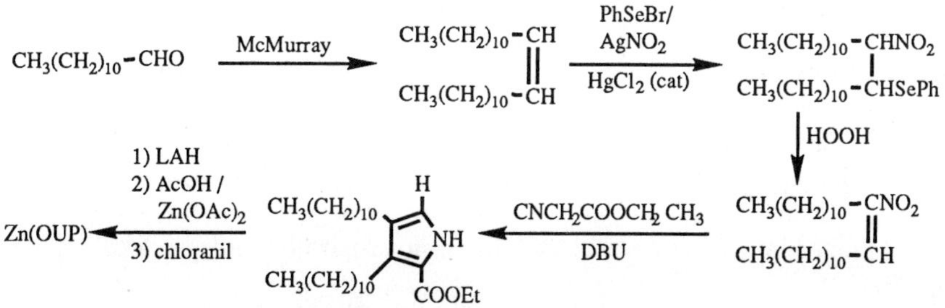

CH3(CH2)10-CHO
McMurray
CH3(CH2)10-CH
CH3(CH2)10-CH
PhSeBr/
AgNO2
HgCl2 (cat)
CH3(CH2)10-CHNO2
CH3(CH2)10-CHSePh
HOOH
CH3(CH2)10-CNO2
CH3(CH2)10-CH
CNCH2COOCH2 CH3
DBU
H
NH
CH3(CH2)10
CH3(CH2)10
COOEt
1) LAH
2) AcOH /
Zn(OAc)2
3) chloranil
Zn(OUP)

Figure 3. Synthesis of Zn(OUP)

Although this new porphyrin, octaundecylporphine (OUP) shows great promise as an ordered material, the phase transitions of the metallated materials take place at higher than ideal temperatures (*vide infra*). Additionally, the first synthetic step is problematic because the McMurray reaction requires 5 equivalents of very finely devided titanium metal; this material is pyrophoric and limits the scale of the reaction. In order to circumvent these problems, we are synthesizing a set of octaalkoxy-substituted porphyrins (see figure 4.)

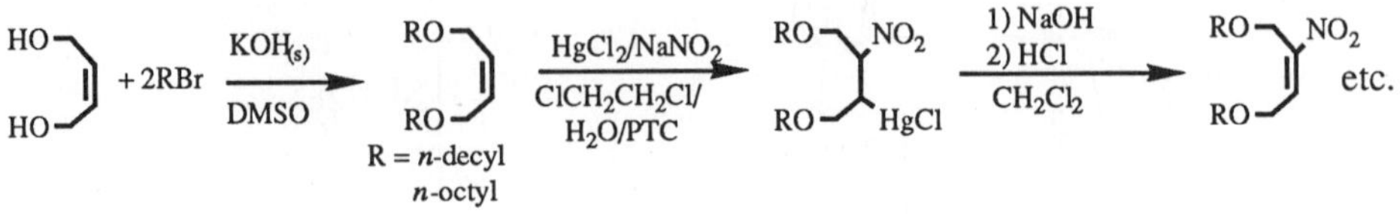

$$\text{HO/HO-alkene} + 2\text{RBr} \xrightarrow[\text{DMSO}]{\text{KOH}_{(s)}} \text{RO/RO-alkene} \xrightarrow[\text{ClCH}_2\text{CH}_2\text{Cl}/\text{H}_2\text{O/PTC}]{\text{HgCl}_2/\text{NaNO}_2} \text{RO, NO}_2\text{, RO, HgCl} \xrightarrow[\text{CH}_2\text{Cl}_2]{\text{1) NaOH; 2) HCl}} \text{RO, NO}_2\text{, RO-alkene etc.}$$

R = *n*-decyl, *n*-octyl

Figure 4: Synthesis of octaalkoxyporphyrins. Subsequent steps as in Figure 3.

Thermal behavior and alignment

Differential scanning calorimetry (DSC) and optical microscopy between -90 and 300 °C show that unmetallated, OUP undergoes a single phase transition from a plastic crystalline solid to an isotropic liquid. However, pure samples of metallated square-planar OUP complexes undergo phase transitions from crystalline (K) through discotic liquid-crysytalline (D) to isotropic liquids (I). Figure 5 shows a typical calorimetric scan for Zn(OUP). Transition temperatures for all of the complexes studied are tabulated in table 1, with the numbers in the headings corresponding to the numbered transitions in figure 5. The phase behaviors of these complexes are consistent with the other known liquid-crystalline porphyrins[6,7]

Table 1. DSC data for M(OUP) complexes

Complex	K - D (1)	D - I (2)	I - D (3)	D - K (4)
H_2(OUP)	None	115°C	None	100°C
Zn(OUP)	140 °C	183°C	178°C	132°C
Ni(OUP)	None	126°C	124°C	123°C
Pd(OUP)	135°C	136°C	132°C	127°C
$[Ru(OUP)]_2$ *	None	55°C	None	50°C

* $Ru(OUP)]_2$ data for 95% pure material

Solubility characteristics of these materials indicate that there is a strong interaction between the macrocycles: the unmetallated and four coordinate Zn, Ni and Pd complexes show limited solubility in all solvents at room temperature, while six-coordinate ruthenium complexes show high solubility even in hexanes. All OUP complexes are very soluble in hydrocarbon, chlorinated and aromatic solvents at temperatures greater than 60 °C. Except for those containing ruthenium, M(OUP) complexes crystallize from toluene/hexanes as fibers that are several millimeters long and several micrometers thick. Such fibers show optical extinction both parallel and perpendicular to the long fiber axis.

For known discotic liquid crystals, complete alignment has only been accomplished by spinning fibers at just below the discotic/isotropic phase-transition temperature.[9]

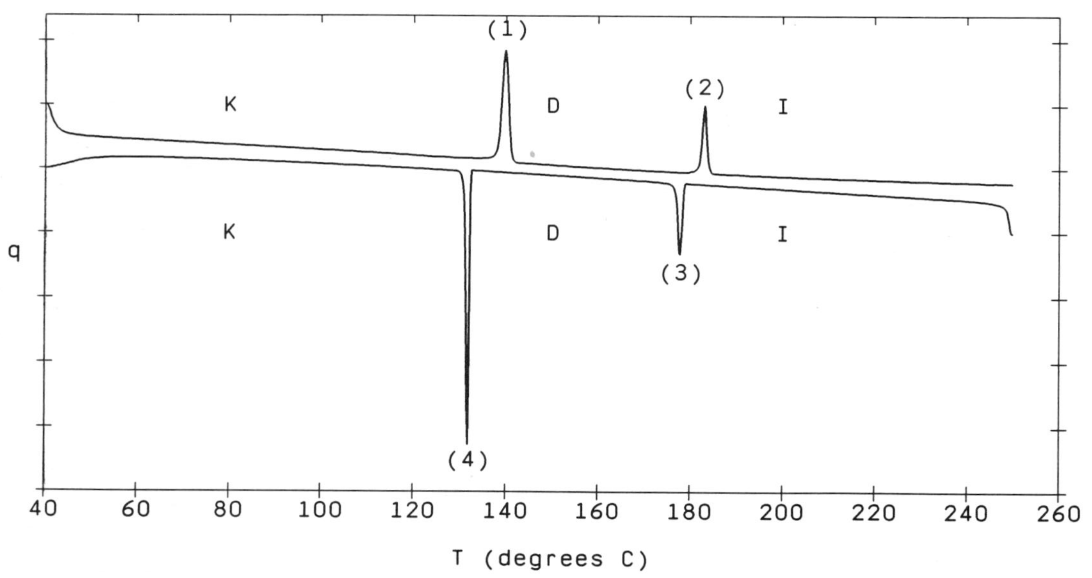

Figure 5. DSC scan of Zn(OUP) K region is crystalline, D region is discotic liquid crystalline and I region is isotropic liquid

Partial alignment of discotic liquid-crystals has been accomplished by slow flow through capillary tubes, and through their interaction with various coatings on glass.[10] These materials may also be aligned on suitable substrates such as pyrolytic graphite and muscovite. Films of the Zn(OUP) appear to be flow aligned between glass slides while in their liquid-crystalline phases. Evidence for this alignment comes from their optical textures in crossed polarized light. After cycling an aligned film through its various phases, it loses its alignment and forms a mosaic pattern characteristic of a D_{ho} or D_{hd} phase.[11] The assignment of the exact discotic phase of these materials will have to await planned X-ray scattering studies.

Electronic and magnetic properties

Our interest in meso-discotic liquid-crystalline porphyrins containing second and third-row transition metals stems from the anisotropic environment in which the metals are placed in such materials. This anisotropy may lead to low-dimensional conductors, as suggested in the introduction to this paper, and this anisotrpy may also lead to materials with other interesting and perhaps unprecedented magnetic and electronic properties. In particular, alignment of metal-metal bonded dimeric ruthenium containing porphyrins could lead to a macroscopic ordering of the unpaired spins in these materials. As illustrated in figure 6, the HOMO in neutral and singly-oxidized metal-metal multiply bonded ruthenium porphyrins is of π symmetry.[12]

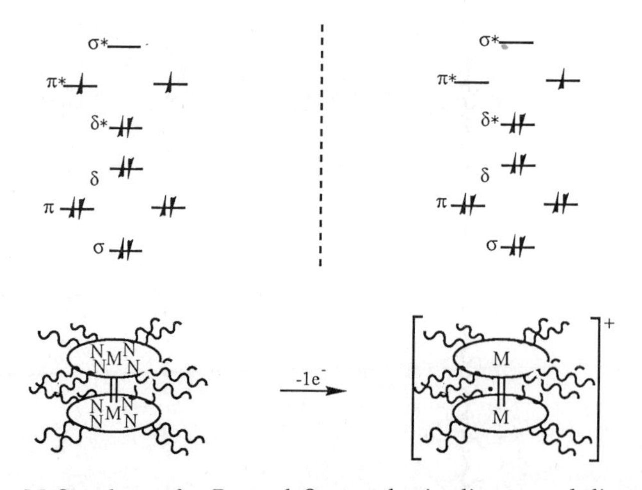

Figure 6. M.O. scheme for Ru and Os porphyrin dimers and dimer cations.

This situation is similar to a model proposed by McConnell for molecular ferromagnetism.[13] Where there is alignment of these dimers within a stack, and with partial oxidation of the stacks, ground-state bulk ferromagnetism may result from configurational-interaction with a stabilized high-spin lowest excited-state (see figure 7.) If there is insufficient exchange between neighboring porphyrin stacks, these materials may exhibit bulk anti-ferromagnetism.

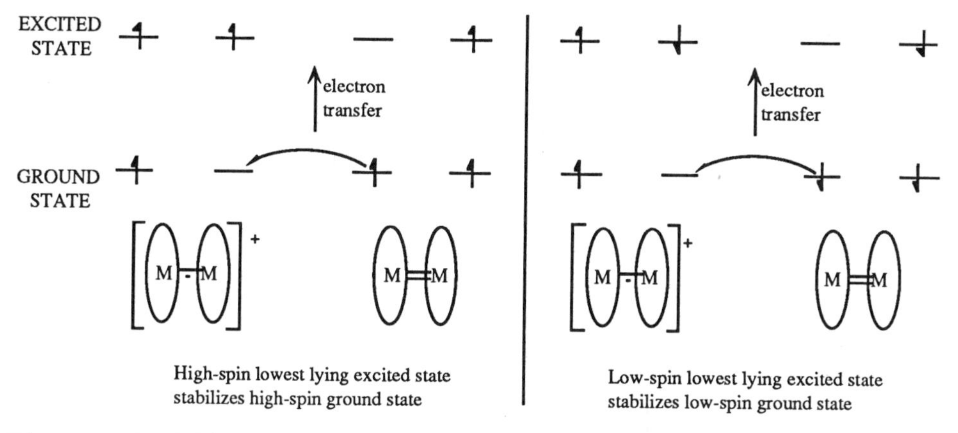

Figure 7. Model for configurational interaction stabilized intra-column ferromagnetic interaction. Hund's rule suggests that the left-hand situation is more stable than the right-hand situation.

Metal-metal bonded species, when aligned in a columnar fashion, may also show the unprecedented structure of an extended polymer of metal atoms chained together with cumulated bonds. This may happen in particular on partial oxidation of these species at an electrode surface (akin to an electrocrystalization; see figure 8.)

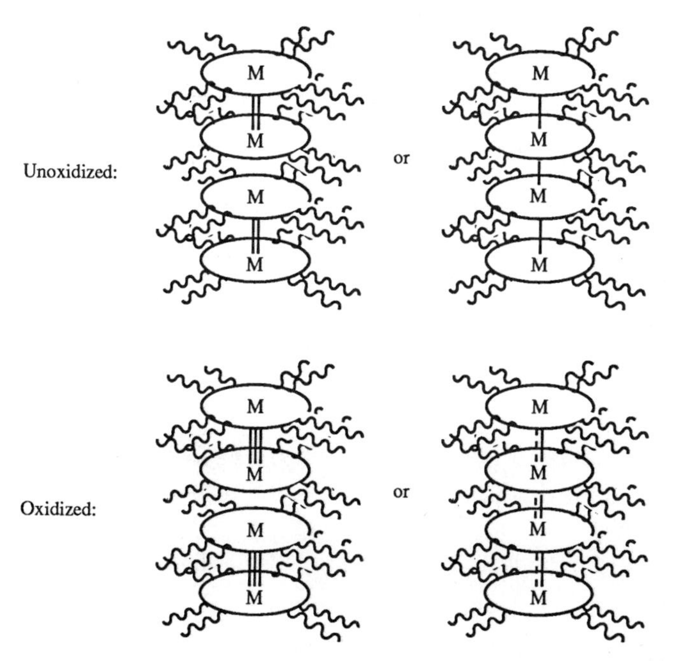

Figure 8. Proposed cumulated polymeric structures for Ru and Os OUP

We plan to introduce bridging axial ligands to these species in the hope of creating quasi-one-dimensional conductors (as has been shown in other organic metals, careful selection of counter ions should provide a small amount of inter-column charge transfer to prevent the formation of a charge density wave and a pinning of charge carriers.) We will synthesize μ-pyrazine polymers and also attempt to form a μ-oxo polymer through the dehydration of a columnar form of HO-Ru(OUP)-O-Ru(OUP)-OH. We hope to introduce these axial bridges by diffusion into an existing columnar phase, but where this is impossible, it is presumed that the characteristics of the paraffin side-chains will permit the alignment of oligomeric species. We will undertake analogous studies of these materials to those already published[5] for the bridged iron, ruthenium and osmium OEP series.

Acknowledgement

The support of the U.S. Department of energy (grant number DE-FG03-86ER45245) and the National Science Foundation (grant number CHE88-14949) is gratefully acknowledged.

References

1. W.A. Little, *Phys. Rev.*, **1964**, *134A*, 1416.

2. Davis, D., Gutfreund, H., and W.A. Little, *Phys. Rev.*, **1976**, *B13*, 4766.

3. K. Krogmann, J. W., *Angew. Chem.*, **1969**, *8*, 35.

4. Collman, J. P., *J. Polym. Chem.*, **1970**, *C29*, 138.

5. Collman, J. P. *et al.*, "A search for New Classes of Conducting Polymers Using 4d and 5d Metal Porphyrin Complexes", *this volume*, and references therein.

6. Piechocki, C., Boulou, J.-C., and J. Simon, *Mol. Cryst. Liq. Cryst.*, **1987**, *149*, 115 and references therein.

7. Gregg, B. A., Foxe, M. A., and A. J. Bard, *J. Am. Chem. Soc.*, **1989**, *111*, 3024; Gregg, B. A., Foxe, M. A. and A. J. Bard, J. C. S. Chem. Commun., **1987**, 1134; Goodby, J. W., *et al., Mol. Cryst. Liq. Cryst. Lett.*,**1980**, *56*, 303.

8. Ono, N. and K. Muruyama, *Chem. Lett.*, **1988** 1511; Sessler, J. *Synthesis, submitted.*

9. Giroud-Godquin, A.-M. *et al., J. Phys. France*, **1989**, *50*, 513.

10. Vauchier, C. *et al., Mol. Cryst. Liq. Cryst.*, **1981**, *66*, 103.

11. Destrade, C. *et al., Mol. Cryst. Liq. Cryst.*, **1984**, *106*, 121.

12. Collman, J. P., Prodelliet, J. W., and C. R. Leidner, *J. Am. Chem. Soc.*, **1986**, *108*, 2916.

13. McConnell, H. M., *J. Chem. Phys.*, **1963**, *39*, 1910

STRUCTURAL AND ELECTRONIC PROPERTIES OF HIGHLY CONDUCTING DICYANOQUINONEDIIMINE (DCNQI) RADICAL ANION SALTS

P. Erk and S. Hünig

Institut für Organische Chemie
Universität Würzburg, Am Hubland
D-8700 Würzburg, FRG

J. U. von Schütz, H.-P. Werner and H. C. Wolf

3. Physikalisches Institut
Universität Stuttgart, Pfaffenwaldring 57
D-7000 Stuttgart 80, FRG

D. Jérome, S. Tomić and R. T. Henriques

Laboratoire de Physique des Solides, Bâtiment 510
F-91405 Orsay-Cédex, France

D. Schmeisser

Institut für Physikalische und Theoretische Chemie
Universität Tübingen, Auf der Morgenstelle 8
D-7400 Tübingen, FRG

INTRODUCTION

The origin of the multidimensional band structure of all donor-based organic superconductors is due to interactions between chalcogen atoms in the periphery of the donors π-systems [1]. In general the anisotropy, usually found in columnar organic conductors is caused by different intra- and inter-π-stack interactions. If Halfchair- or envelope-like rings are annelated to TTF-like π-systems, a more isotropic behaviour is observed, as is demonstrated by studies of superconducting BEDT-TTF radical cation salts [2].

The discovery of unprecedented high conductivity and a stable metallic behaviour in [2,5-Me_2-DCNQI]$_2$Cu [3] lead to the development of a new approach for creating multidimensional highly conducting radical salts derived from *acceptor molecules* with powerful ligating abilities [4, 5, 6, 7].

The recently discovered facile one-pot synthesis of DCNQIs [8, 9] opens up the possibility of modifying their redox properties [10] as well as their steric properties by introducing a widespread range of substituents [11, 12, 13]. Furthermore, all salts derived from 2,5-substituted DCNQIs (Scheme 1) with various cations of the alkaline [12, 14] or coinage metals [15, 4, 5, 12] are isostructural and this permits the investigation of the subtle influence that the substituents have on their electronic properties. This paper illustrates the state of the

Organic Superconductivity, Edited by V.Z. Kresin and
W.A. Little, Plenum Press, New York, 1990

Scheme 1

R^1	R^3
MeO	MeO
MeO	Cl
MeO	Br
MeO	I
Cl	Cl
Br	Br
I	I
Br	I

$[\text{2-}R^1\text{-5-}R^3\text{-DCNQI}]_2M$ (NC–N=, R¹, R³, N–CN)

M	
Li, Ag, Cu	$I\,4_1/a$
Na	$C\,2/c$
K, Rb, NH_4, Tl	$P\,4/n$

R^1	R^3
Me	Me
Me	Cl
Me	Br
Me	I
Me	Et
Me	MeO
Cl	Br
Cl	I

art in the area and provides a background understanding of this new concept for the design of molecular organic conductors.

STRUCTURES OF $[2\text{-}R^1\text{-}5\text{-}R^3\text{-DCNQI}]_2M$

The copper and silver salts of 2,5-substituted DCNQIs with the substituents methyl, methoxy and the heavier halogens are all isomorphic [4, 12] and crystallize in the centrosymmetric tetragonal space group $I\,4_1/a$. The a,b-projection of the unit cell shows that the cations are aligned in rows surrounded by four DCNQI stacks (Figure 1 A). The cyano groups point toward the cations and the substituents project into channels between the DCNQI stacks. The DCNQI molecules lie on the crystallographic center of inversion and consequently the substituents of unsymmetrically substituted DCNQIs are statistically disordered.

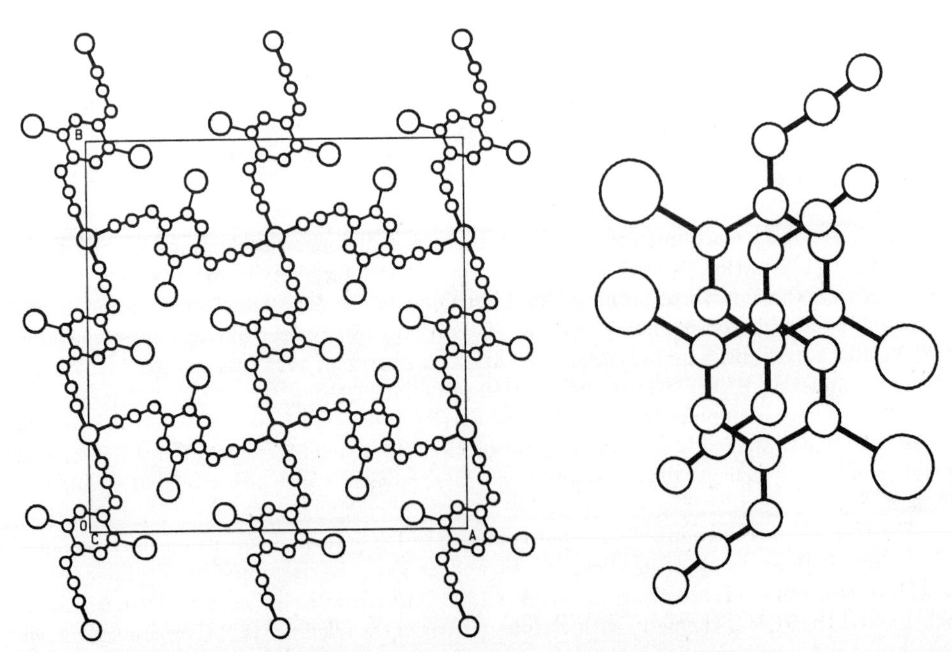

Figure 1: $[2\text{-I-5-Me-DCNQI}]_2Cu$, space group $I\,4_1/a$: a,b-projection of the unit cell (A) and view perpendicular to the molecular plane of stacking DCNQIs (B)

The molecular planes of the DCNQIs are skewed towards the stacking direction (c-axis) at an angle of $\approx 34°$ (Table 1). They form the well known "ring over bond" stacking arrangement which is usually observed in stacks of quinoid acceptors (Figure 1 B).

Table 1: $[2\text{-X-5-Me-DCNQI}]_2Cu$: c-axis, van der Waals radii, $r_{vdW}(X)$ [16], of the substituent X, skewing angle β (see Figure below) and interplanar distance $d(\pi-\pi)$ between the DCNQIs; shortest interstack distance of substituents d(X–Me) and sum of the van der Waals radii of the substituents, $\sum r_{vdW}$; copper-nitrogen distance, d(Cu–N) and distortion angle α (see Figure below). All distances in [pm], all angles in [°].

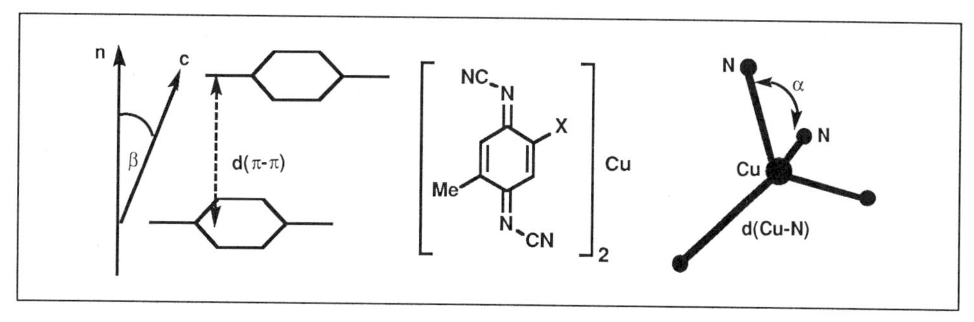

X	c-axis	$r_{vdW}(X)$	β	$d(\pi-\pi)$	d(X–Me)	$\sum r_{vdW}$	d(Cu–N)	α
Cl	383.0	180	33.9	317.9	382.4	380	198.4	125.8
Br	387.1	195	34.9	317.4	369.4	395	199.1	126.7
Me	388.7	200	34.4	320.7	397.1	400	198.3	125.0
I	397.3	215	35.8	322.4	353.0	415	199.8	123.2

A careful comparison of structural features indicates several trends in the arrangement of the acceptor stacks that are associated with the size of the substituents. They are exemplifyed by the four copper salts in Table 1. All known DCNQI copper salts show the same dependencies. The elongation of the c-axis correlates with the size of the substituents in the order Cl $<$ Br $\leq$ Me $\ll$ I. Increasing size of the substituents also causes an increase of the angle β between the normal vector of the DCNQI planes and the c-axis, maintaining the distance between the DCNQIs π-systems near a value of 320 pm.

The channels between the DCNQI stacks are tightly packed with the substituents. Contacts between the substituents are close to, or even significantly shorter than, the sum of the van der Waals radii. The closest distances, d(X–Me), between substituents of different stacks are found in the iodo/methyl-substituted compound.

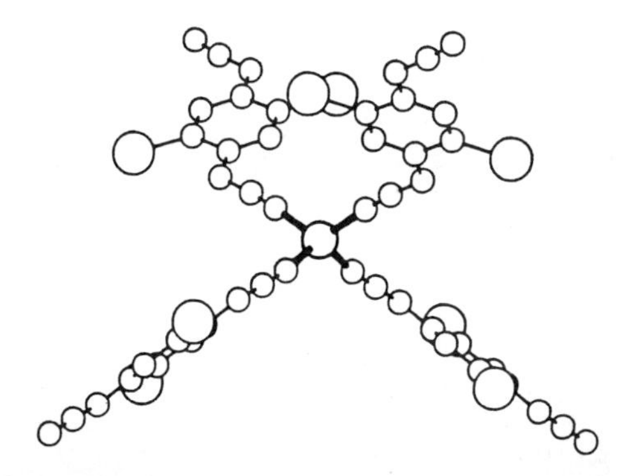

Figure 2: $[2\text{-I-5-Me-DCNQI}]_2Cu$: coordination sphere of the copper.

The copper atoms are coordinated by four DCNQI molecules in a distorted tetrahedral arrangement (Figure 2). The metal-nitrogen distances are within the range of the bond distance found in the tetrakis(acetonitrile)copper complex [17] (Table 1). The cyano groups point towards the copper, forming a bond angle of $\approx 170°$ at the cyanonitrogen. The distortion of the tetrahedral environment around the copper is best described by the angle α, which indicates that the coordination sphere of the copper is distorted towards square planar (Table 1). The angle α in $[2\text{-I-5-Me-DCNQI}]_2Cu$ comes closest to the ideal value of 109.5°. The reason for this may be seen in the skewing of the DCNQI molecules towards the stacking axes caused by the size of the iodo substituent.

ELECTRONIC PROPERTIES OF DCNQI RADICAL ANION SALTS

DCNQI non-copper salts

All salts of the type $[2\text{-R}^1\text{-5-R}^3\text{-DCNQI}]_2M$ (R = Me, MeO, Cl, Br, I; M = Li, Na, K, Rb, NH_4, Tl, Ag) show well defined 1d character [18, 19, 20]. Their room temperature conductivity ranges between 100 and 300 Scm^{-1} and since the temperature dependence of their conductivity can be described by Equation 1, they behave like metal-like semiconductors with activation energies between 30 and 70 meV (Figure 3) [21].

$$\sigma(\alpha, T) = AT^{-\alpha} \exp^{\frac{-E_g}{k_b T}} \quad (1)$$

Due to the low dimensionality of the DCNQI non-copper salts the system is susceptible to CDW formation, which has been observed in the silver salts by x-ray scattering [22]. At 100 K a $4k_F$ phase transition pins the electrons in $[2,5\text{-Me}_2\text{-DCNQI}]_2Ag$, followed by a $2k_F$ (spin Peierls) transition at 83 K, which is indicated by a drop in the ESR susceptibility.

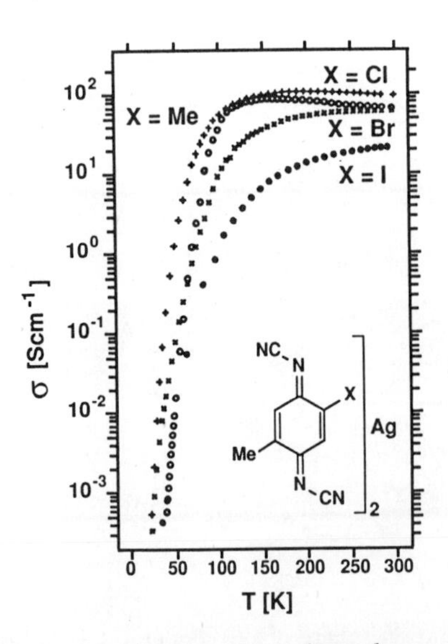

Figure 3: Temperature dependence of the conductivity of the salts $[2\text{-X-5-Me-DCNQI}]_2Ag$.

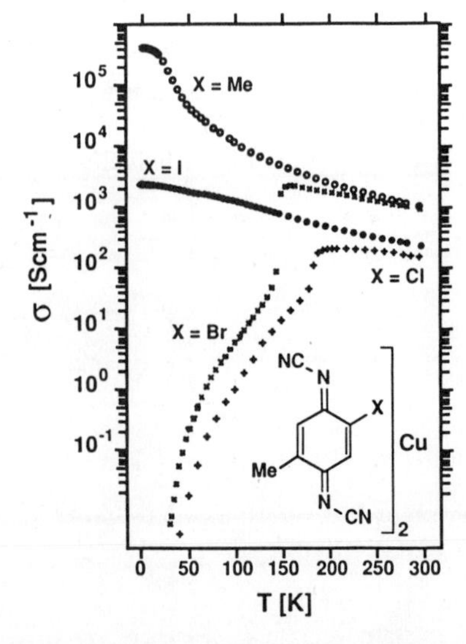

Figure 4: Temperature dependence of the conductivity of the salts $[2\text{-X-5-Me-DCNQI}]_2Cu$.

DCNQI copper salts

All copper salts of the 2,5-disubstituted DCNQIs depicted in Scheme 1 exhibit high room temperature conductivities of 700–1000 Scm^{-1}. The temperature dependence of their conductivity follows a power law, indicating metallic behaviour [4, 5] (Figure 4).

For the salts $[2\text{-Cl-5-Me-DCNQI}]_2Cu$ and $[2\text{-Br-5-Me-DCNQI}]_2Cu$ phase transitions into a semiconducting state are observed at 210 and 150 K, respectively. In both cases the semiconducting states show the same activation barriers as the corresponding silver salts [18].

The salts $[2,5\text{-Me}_2\text{-DCNQI}]_2Cu$ [3] and $[2\text{-I-5-Me-DCNQI}]_2Cu$ [23] retain metallic behaviour down to 0.45 K, without any indications of a superconducting state. They reach a minimum resistivity at about 30 K and subsequently their behaviour can be described by Matthiessen's rule (Equation 2), as composed of a temperature dependent resistivity, $\rho_P(T)$, due to electron-phonon scattering and a temperature independent residual resistivity, ρ_D, where impurities or imperfections in the crystal limit the conductivity [24].

$$1/\sigma(T) = \rho(T) = \rho_D + \rho_P(T) \qquad (2)$$

The metallic behaviour of the DCNQI copper salts is characterized by the occurrence of a Fermi edge in the UPS spectra at room temperature. At 100 K, i. e. below T_{M-I} of $[2\text{-Cl-5-Me-DCNQI}]_2Cu$, the DOS at the Fermi energy is drastically reduced and, as expected, a gap opens up. In contrast, the salt $[2,5\text{-Me}_2\text{-DCNQI}]_2Cu$ shows a Fermi edge down to 100 K due to the stability of the metallic state of this compound [25].

The pressure-temperature phase diagrams of the salts $[2\text{-Cl-5-Me-DCNQI}]_2Cu$ [26] and $[2,5\text{-Me}_2\text{-DCNQI}]_2Cu$ [27] indicate clearly that the metallic states in both salts are pressure sensitive (Figure 5). Upon application of 9 kbar pressure even $[2,5\text{-Me}_2\text{-DCNQI}]_2Cu$ is forced into an insulating state at room temperature. In the low pressure region (40 bar $\leq$ p $\leq$ 500 bar) $[2,5\text{-Me}_2\text{-DCNQI}]_2Cu$ shows a metal-insulator transition between 80 and 50 K (depending on the pressure) and reenters a metallic state below 50–10 K [28].

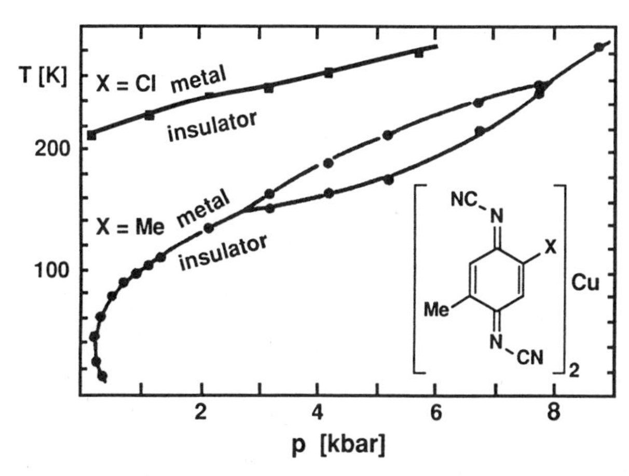

Figure 5: Pressure-temperature phase diagrams of the salts $[2\text{-Cl-5-Me-DCNQI}]_2Cu$ and $[2,5\text{-Me}_2\text{-DCNQI}]_2Cu$

ESR lines are extremely broad in the metallic state of the salts, due to strong spin-orbit coupling of the conducting electrons with the copper ions [19]. In the ^{63}Cu-NMR a Knight shift close to that for metallic copper is observed and a Korringa-like temperature dependence of the relaxation time [29]. Both experiments indicate that copper participates in the conduction mechanism, leading to conductivity in more than one dimension, which suggests that the high room temperature conductivity is due to screening of defects by copper bridges. The results of UPS spectra confirm this suggestion. The participation of Cu-d states in the DOS at the Fermi level is clearly demonstrated by the occurence of a strong peak at -4 eV relative to the Fermi energy [25], which can be assigned to the Cu-3d states as shown by resonance experiments [30].

X-ray investigations of the nature of the phase transitions have revealed that they are correlated with a discontinuity in the shortening of the c-axis [5], which also seems to be the case under applied pressure [31]. In the stable metallic salt $[2,5\text{-Me}_2\text{-DCNQI}]_2Cu$ no satellite reflections are observed over the whole temperature range. A superstructure is observed in the salts $[2\text{-Cl-5-Me-DCNQI}]_2Cu$ [32] and $[2\text{-Br-5-Me-DCNQI}]_2Cu$ [5] below T_{M-I}, which is due to an approximate trimerisation of the c-axis ($c^* \rightarrow c^*/3\pm\varepsilon$). The trimerisation of the unit cell implies an oxidation state of the copper of +4/3, consistent with the observation of approximately 30 % Cu^{2+} in the ESR susceptibility [19] and from XPS measurements [32]. However, samples of $[2\text{-Cl-5-Me-DCNQI}]_2Cu$ carefully grown under the exclusion of oxygen show a much lower content of Cu^{2+} (≤ 5 %) as revealed from XPS spectra [25].

DISCUSSION

The unique properties of the DCNQI copper salts can be attributed to the participation of Cu-3d states in the conduction band of the π^*-orbitals of the organic acceptors. Although most band structure calculations based on the extended Hückel approximation do not predict an admixing of copper states at the Fermi level [33, 34], this is clearly indicated by the experimental facts. Kobayashi et al. found, that the dicontinuity in the c-axis at the phase transitions leads to a flattening of the distorted tetrahedral environment around the copper ions [35, 5]. Thus a close relation between the coordination geometry of the copper and the electronic properties of the system can be assumed.

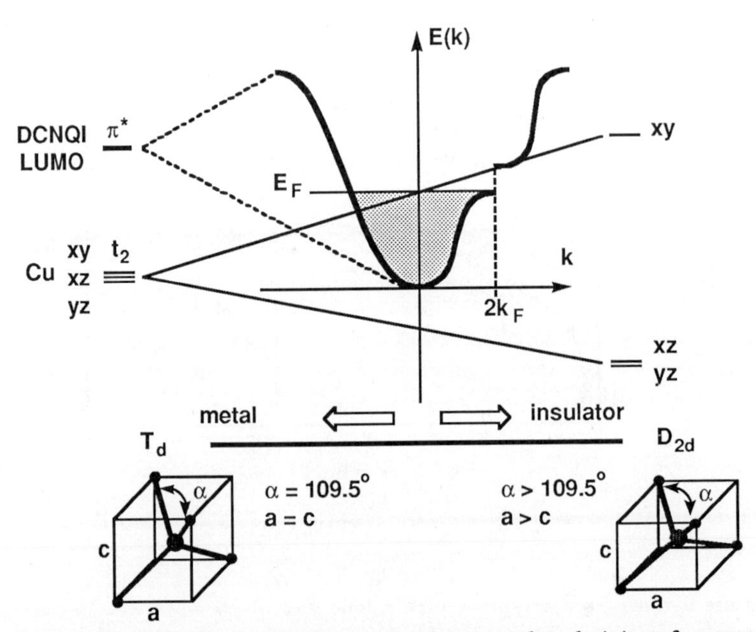

Figure 6: Qualitative description of the connection between the admixing of copper d-orbitals with the π^*-conduction band and the split of the d-orbitals caused by the distortion of the T_d symmetry of the coordination sphere of the copper.

In a tetrahedral transition metal complex the d-orbitals are split into t_2 and e_g subgroups. Under a distortion towards a square planar geometry the xz- and yz-orbital of the t_2 set are lowered in energy while the xy-orbital is raised (Figure 6).

As long as the xy-orbital lies energetically within the range of the DCNQIs π^*-band a highly conductive stable metallic state is preserved. As the temperature is lowered the c-axis starts to shorten, the tetrahedron around the copper is distorted and subsequently the xy-orbital raises further in energy. It is expected that the rise of the xy-orbital causes electrons to drop from the copper states into the π^*-conduction band and at a certain level the system becomes unstable towards a localised state and opens a gap at the Fermi energy. This phenomenon is chemically well documented, since solutions of Cu^{2+} (e. g. $CuBr_2$) in acetonitrile show a tremendous oxidation power caused by the driving force to reach the stable tetrahedrally coordinated tetrakis(acetonitrile)copper(I) state [36, 37]. The reverse of this process is seen during the phase transition of the copper salts. Kobayashi et al. calculated bandstructures of the DCNQI copper salts at the extended Hückel level [32, 5]. For a small splitting of the t_2 set of the copper d-orbitals they find a "multi-Fermi surface" system, unable to lower its energy by a simple lattice distortion. Furthermore, a correlation between the energies for the crystal field splitting of the copper orbitals and T_{M-I} provides evidence that the nature of the ligand sphere around the copper is related to the conducting properties of the copper salts [5].

In is still an open question, why, and in which way, sustituents on the DCNQI moiety affect the properties of the DCNQI copper salts so dramatically. The early suggestion, that the stronger electronaffinity of the halogen-substituted DCNQIs may be responsible for the the breakdown of the metallic state is no longer valid, since $[\text{2-I-5-Me-DCNQI}]_2Cu$ [23] and other salts derived from DCNQIs with high electronaffinities [38] retain their metallic behaviour down to the lowest temperatures.

According to a qualitative description of the MO diagram of tetrahedral complexes [39], the metal t_2 set of d-orbitals interacts predominantly with the donating σ-orbital/lone pair of the ligand, whereas its overlap with the π- and π^*-levels of the ligand are weaker. The bond strength between copper and the DCNQI should therefore be strongly dependent on the donating properties of the ligand, i. e. the energy level of the terminal nitrogens lone pair. The latter is also suggested by the correlation of the crystal field splitting with T_{M-I} [5].

As the correlation with photo electron spectra [40] and the reduction potentials [12] of the DCNQI demonstrates, AM1 calculations [41] are an appropriate means to illustrate the electronic properties of quinoid acceptors, especially the ligating abilities of the DCNQIs. As expected AM1 calculations reveal that the DCNQIs posess a σ-MO, which is predominantly located on the cyano nitrogen. The energy level of this orbital in the unsubstituted DCNQI lies ca. 0.8 eV above the comparable orbital of the TCNQ and thereby accounts for the differences in reactivity of both polycano acceptors.

The introduction of substituents affects the energy levels of the donating σ-orbital and the π^*-levels of the DCNQIs as well as the charge density on the nitrile nitrogen in the radical anions (Figure 7). The LUMO energies correspond well to the redox potentials of the DCNQIs and suggests their classification as 1) a group of weak acceptors with $\varepsilon_{LUMO} \geq -2.4$ eV represented by the dimethyl- and dimethoxysubstituted derivative, and 2) a group of halogen/methyl-substituted moderate acceptors with $-2.8 \leq \varepsilon_{LUMO} \leq -2.4$ eV; 3) a group of dihalogenated strong acceptors with $\varepsilon_{LUMO} \leq -2.8$ eV. The same classification is suggested by the charge densities on the terminal nitrogen of the DCNQI. DCNQI radical anions with only donating substituents show a fractional charge of –0.208. By the introduction of halogens the value is lowered by ≈ 0.013 per halogen.

The highest σ-orbital almost exclusively located on the terminal nitrogen shows up at energies between –14.5 and –14.9 eV. All the DCNQI radical anion salts with stable metallic properties are derived from acceptors with a lone pair above –14.6 eV. In contrast to the LUMO energies the nitriles lone pair orbital energies mirror the differences in the electron withdrawing character of the halogens, thus lowering in energy according to the order I > Br > Cl.

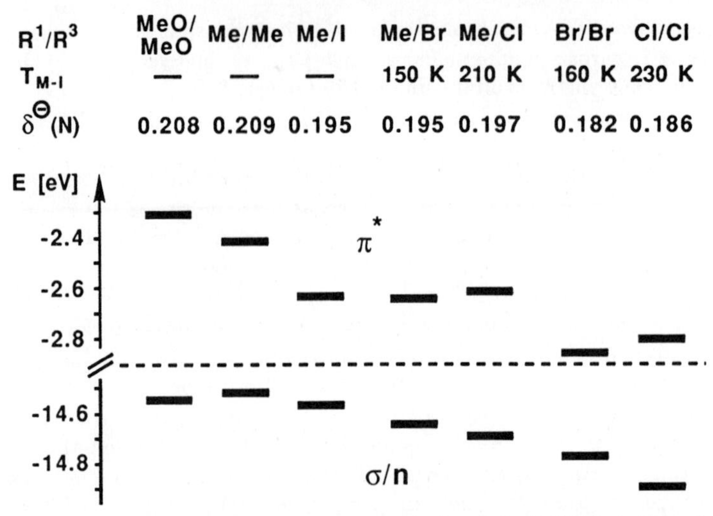

Figure 7: Substituent dependence of the energy levels of the LUMO (π^*) and the lone pair orbital of the cyano nitrogen (σ/n) versus T_{M-I} [5] and the partial charge on the cyano nitrogen, $\delta^{\ominus}$(N), in the radical anion, derived from AM1 calculations.

In conclusion we find, that DCNQI copper salts with stable metallic states are based on DCNQIs with moderate or weak π-acceptor character and good σ-donor properties. The last should dominate the bond strength in tetrahedral copper complexes and may therefore be held responsible for the stabilization of a low valent copper ion in a tetrahedral ligand environment, which is associated with the metallic character of the whole system.

Further experimental and theoretical attempts to investigate the connections between the electronic structure of the DCNQIs and the properties of their copper salts are under active investigation.

CONCLUSIONS

DCNQI radical anion salts give rise to a nearly unlimited series of quasi 1d metal-like semiconductors. However, the copper salts of 2,5-disubstituted DCNQIs are characterized by their metallic behavior due to an admixing of Cu-d orbitals and the π^* conduction band of the DCNQIs. The key to this unprecedented behaviour is the ligating power of the DCNQI radical anions, which also dominates their chemical behaviour.

Some of the DCNQI copper salts undergo phase transitions into a semiconducting state induced by cooling or under applied pressure. These are accompanied by a distortions in the coordination sphere around the copper. What finally determines whether DCNQI copper salts will exhibit a stable metallic state or not is still unclear. It seems to be related to the volume of the substituents and its steric needs and also to the electronic properies of the DCNQI to form dative and accepting bonds to the metal core.

Further investigations on the properties of DCNQI radical anion salts will lead the way to a deeper understanding of the concept for the design of metallic molecular conductors demonstrated in this paper. The stabilization of 1d conducting systems by metal-ligand interactions may open up possibilities to new highly conducting and superconducting materials.

ACKNOWLEDGEMENT

We are grateful to a large number of enthusiastic coworkers. Their names are given in the references. This work was supported by the Stiftung Volkswagenwerk, the Fonds der Chemischen Industrie and the BASF AG Ludwigshafen/Rhein.

REFERENCES

[1] H. Inokuchi, *Angew. Chem. Adv. Mater.* **100** (1988) 1817.

[2] J. M. Williams, H. H. Wang, T. J. Emge, U. Geiser, M. A. Beno, P. C. W. Leung, K. D. Carlson, R. T. Thorn, A. J. Schultz, M.-H. Whangbo, *Prog. Inorg. Chem.* **35** (1987) 51.

[3] A. Aumüller, P. Erk, G. Klebe, S. Hünig, J. U. von Schütz, H.-P. Werner, *Angew. Chem.* **98** (1986) 759.

[4] S. Hünig, A. Aumüller, P. Erk, H. Meixner, J. U. von Schütz, H.-J. Groß, U. Langohr, H.-P. Werner, H. C. Wolf, C. Burschka, G. Klebe, K. Peters, H.-G. von Schnering, *Synth. Metals* **27** (1988) B181 and literature cited therein.

[5] R. Kato, H. Kobayashi, A. Kobayashi, *J. Am. Chem. Soc.* **111** (1989) 5224 and literature cited therein.

[6] S. Hünig, *Pure Appl. Chem.* **62** (1990) 395.

[7] S. Hünig, P. Erk, *Angew. Chem. Adv. Mater.* **102** (1990) in press.

[8] A. Aumüller, S. Hünig, *Angew. Chem.* **96** (1984) 437.

[9] A. Aumüller, S. Hünig, *Liebigs Ann. Chem.* **1986** 142.

[10] A. Aumüller, S. Hünig, *Liebigs Ann. Chem.* **1986** 165.

[11] A. Aumüller, P. Erk, S. Hünig, H. Meixner, J.-U. von Schütz, H.-P. Werner, *Liebigs. Ann. Chem.* **1987** 997.

[12] P. Erk, *Dissertation*, Universität Würzburg, 1989.

[13] H. Meixner, *Dissertation*, Universität Würzburg, 1991.

[14] R. Kato, H. Kobayashi, A. Kobayashi, T. Mori, H. Inokuchi, *Synth. Metals*, **27** (1988) B263.

[15] A. Aumüller, P. Erk, S. Hünig, J. U. von Schütz, H.-P. Werner, H. C. Wolf, *Mol. Cryst. Liq. Cryst. Inc. Nonlin. Opt.* **156** (1988) 215.

[16] L. Pauling, *Die Natur der Chemischen Bindung*, Verlag Chemie, Weinheim 1960.

[17] I. Csoregh, P. Kierkegaard, R. Norrestam, *Acta Crystallogr. Sec. B* **B31** (1975) 314.

[18] H.-P. Werner, Dissertation, Universität Stuttgart 1988.

[19] H.-P. Werner, J. U. von Schütz, H. C. Wolf, R. K. Kremer, M. Gehrke, A. Aumüller, P. Erk, S. Hünig, *Solid State Comm.* **65** (1988) 809.

[20] R. T. Henriques, W. Kang, S. Tomic, D. Jérome, P. Erk, S. Hünig, J. U. von Schütz, *Solid State Comm.* **68** (1988) 909.

[21] J. U. von Schütz, M. Bair, H.-J. Groß, U. Langohr, H.-P. Werner, H. C. Wolf, D. Schmeisser, K. Graf, W. Göpel, P. Erk, H. Meixner, S. Hünig, *Synth. Metals*, **27** (1988) B249.

[22] R. Moret, P. Erk, S. Hünig, J. U. von Schütz, *J. de Physique* **49** (1988) 1925.

[23] P. Erk, S. Hünig, J. U. von Schütz, H.-P. Werner, H. C. Wolf, *Angew. Chem.* **100** (1988) 286; *Angew. Chem. Int. Ed. Engl.* **27** (1988) 267.

[24] H.-J. Gross, U. Langohr, J. U. von Schütz, H.-P. Werner, H. C. Wolf, S. Tomic, D. Jérome, P. Erk, H. Meixner, S. Hünig, *J. Phys. France* **50** (1989) 2347.

[25] D. Schmeisser, K. Graf, W. Göpel, J. U. von Schütz, P. Erk, S. Hünig, *Chem. Phys. Lett.* **148** (1988) 423.

[26] R. T. Henriques, D. Jérome, P. Erk, S. Hünig, J. U. von Schütz, *J. Phys. C: Solid State Phys.* **23** (1988) L1205.

[27] S. Tomic, D. Jérome, A. Aumüller, P. Erk, S. Hünig, J. U. von Schütz, *J. Phys. C: Solid State Phys.* **21** (1988) L203.

[28] S. Tomic, D. Jérome, A. Aumüller, P. Erk, S. Hünig, J. U. von Schütz, *Europhys. Lett.* **5** (1988) 553.

[29] D. Köngeter, F. Hentsch, H. Seidel, M. Mehring, J. U. von Schütz, H. C. Wolf, P. Erk, S. Hünig, *Solid State Comm.* **65** (1988) 453.

[30] D. Schmeisser, to be published.

[31] R. Moret, *Synth. Metals* **27** (1988) B301.

[32] A. Kobayashi, R. Kato, H. Kobayashi, T. Mori, H. Inokuchi, *Solid State Comm.* **64** (1987) 45.

[33] W. Koch, F. F. Seelig, *Z. Naturforsch.* **42** (1987) 875.

[34] W. Koch, private communication, 1989.

[35] A. Kobayashi, T. Mori, H. Inokuchi, R. Kato, H. Kobayashi, *Synth. Metals*, **27** (1988) B275.

[36] P. Erk, unpublished results.

[37] V. A. Pleskov, *Zurnal fiz. Chim.*, **22** (1948) 351; C. A. **1948** 6249.

[38] S. Hünig, H. Meixner, to be published.

[39] H. B. Gray, *J. Chem. Educ.* **41** (1964) 2.

[40] P. Erk, B. Mayer, to be published.

[41] M. J. S. Dewar Research Group, AMPAC, *Austin Method 1 Package*, QCPE # 527, University of Texas, Austin, 1986.

CONJUGATED IONIC POLYACETYLENES I: NOVEL STRUCTURES AND MODEL FOR A POLYMERIC HIGH T_c SUPERCONDUCTOR

Alexandre Blumstein and Sundar Subramanyam

University of Lowell
Department of Chemistry, Polymer Program
Lowell, MA 01854

INTRODUCTION

In the quest for an organic polymer high temperature superconductor, W.A. Little[1] proposed a model wherein an extensively conjugated and substituted polyacetylenic chain is endowed with a repeating unit in which one of the hydrogen atoms is substituted with a positively charged (quaternized) side group. The structure of the side group is such that the charge located in the alpha position to the conjugated chain can be delocalized and can resonate between quaternizable sites in the side group (Fig. 1). To our knowledge, such structures have not been previously synthesized.[2] The polymers obtained by us (Fig. 1b) are structurally similar to the model (Fig. 1a). Details of their synthesis are reported elsewhere.[3,4,5] Typical examples of such structures are shown in Figure 2.

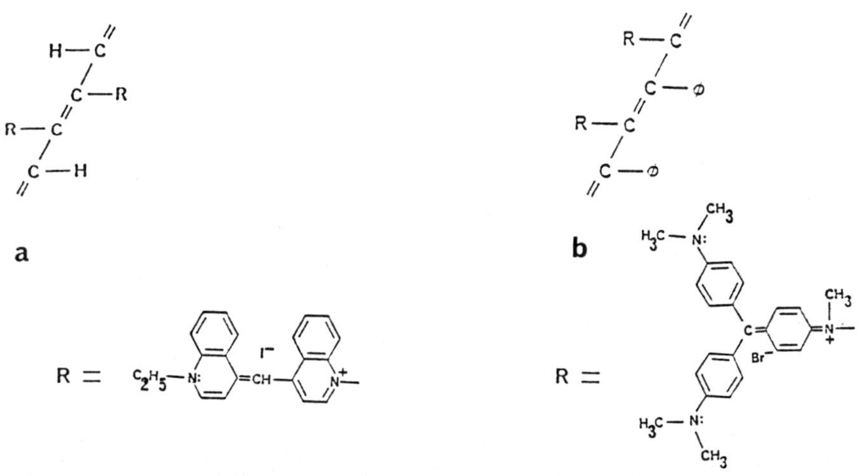

Figure 1. (a) Model for a high T_c superconductor proposed by Little (b) Structure of synthesized polymer

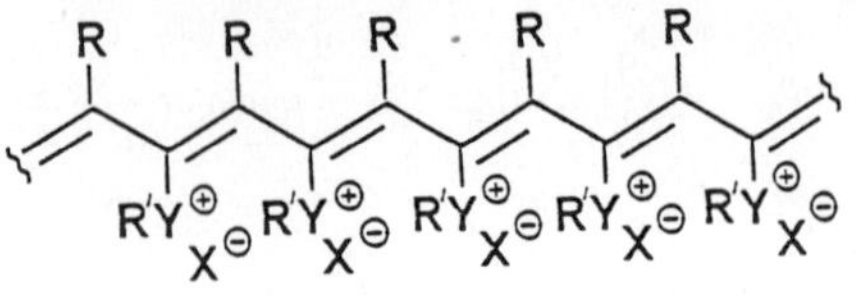

Figure 2. General structures of polyacetylenes obtained by present method

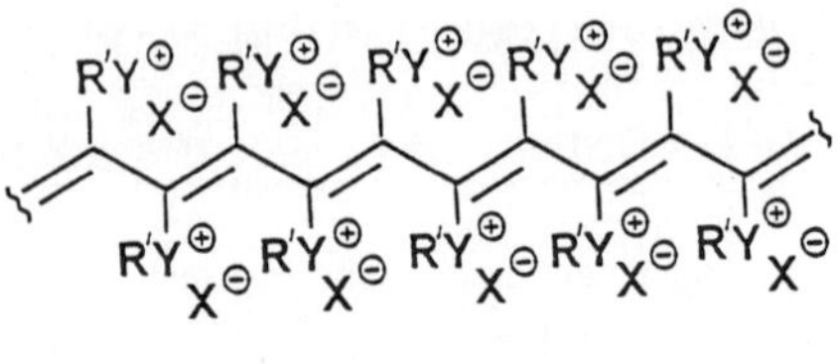

RESULTS AND DISCUSSION

All polyenes prepared by our method are black or purple-black solids and display large bathochromic shifts in the visible range of the UV-visible spectrum, (Table 1), in comparison to the starting monomer. These polymers are novel examples of disubstituted polyacetylenes with an extensively conjugated backbone. Their infrared spectra reveal absence of the absorbtion bands at 3128 $cm.^{-1}$ and 2098 $cm.^{-1}$ ($C\equiv CH$ and $C\equiv C$ stretch) present in the monomer which are replaced by a new, intense absorption at 1633 $cm.^{-1}$ ($C=C$ stretch), indicative of extensive conjugation. This band was shifted to as much as 1560 $cm.^{-1}$ for some of the polymers. Table 1 gives the typical UV spectral characteristics (λ_{max} and ε_{max}) and intrinsic viscosities $[\eta]$ of polymers obtained by the present method, together with those of substituted polyacetylenes obtained by polymerization with transition metal coordination catalysts.[6]

Figure 3 shows the relationship between the absorbtion maximum (λ_{max}) and the number of $C=C$ double bonds in conjugation, *n*, for polyenes.[7] Assuming this relationship is valid for higher values of *n*, these polyacetylenes have *n* ranging from 10 to 16 which is much larger than those for substituted polyacetylenes obtained by using coordination catalysts where n is typically 2 to 3.[6] We have also found that the absorption maxima in the UV spectrum for these polymers are uninfluenced by both the nature of the side group and by the type of counterion (for Br^-, I^-, and $CH_3SO_3^-$). Figure 4 gives the ^1H NMR spectrum for polyacetylene Ip. The sharp peak at 7.2 ppm is assigned to the $C=C-H$ proton of a highly conjugated conformer. The exact nature of this conformer cannot be identified with certainty at this point, but based on previous studies on poly(phenylacetylenes)[8], it may be attributed to either a trans-transoidal or a trans-cissoidal conformer. Figure 5 gives a typical DSC thermogram for polymer IIp. The first heating shows a distinct transition at 342° K (69° C), but no significant heat evolution is associated with it. These changes are reversible, but can be "locked in" by rapid cooling. The original conformer is restituted by annealing as is evident from figures 5b, c and d. This phenomenon may be attributed to a reversible conformational change.

Table 1. Physical properties of polymer $-(\underset{R}{C}=\underset{R'}{C})_n-$

	R	R'	color	λ_{max} (nm)	ε_{max}	$[\eta]$ (dL / g)	ref.
	H	t-Bu	colorless	283	1500	0.97	6
	Me	Ph	colorless	283	2500	2.7	6
Ip:	H	a	black	450	22000	0.91*	this work
IIp:	a	a	black	599	82000	0.93*	"
IIIp:	H	b	purple-black	443	142000	0.90*	"
IVp:	Ph	c	purple-black	584	25800	0.10*	"

a = 4-N-dodecylpyridiniumbromide b = 4-N-ethylpyridiniumiodide c = see fig. 1b

*Inherent viscosity in absolute methanol

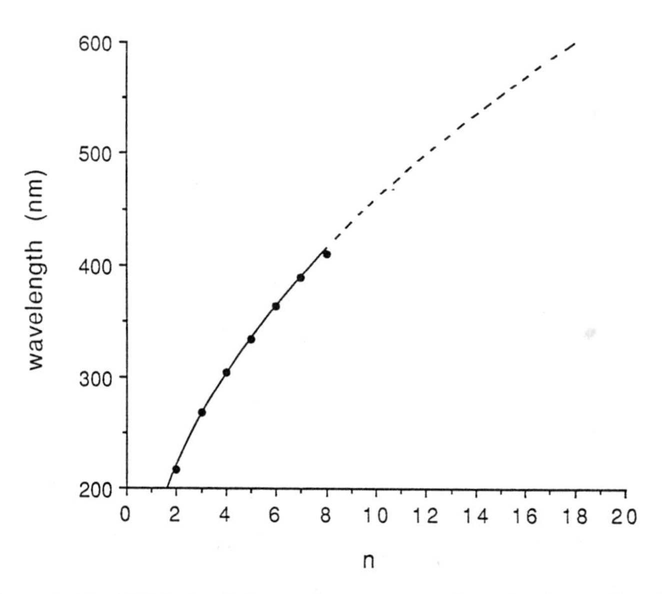

Figure 3. Plot of UV absorbtion maxima versus number of conjugated bonds '*n*' for polyenes

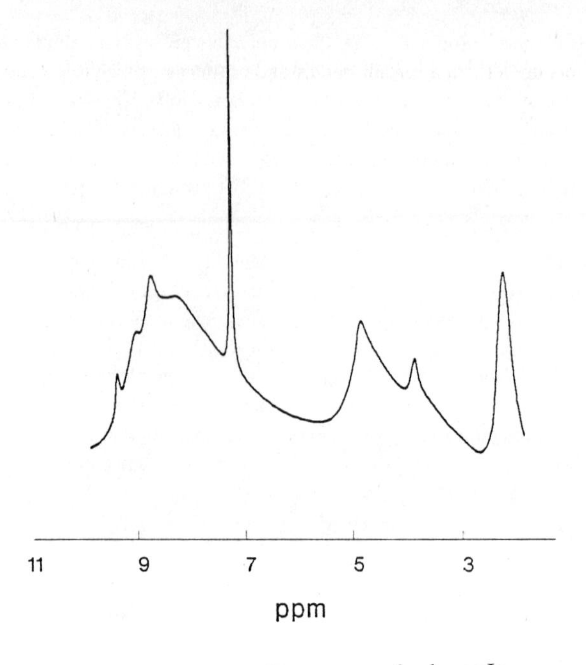

Figure 4. Proton NMR spectrum of polymer Ip

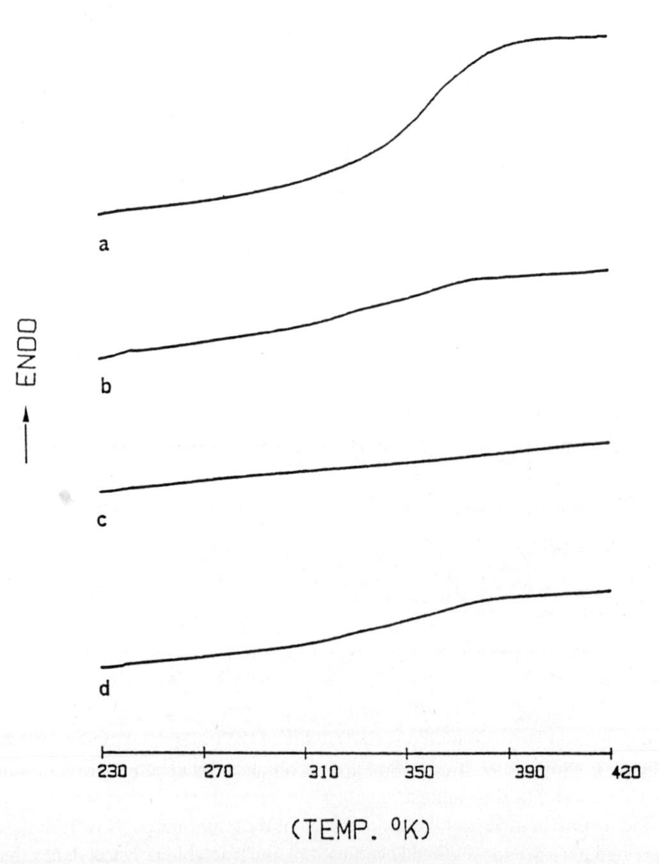

Figure 5. DSC thermogram of polymer IIp : (a) first heating (b) second heating (c) third heating (d) fourth heating after annealing sample at 23° C for 170 hours

It is evident from the spectroscopic data that these polymers possess extensively conjugated structures. This confers upon them a certain rigidity and brittleness, which is in contrast to mono and disubstituted polyacetylenes obtained with transition metal catalysts. All polymers prepared by us are readily soluble in polar solvents such as methanol, ethanol, dimethylsulfoxide, dimethylformamide.
The solubility in water is dependant on molecular mass and on the counterion. Those polymers with lower M were found to be soluble in chlorinated hydrocarbon solvents. Despite the long sequences of conjugated double bonds, these polymers appear quite stable up to temperatures of 250^{0} C. Figure 6 shows a typical TGA run performed on polymer Ip. A sharp decrease in weight occurs at 250^{0} C indicating rapid decomposition. No significant differences were observed between heating runs in air and in nitrogen. Thus, in contrast to polyacetylene, polymers obtained by us are more stable to oxidation. The molecular mass distribution for these polyacetylenes is difficult to determine due to their ionic nature and their strongly colored solutions. Nevertheless, with the exception of IVp, measurements of inherent viscosity are indicative of molecular masses in the high polymer range, (Table 1). X-ray diffraction patterns of polymers Ip and IIp indicate that their morphology "in situ" is amorphous and DSC thermograms of these polymers confirm the absence of crystallinity. This aspect is being currently investigated.

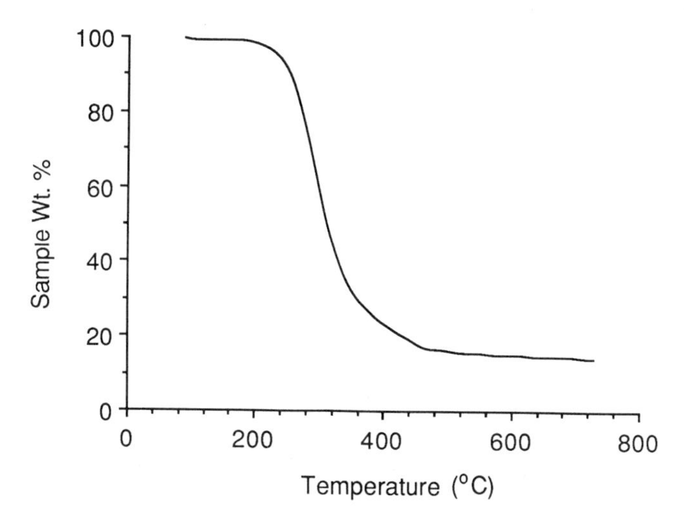

Figure 6. TGA of polymer Ip in nitrogen atmosphere

Extensive studies remain to be done to test the validity of the theoretical model for a polymeric high temperature superconductor and structures similar to it obtained by us as illustrated in Figure 1. Such studies would include control of the conjugation length, nature and size of the side group, the molecular mass and its distribution and the morphology of these polymers. Neverthless, such structures as described here are novel, highly conjugated and tractable polyacetylenes that may have potential for applications in the areas of energy storage, conductive polymer devices and permselective membrane technology.

ACKNOWLEDGEMENT

The authors wish to acknowledge the support of the National Science Foundation's Polymer Program under Grant DMR 8823084.

REFERENCES

1. Little, W.A. *Phys. Rev.* A**134,** 1416-1424 (1964).
2. Ku, C.C.; Liepins, R. *Electrical Properties of Polymers*, (Hauser Publishers, New York, 1987) 290.
3. Blumstein, A.; Subramanyam, S. *U. S . Patent* pending.
4. Subramanyam, S; Blumstein, A. *Macromolecules*, in preparation.
5. Blumstein, A.; Subramanyam, S.; *Polymer Bulletin*, in preparation.
6. Masuda, T.; Higashimura, T. *Adv. Polymer Sci.* **81**, 121-165, (1987).
7. Sondheimer, F.; Ben-Efraim, D.A.; Wolovsky, R. *J. Am. Chem. Soc.*, **83**, 1675-1681, (1961).
8. Berlin, A.A.; Tcherkashin, M.I. *Vysokomol. Soed.*, A**13**, 2298-2307, (1971).

PROSPECTS FOR ORGANIC POLYMERIC SUPERCONDUCTORS AND NEW ORGANIC CRYSTALLINE SUPERCONDUCTORS

Ronald L. Elsenbaumer

Corporate Research and Technology
Allied-Signal, Inc.
P.O. Box 1021R
Morristown, NJ 07962

INTRODUCTION

Although at this time there are no known (confirmed) organic conductive polymers that exhibit a metal to superconductor transition, the fact that a polymeric system, $(SN)_x$, does become superconductive at low temperature, and given the enormous progress made in recent years on the improvement in conductive and transport properties in organic conductive polymer systems, it is felt that the prospects are now quite good for observing superconductivity in an organic polymer system. It is no longer a question of if, rather a question of when and what polymer system it will be that exhibits superconductivity. The purpose of the discussions by the synthesis panel at this conference, and in particular this paper, is to explore highly probable organic polymer candidates for exhibiting superconducting phenomenon. In addition, some suggestions for new types of organic crystalline superconductor compositions, and unusual preparative methods will be presented. Lastly, researchers are urged to look for important technological applications beyond superconductivity for many of these new types of materials. Scientifically interesting and perhaps technologically important surprises are bound to be byproducts of the important research in this field.

PROSPECTIVE ORGANIC POLYMERIC SUPERCONDUCTORS

Today, there are two classes of conductive organic polymers that appear to have high probabilities of displaying superconductive phenomena, new polyacetylene and poly(arylene vinylenes). As synthetic techniques have improved, quite substantial improvements have been realized in terms of the conductive properties of organic conjugated polymers. These two polymer systems are especially noteworthy in this respect.

Organic Superconductivity, Edited by V.Z. Kresin and W.A. Little, Plenum Press, New York, 1990

Polyacetylene

The first modern organic polymer to be discovered to go highly electrically conductive is perhaps the one that also has the best apparent prospects for being superconductive as well. Although acceptor doped compositions are highly conductive, they tend not to be highly crystalline or inherently stable materials. On the other hand, it has been shown that donor doped polyacetylenes, especially potassium and rubidium doped ones, display unprecedented levels of crystallinity and very good inherent stability (under high vacuum or argon).[1]

The temperature dependence of the conductivity of *cis*-polyacetylene doped with potassium is shown in figure 1. Surprising behavior is noted at temperatures above ambient. Upon heating up to 200°C, a large permanent conductivity increase is observed. The linearity over this range corresponds to an apparent activation energy of 0.005 eV. This effect is not a general property of all alkali-metal doped polyacetylene complexes. It has been observed only with highly crystalline complexes of potassium and rubidium doped polyacetylenes, where there is a good match between the ionic diameters of the alkali-metal ions and the cavity size in the observed channel structure (see figure 2).[2] Each column of alkali-metal ions is surrounded by four polyacetylene chains, and each chain is nearest neighbor to two alkali metal columns.

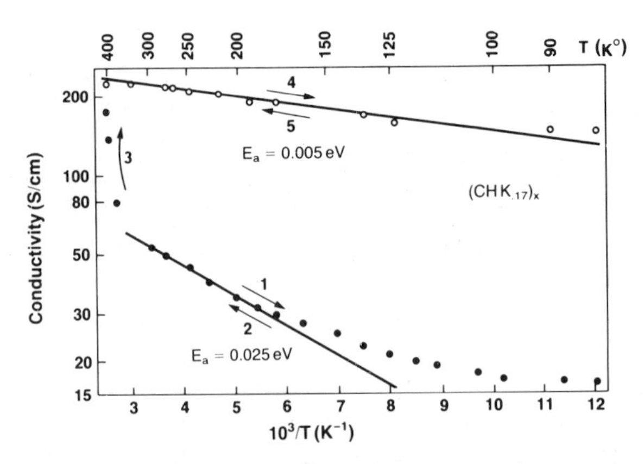

Fig. 1. Conductivity vs. T^{-1} for potassium doped polyacetylene. The numbers indicate the sequence in which the data were taken.

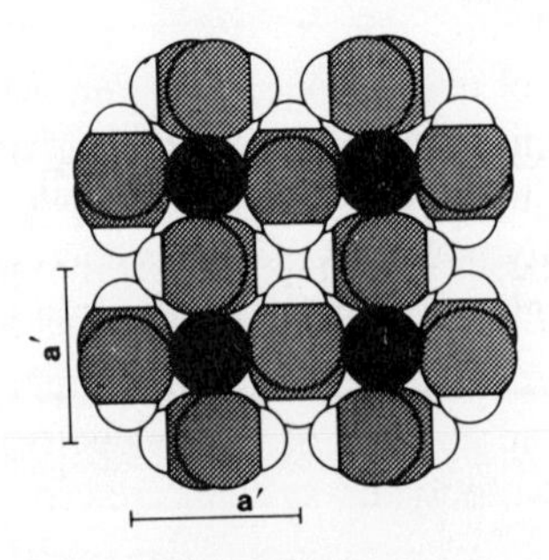

Fig. 2. Crystalline structure of heavily potassium doped polyacetylene. View is along the chain axis direction. The potassium ions (black circles) are arranged in columns.

Thermal annealing of doped *cis*-polyacetylene causes a contraction of the crystalline lattice giving rise to an increase in hybridization between the p_z orbitals on the polymer chain and the s orbitals on the metal ion. It is believed that annealing removes polymer chain configurational and conformational structural defects trapped in the lattice (see figure 3). The conductivity enhancements arise from both an increased effective conjugation length and improved electron transport orthogonal to the chain-axis direction. Hybridization between carbon orbitals and alkali metal s orbitals has been used to explain the decreased conductivity anisotropy in potassium doped graphite as compared to that of undoped and acceptor doped graphite. [3]

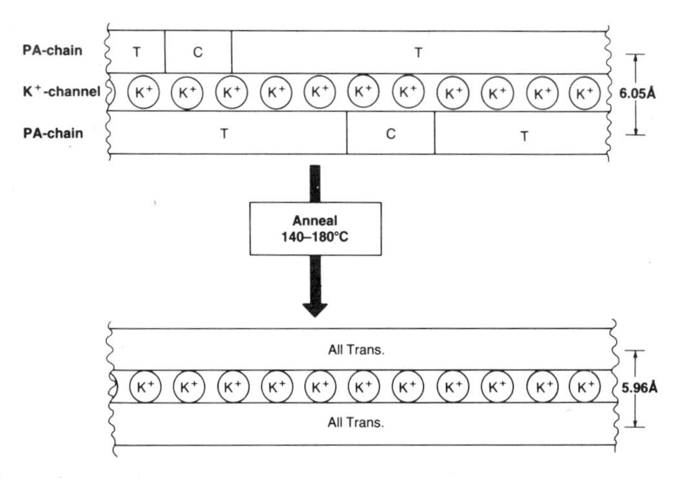

Fig. 3. Schematic representation of the effects of annealing potassium doped *cis*-polyacetylene. The view is normal to the chain-axis and the alkali metal column direction. The symbols used are as follows: T denotes *trans-transoid* chain segments and C denotes *cis-transoid* or *trans-cisoid* chain segments.

Recent improvements in the synthetic preparation of polyacetylene now provide acceptor doped, highly oriented conductive polymer samples that reportedly have conductivities approaching that of copper metal at (room temperature) on a per weight basis.[4] It is believed that these new synthetic techniques provide greater degrees of polymer perfection (longer conjugation lengths) than previously obtainable. Given these new forms of higher conductivity polyacetylene, it is conceivable that annealed potassium doped samples would have a high probably of displaying a low temperature metal-superconductor transition.

Poly(arylene vinylenes)

Poly(arylene vinylene) polymers are a relatively new class of conductive materials that are conveniently prepared by precursor polymer routes. They are typically prepared with high molecular weights, can be highly stretch oriented, and display high conductivities, rivaling those of doped polyacetylenes.[5] Although not much is known about their low temperature properties, the relatively high conductivities obtainable with these materials, and their ease of synthesis, suggests that significant improvements are yet to be made. As with polyacetylene, as synthetic methods improve, so will the degree of polymer perfection, the degree of crystallinity and, hence, the probability of observing a metal-superconductor transition. A key feature will be identifying dopants that provide conductive complexes with high degrees of crystallinity.

From a technological stand point, organic polymeric superconductors can offer significant advantages over the known high temperature ceramic superconductors and the organic crystalline complexes. Compared with these latter type materials, the organic polymers promise significant advantages in terms of processibility and design flexibility. Superconductive fibers (wires), films, and coatings are possibilities, and judging from what we know today about the high degree of processibility for conductive polymers, such forms most likely would be obtainable by conventional polymer processing technologies.

NEW OPPORTUNITIES FOR SUPERCONDUCTIVE ORGANIC CRYSTALS

New Types Of Superconductive Complexes

Rapid advances have been made in the area of superconductive organic charge transfer complexes in the last several years. Somewhat surprising is the observation that the vast majority of superconductive charge transfer complexes are derived from organic donors. Essentially no organic acceptor (only) superconductive complexes are known. One might ask if there is some fundamental reason why acceptor-inorganic complexes fail to display superconductive behavior. A key synthetic challenge with the acceptor complexes is identifying appropriate counter ions of the appropriate size and shape to allow for crystal packing conducive to superconductivity.

New Technologies for Preparing Superconductive Crystalline Complexes

Superconductivity in organic crystalline complexes is critically dependent upon the very nature of the crystalline complex - the way the organic structures are arranged in the lattice, their periodicity, spacing, and interactions with the counter ions being all important. Typically, these organic crystals are grown electrochemically (electrochemical crystallization) at reasonably slow rates. Under these conditions, one would expect to grow crystals which are essentially in a thermodynamically low energy phase. It is conceivable that such low energy phases are not necessarily the most appropriate for exhibiting metal to superconducting transitions. Further, high degrees of crystalline perfection may not always be desirable (critical current concerns). As such, a synthetic challenge lies in attempts to prepare possibly metastable crystalline phases that might display unusual behavior.

Several approaches to metastable complexes are evident. Since electrochemical techniques are prevalent, some techniques for preparing highly conductive forms of certain types of conductive polymers may be useful in the growth of crystalline complexes. Alternate polling of the working electrode at set intervals, and the use of high overpotentials have been used to prepare highly conductive polythiophenes, where constant low current techniques give inferior products. Aside from electrochemical techniques, simple thermal treatments followed by quenching may prove interesting. Admittedly, such techniques may be difficult to reproduce, but the observation of interesting phenomenon may provide clues as to what crystalline features are important for higher temperature metal to superconducting transitions.

Novel Properties Beyond Superconductivity

The highly unusual nature of the conductive organic crystalline complexes and polymeric materials provides opportunities for uses for these materials in a variety of other important technological applications. Already, the highly conjugated organic polymers are attractive materials for nonlinear optical applications, solar energy absorbing window treatments, and possibly as micromechanical actuators. It is conceivable that unusual piezoelectric phenomenon might be observed with certain forms of the polymers and crystals.

These exciting new materials have not yet come close to realizing their full applications potentials. It is clear that the science and technology surrounding conductive organic materials remains in its infancy, with many promising commercial opportunities and exciting discoveries yet to be made.

REFERENCES

1. R.L. Elsenbaumer, P. Delannoy, G.G. Miller, C.E. Forbes, N.S. Murthy, H. Eckhardt, alnd R.H. Baughman, Synthetic Metals, 11:251 (**1985**).

2. R.H. Baughman, N.S. Murthy and G.G. Miller, J. Chem. Phys., 79:515 (**1983**).

3. J.E. Fischer in: "Intercalated Layered Materials," F. Levy, ed., Reidel, Dordrecht, Holland (**1979**), p. 481.

4. Th. Schimmel, G. Denninger, W. Riess, J. Voit, M. Schwoerer, W. Schoepe, and H. Naarmann, Synthetic Metals, 28:D11 (**1989**). N. Theophilou, D.B. Swanson, A.G. MacDiarmid, A. Chakraborty, H.H.S. Javadi, R.P. McCall, S.P. Treat, F. Zuo, and A.J. Epstein, Synthetic Metals, 28:D35 (**1989**).

5. K.Y. Jen, L.W. Shacklette, and R.L. Elsenbaumer, Synthetic Metals, 22:179 (**1987**).

MULTIDECKER ORGANOMETALLIC SANDWICH POLYMERS; PROSPECTIVE CONDUCTIVE BACKBONES FOR THE FORMATION OF MACROMOLECULAR ARCHITECTURES CONFORMING TO THE EXCITONIC SUPERCONDUCTOR TOPOLOGY PROPOSED BY LITTLE

Ben Bush and J. J. Lagowski
Department of Chemistry
The University of Texas
Austin, TX 78712

Introduction

The recent interest in the high temperature superconductivity of certain materials exhibiting the perovskite structure was initiated by the report of Bednorz and Müller.[1] This discovery brought into sharp focus the inability of the BCS theory[2] and its phonon mediated Cooper pair formation mechanism to adequately describe high temperature superconductivity. Past theoretical interest in superconductors incorporating an excitonic interaction as the Cooper pair stabilizer has diverged into two related schools of thought: one emphasizing surface effects between two dimensional thin films of metallic conductors and semiconductors,[3,4] and the other related to one dimensional polymeric conductors based on the existence of certain features in macromolecular architecture.[5] The multilayer structure of perovskites has been described[6] as a factor in enhancing an excitonic interaction between conductive planes mediated through an intervening Cu-O chain (Figure 1).

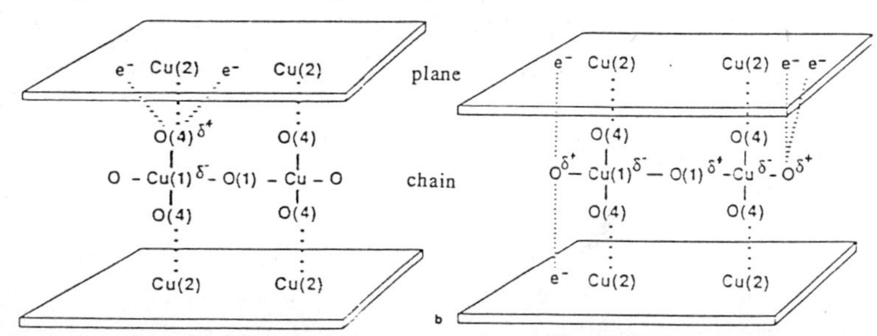

Figure 1: Adapted from reference 6: a) depicts intra-plane paring mediated by transverse excitation b) depicts both intra- and inter-plane pairing mediated by a longitudinal excitation.

The non-stoichiometric content of oxygen is thought to permit adjustment of polarizability to match the coherence length. Furthermore, the one dimensional nature of the Cu-O chain is regarded as particularly conducive in the stabilization of Cooper pair formation. Experimental support for modeling the perovskites as excitonic superconductors is derived from the lack of a significant isotope effect.[7]

Early interest in high-temperature superconductivity in molecular systems centered on the proposal by Little[5] of the excitonic stabilization of the Cooper pair. Attempts to produce macromolecular arrays conforming to the topology suggested by Little (Figure 2) rapidly diversified from a focus on purely organic systems to organometallic systems.[8] These experiments produced a steady shift of interest away from the unsuccessful organic systems[9] toward organometallic polymers.[10]

Synthetic convenience has largely constrained efforts along this line to ferrocene polymers containing organic connecting groups, despite the observed lack of conjugation between the two rings through the metal atom.[11,12,13] The organic connecting groups can disrupt delocalization, since the band structural properties of the system depend critically on the mutual orientations between neighboring unit cells in the case of bridging ligands with low spatial symmetry, e.g., where there is the possibility of rotation around the organic connecting group in the polymer.[14] By contrast, rigidity between metal sites can be assured by employing μ-arene ligands; further, Elschenbroich et al. found evidence of delocalization through the metal moiety in bis(fused-polycyclic arene)chromium compounds.[15]

Multidecker organometallic sandwich compounds wherein the organic moiety contains no hetero-atoms are relatively uncommon.[16] True multidecker sandwich compounds do not contain purely organic connecting moieties (e.g., polyferroceneolenes[11]), and the μ-cyclic moieties do not deviate greatly from aromatic planarity.[17] Many of these multidecker compounds (triple-deckers specifically) conform to the theoretically stable 30 and 34 electron systems of Hoffmann et al.[18] We are attempting to employ poly[(μ-η^6,η^6-naphthalene)chromium] as a component in Little-like structures to probe the phenomenon of excitonic superconductivity.

Theoretical interest in macromolecular conductors has centered on (the then synthetically unavailable) polydecker sandwich compounds (I and II)[14] which could form the conductive spine component (feature A in Figure 2) in the superconductor topology suggested by Little. A recent theoretical treatment of the slipped polydecker sandwich compound poly[(μ-η^6,η^6-naphthalene)chromium] (III) by Burdett,[19] focused upon the electrical conductivity of this (at that time, unknown)

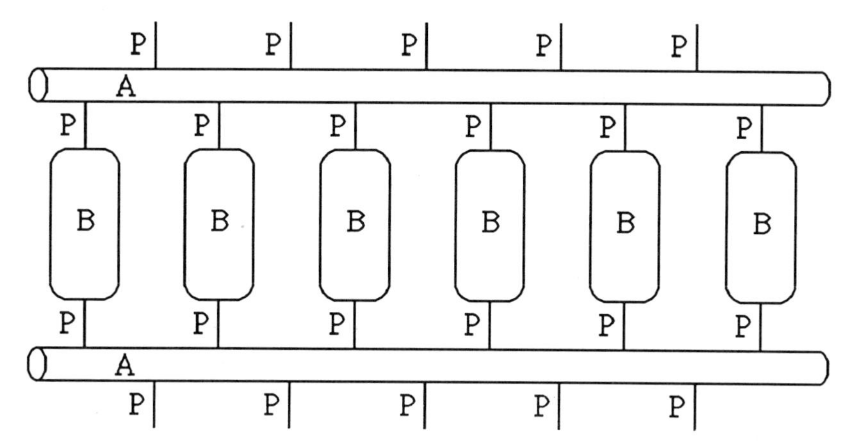

Figure 2. Schematic diagram of a proposed two dimensional excitonic superconductor. A is the conductive element; B is the polarizable element; P is the insulative connector between A and B.

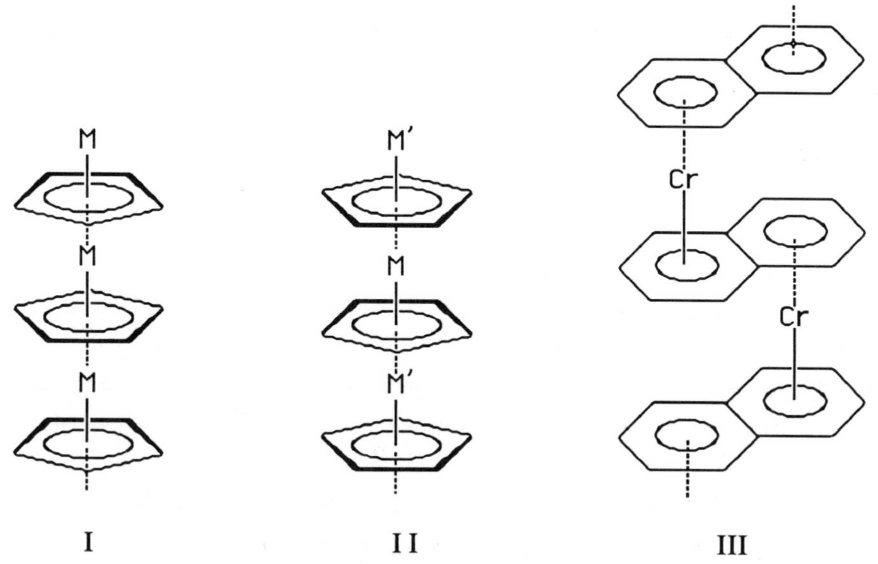

compound. Our recent synthesis and characterization of poly[(μ-η^6,η^6-naphthalene)chromium] (III) is the first known example of this new class of macromolecular organometallic compounds.[20]

The properties to be expected of the polymer (III), can be approximated from the chemistry and properties of the monomeric precursor, bis(η^6-naphthalene)chromium (IV), of its derivatives (V),[21] and of oligomers of the type VI,[22] all species being known.

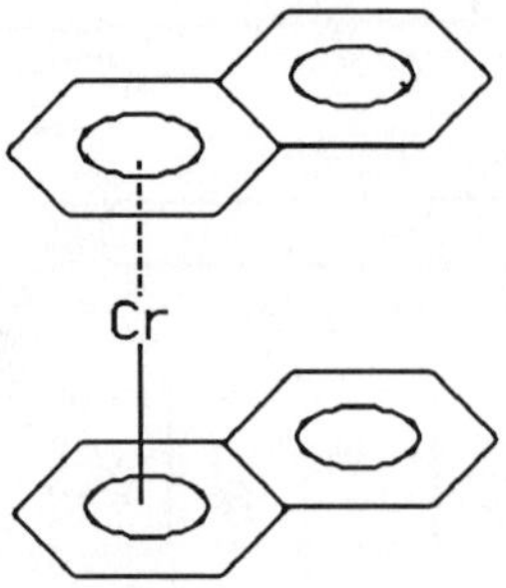

IV 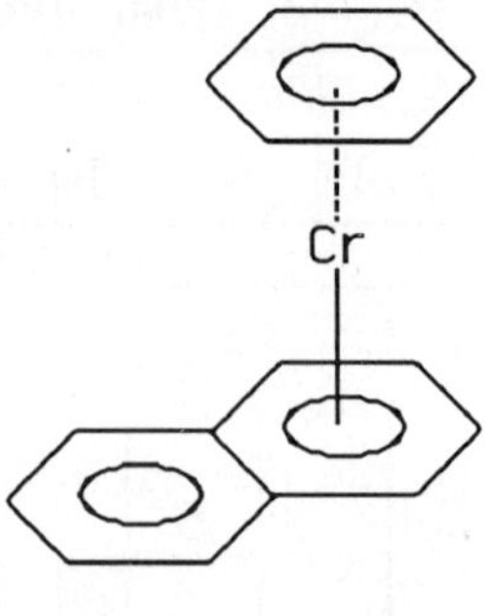 V 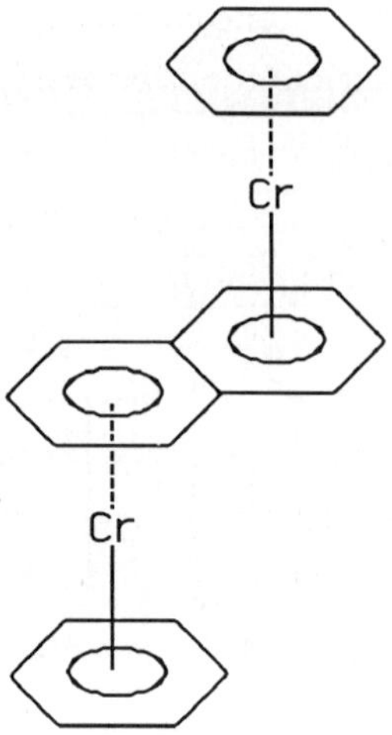 VI

Polymers

The ultimate goal of our efforts is a generalized macromolecular organometallic architecture optimized toward the realization of high temperature superconductors based upon the excitonic stabilization of the Cooper pair as proposed by Little (Figure 2).[5] To this end, we seek other examples of μ-polyarene structures to complement our discovery of poly[(μ-η^6,η^6-naphthalene)chromium] that can act as the conductive element. The ability of bis(η^6-naphthalene)chromium to undergo ligand exchange to form poly[(μ-η^6,η^6-naphthalene)chromium][20] exemplifies some of the reaction chemistry of the bis(polycyclic-fused-arene)chromium organometallics,[23] and the corresponding oligomer chemistry as well.[13,22] The macromolecules and the monodentate ligand complexes thereof conform to the 1-dimensional topology envisioned by Little,[5] with the advantage of an enlarged unit cell[25] and a correspondingly greater coherence volume. However, the use of a bidentate ligand to form the 2-dimensional complexes is preferred.

The failure of the concerted efforts to form purely organic macromolecular superconductors[25] can be traced to the the formidable constraints implied in the synthetic pathways. The constraints imposed by steric hindrance and torsional randomization of the polymeric structure are largely overcome by the use of polydecker sandwich organometallics.[20] Contemporary theoretical interest in these compounds[14,26] centers on bulk conductivity, but examples have been lacking until our recent synthesis and characterization of poly[(μ-η^6,η^6-naphthalene)chromium];[20] this substance represents an important first example of an organometallic structure for which explicit theoretical interest has been indicated.[19]

π-Naphthalene Chemistry, The existing Example and Its Implications

We present here a brief synopsis of the synthesis and characterization of poly[(μ-η^6,η^6-naphthalene)chromium] which serves to provide detailed information on the complexation properties of this polymer and is the basis for the general strategy employed in this work. Poly[(μ-η^6,η^6-naphthalene)chromium] is important because it is the model compound on which to base the Little-structures, and its synthesis represents the general strategy for our synthesis of other back-bone systems (Figure 1) incorporating μ-arene moieties. The key compound in the synthesis of poly[(μ-η^6,η^6-naphthalene)chromium] is bis(η^6-naphthalene)chromium which is prepared by metal atom techniques;[27-31] the self-exchange of this compound produces poly[(μ-η^6,η^6-naphthalene)chromium].[20] An understanding of the constitution of poly[(μ-η^6,η^6-naphthalene)chromium] starts with the NMR spectroscopy of a series of compounds containing the bound naphthalene moiety (Table I).

The ^1H NMR spectrum of bis(η^6-naphthalene)chromium] obtained in ND_3 (Table I) exhibits a marked change when compared to the spectrum of this compound in C_6D_6;[20] the implied change in anisotropy is ascribed to a solvent effect which constrained the naphthalene moieties to an eclipsed configuration (also see reference 31). The ^1H NMR spectrum of poly[(μ-η^6,η^6-naphthalene)chromium]

Table I. Comparison of Proton NMR Spectra Between C_6D_6 and ND_3 for selected μ-η^6,η^6-naphthalene compounds and bis(η^6-naphthalene)chromium in ppm.

		naphthalene		
		unbound	η^6-bound	η^6-benzene
bis(η^6-naphthalene) chromium:	ND_3	3.42	5.06, 5.74	
	C_6D_6	6.92	4.37, 5.29	
(μ-η^6,η^6-naphthalene) bis(η^6-benzene) dichromium:	ND_3		5.04, 5.73	3.38
	C_6D_6		3.97, 5.67	3.89
poly[(μ-η^6,η^6-naphthalene) chromium]:	ND_3		5.13	
	C_6D_6		INSOLUBLE	

is a remarkably sharp singlet in liquid ND3, suggesting that this macromolecule is similarly constrained to the eclipsed configuration in this solvent.[20] Recent attempts to grow crystals of poly[(μ-η^6,η^6-naphthalene)chromium] for X-ray analysis by the slow cooling of ethylenediamine solutions has resulted in extremely fine whisker like needles which are to small for analysis.

The electrochemistry of poly[(μ-η^6,η^6-naphthalene)chromium] (Figure 3) shows both oxidation of the chromium moiety and reduction of the naphthalene moiety.[32] The apparent equivalence of redox potential between the chromium $Cr^{o/+}$ and naphthalene (o/-) redox couples can be interpreted as a delocalization process through the Cr moiety.[32] Burdett's[19] theoretical analysis of poly[(μ-η^6,η^6-naphthalene)chromium] predicts it to be a conductor in the oxidized state wherein "holes" would be formed in the valence band; presumably reduction of the polymer

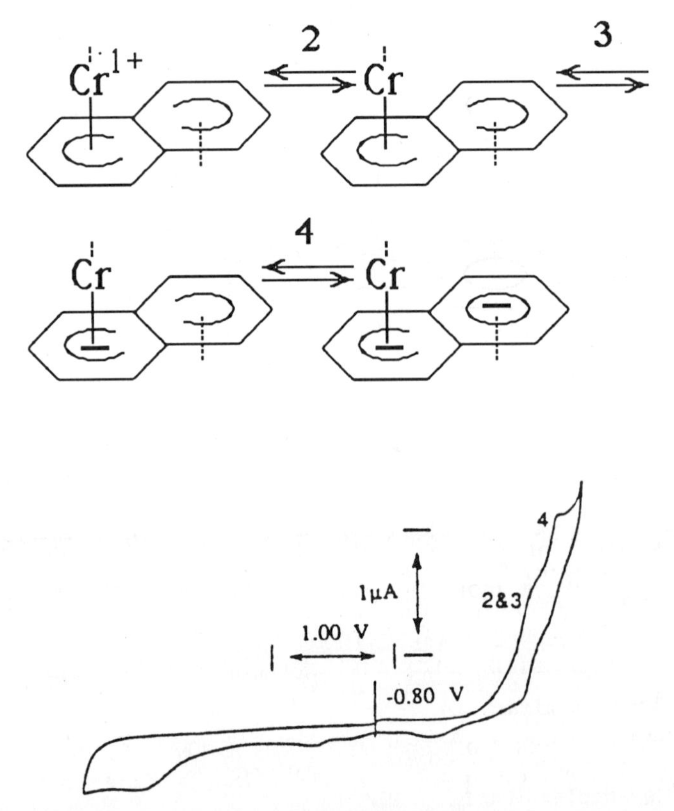

Figure 3: 50 mV/s Cyclic voltammogram of poly[(μ-η^6,η^6-naphthalene)chromium] in ethylenediamine. The $Cr^{o/+}$ [2] couple overlaps with the naphthalene (o/-) [3] couple at -2.04 volts versus Ag/Ag^+, and the naphthalene (-/=) [4] couple appears at -2.24 volts.

would populate the conduction band with itinerant electrons, likewise resulting in conductivity. Thus, both oxidative and reductive doping of the polymer should produce a conducting substrate. This type of compound appears to form complexes with strongly coordinative solvents, as indicated by its solubility characteristics; the polymer is insoluble in ethers but soluble in ethylenediamine. The observation that the naphthyl moiety can act as the excitonic stabilizer of the Little-type superconductor as discussed by Wiley[33] and Paulus[34] suggests the use of 1,5-naphthalenediamine as the complexing excitonic stabilizer for these proposed organometallic complex superconductors (VII); this strategy is especially

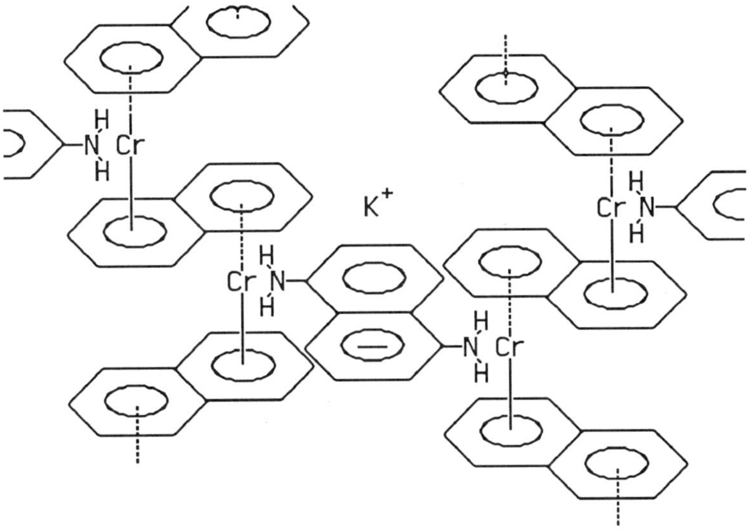

VII

interesting in view of the discussion of Salem[35] who suggested the use of terminal amine moieties in the excitonic stabilizer.

Multidecker Sandwich Polymers incorporating μ-monocyclic ligands

The synthesis of polymers represented by topologies I and II was recently started.[36] Rather than developing an exchange chemistry, metals were evaporated into existing organometallic sandwich compounds in order to determine whether relatively stable polymers might result. The synthesis of tris(mesitylene)dichromium by Lamanna[37] during the co-condensation of mesitylene with chromium under high metal loading, indicated that this would be a rational approach, because the synthesis may be regarded as the evaporation of chromium into bis(mesitylene)chromium.

Thus we evaporated chromium into bis(η^6-mesitylene)chromium, and bis(η^6-benzene)molybdenum to produce VIII and IX respectively; iron into nickelocene; and nickel into ferrocene to produce X and XI respectively. The bulk of the organometallic precursor is unreacted, but we believe some oligomeric species are made. Reactions with mesitylene showed a low yield of tris(mesitylene)dichromium by mass spectroscopy and NMR. Likewise, bis(η^6-benzene)chromium was formed from bis(η^6-benzene)molybdenum, ferrocene from nickelocene, and nickelocene from ferrocene; and some higher oligomeric fragments were tentatively observed, by low resolution chemical ionization mass spectroscopy. Observation of mixed metal species militates against spurious formation in the mass spectrometer. These preliminary results which were obtained on unfractionated reaction mixtures suggest that such polymers are thermodynamically stable. High resolution mass spectroscopy will be necessary to provide definitive identification of fragments. Efficient methods of synthesis and

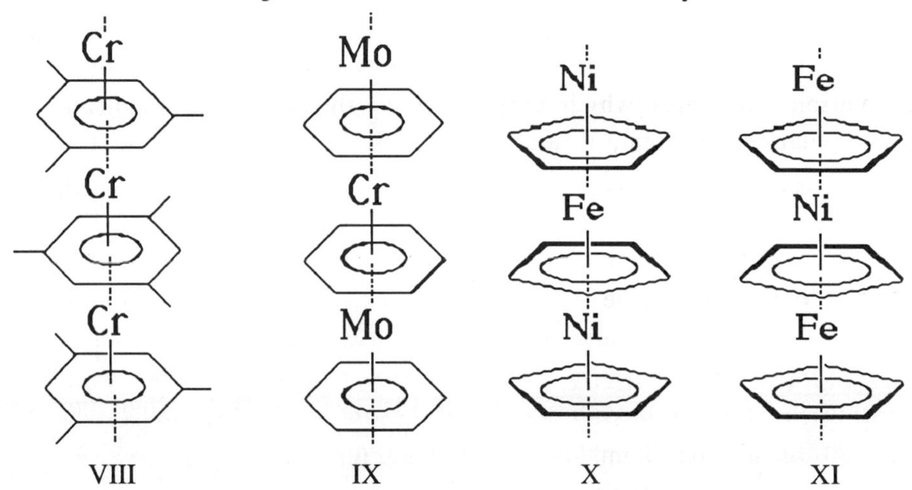

isolation will need to be developed, and combined with mass spectrometry methods that overcome polymer involatility (e.g. laser desorption).

CHARGE TRANSFER PROCESS:

Our interest in charge transfer phenomena in species containing π-arene moieties is driven by a need to understand the optimum geometry of π-complexed arenes to produce maximum interactions between metal sites. Such measurements can be made readily on smaller molecules incorporating the structural units of interest, e.g., IV.

The study of mixed-valence charge transfer processes using the spectroscopic methods delineated by Hush[38] and Day[39] and refined by Meyer[40] provide an effective probe into the degree of delocalization present between intramolecular sites of differing oxidation states. Briefly, a measure of delocalization is afforded by the electron residency α, as established by the expression,

$$\alpha^2 = ((4.2 \text{ X } 10^{-4})\varepsilon_{max}\Delta\upsilon_{1/2})/(\upsilon_{max}d^2)$$

where ε_{max} is the extinction coefficient, $\Delta\upsilon_{1/2}$ is the observed bandwidth at half-height, υ_{max} is the energy of the band at λ_{max},[40] and d is the internuclear separation between redox sites. The extent of delocalization of electrons on the basis of α values has been established by the classification scheme of Day:[39] that is, class I, $\alpha \sim 0$ implies no mixed valence charge transfer; II, $0.707 > \alpha > 0.0$ implies some delocalization with the onset of new optical and electronic properties occurring for $\alpha \geq 0.25$ and; III $\alpha \sim 0.707$ which implies complete electronic delocalization. Using this general approach the effect of the addition of specialized complexing agents (e.g., various diamines which correspond to the insulating connector and the polarized element, P and B, respectively, in Figure 1) to μ-arene systems (VI) should yield information on the electron delocalization in these systems. Hence, the study of oligomers[22] by this technique should provide information related to the expected electronic behavior of the eventual polymers.

The polydecker sandwich polymer chemistry is formidable because of the air-sensitivity, involatility, and insolubility of these systems; however, promising methods of analysis are becoming available. Fast atom bombardment (FAB), and laser desorption methods might be adapted to provide mass spectral measurements of the average molecular weight and molecular weight distributions of both the conductive polymers and the stabilized complexes of the polymers. UV-Vis spectroscopy may be adapted to show the HOMO-LUMO transition[31] (i.e., the energy difference between the conductance and valence bands) in the precursors, oligomers and polymers. The complexed excitonic stabilizer ligand, being essentially a dye, can also be probed spectroscopically to establish its electronic state.[41] 1,5-Naphthalenediamine for example is purple in the air, green when reduced with an alkali metal, and colorless in its neutral state; when complexed between poly[(μ-η^6,η^6-naphthalene)chromium] molecules, 1,5-naphthalenediamine is expected to be reduced and in the green state. Photoacoustic spectroscopy in the UV-Vis. region and near IR should prove particularly useful in studying the extent of delocalization and induced oxidation effected in the polymer by doping or density of state mixing with the excitonic stabilizer ligand, by observing the intervalence charge transfer bands without solvent effects. Comparison between oligomer and polymer optical spectra should provide particular insight. NMR analysis may be possible in deuterated ammonia which is expected to solubilize the new polymers, as it did in the case of poly[(μ-η^6,η^6-naphthalene)chromium]. Combination of the preceding analyses with bulk elemental analysis[42] should provide definitive characterization.[20] Ultimately we are interested in the macroscopic properties of the produced polymer matrices; screening with a SQUID magnetometer[43] will quickly establish superconductivity if it occurs in the systems we synthesize.

SUMMARY

With our development of synthetic and analytical technologies sufficient to execute studies of macromolecular organometallic architecture, the scientifically intriguing possibility of effecting the long-sought high-temperature excitonic superconductor topology of Little[5] (Figure 2) on more promising substrates may be realized. The formation of polydecker sandwich complexes by both direct reaction and ligand exchange offers new large unit cell conductor chains[24] that should readily accommodate complexation to potential excitonic stabilizers (dye-like molecules). Derivatization of the homo- or hetero-cyclic portion of the conductor chain offers direct chemical bonding of the excitonic stabilizer to the chain. Once the Little superconductor topology is realized (preferably in two dimensions), beyond the effect of shared density of states, the controlled oxidation and reduction of the macromolecular complex can be utilized to depopulate the valence band or populate the conduction band to afford conductivity, which can be followed spectroscopically. Parallel spectroscopic studies of mixed-valence charge transfer processes in the corresponding oligomers will afford an understanding of the scale of the delocalization present, in both oligomer and polymer and complexes thereof.

Acknowledgment

We acknowledge the generous support of the Robert A. Welch Foundation, and the Texas Advanced Technology Program.

References

(1) Bednorz, J. G.; Müller, K. A., *Z. fur Physik*, **1986**, *B64*, 189.
(2) Bardeen, J., Cooper, L. N., Schrieffer, J. R. *Phys. Rev.* **1957**, *108*, 1175.
(3) Ginzburg, V. L.; Kirzhnits, D. A., *JETP*, **1964**, *46*, 397. Lett. p.269.
(4) a) Douglass, D. H. (Ed.), Bardeen, J., *Superconductivity in d- and f- band metals* , **1976,** Plenum Press, New York, p.1.
b) Allender, D.; Bray, J.; Bardeen, J., *Phys. Rev. B*, **1973**, *7*, 3, 1020.
(5) Little, W. A. *Phys. Rev.* , **1964**, *134*, 6A, A1416.
(6) Martin, R. L.; Bishop, A. R.; Tesanovic, Z., *Chemistry of High Temperature Superconductors II*, American Chemical Society, **1988**, p. 130.
(7) Bourne, L. C.; et. al., *Phys. Rev. Lett.* **1987**, *58*, 22, 2337.
(8) Collman, J. P.; Little, W. A., NTIS AD-769 630.
(9) Wiley, R. H.; Lee, J., NTIS AD-721 916.
(10) Schneider, R. S. *J. Polymer Sci.: Part C* **1970**, *29*, 27.
(11) Neuse, E. W. J. *Macromol. Sci.-Chem.* **1981**, A16 (1) 3.
(12) Rosenblum, M., Santer, J. O., Howells, W. G. *J. Am. Chem. Soc.* **1963**, *85*, 1450.
(13) Lundquist, R. T., Cais, M. *J. Org. Chem.* **1962**, *27*, 1167.
(14) Böhm, M. C. *Z. Naturforsch.* **1984**, *39a*, 223.
(15) Elschenbroich, Ch.; Möckel, R.; Bilger, E., 29th IUPAC Congress, Cologne, 1983.
(16) Geiger, W. E., Connelly, N. G. *Adv. in Organomet. Chem.* **1985**, *24*, 87.
(17) Edwin, J., Geiger, W. E., Rheingold, A. L. *J. Am. Chem. Soc.* **1984**, *106*, 3052.
(18) Albright, T. A., Hofmann, P., Hoffmann, R., Lillya, P. C., Dobosch, P. A. *J. Am. Chem. Soc.* **1983**, *105*, 3396.
(19) Burdett, J. K.; Canadell, E. *Organometallics* **1985**, *4*, 805.
(20) Bush, B.; Lagowski, J. J. *Organometallics* **1988**, *7*, 1945.
(21) Kündig, E. P.; Perret, C.; Spichiger, S.; Bernardinelli, G. *J. Organomet. Chem.* **1985**, *286* 1983.

(22) Bush, B.; Lynch, V.; Lagowski, J. J. *Organometallics* **1987**, *6*, 1267.
(23) Elschenbroich, C.; Möckel, R.; Bilger, E. Z. *Naturforsch.* **1984**, *39b*, 375.
(24) Little, W. A. *J. Less Comm. Metals* **1978**, *62*, 361.
(25) Wiley, R. H.; Lee, J.; Raffo; J.; Chen, V., NTIS AD-677 513.
(26) Lauher, J. W.; Elian, M.; Summerville, R. H.; Hoffmann, R. *J. Am. Chem. Soc.* **1976**, *98*, 3219.
(27) Kündig, E. P.; Timms, P. L. *J. Chem. Soc., Dalton Trans.* **1980**, 991.
(28) Elschenbroich, C.; Möckel, R. *Angew. Chem., Int. Ed. Engl.* **1977**, *16*, 870.
(29) Bard, R. A.; Lagowski, J. J. Presented at the 175[th]National of the American Chemical Society, Meeting Anaheim, CA March 1978; Bis(naphthalene)chromium: Preparation and Properties.
(30) The synthesis of bis(η^6-naphthalene)chromium has been accomplished by the method described by: Markle, R. J.; Pettijohn, T. M.; Lagowski, J. J. *Organometallics* **1985**, *4*, 1529.
(31) Osborne, J. H., Trogler, W. C., Pascale, D. M., Francis, C. G. *Organometallics* **1987**, *6*, 94.
(32) Bush, B.; Lagowski, J. J., *J. Organomet. Chem.* , in press.
(33) Wiley, R. H.; Sapse, A. M.; Lee, J. Y.; *J. Macromol. Sci.-Chem.* **1970**, *A4*(1), 203.
(34) Paulus, K. F. G. *Mol. Phys.* **1966**, *10*, 381.
(35) Salem, L. *Mol. Phys.* **1966,** *11*, 499.
(36) Dr. T. M. Pettijohn of this lab.
(37) Lamanna, W. M. *J. Am. Chem. Soc.* **1986**, *108*, 2096.
(38) Hush, H. S. *Prog. Inorg. Chem.* **1967**, *8*, 391.
(39) Robin, M. B.; Day, P. *Adv. in Inorg. Chem. Radiochem.* **1967**, *10*, 247.
(40) Powers, M. J.; Salmon, D. J.; Callahan, R. W.; Meyer, T. J. *J. Am. Chem. Soc.* **1976**, *98*, 6731.
(41) J. Fabian & H. Hartmann, *Light Absorption of Organic Colorants*, Springer-Verlag New York, **1980**.
(42) Schwarzkopf Microanalytical Lab, 56-19 37th Avenue, Woodside, N.Y., 11377.
(43) In collaboration with Prof. James L. Dye of Michigan State University.

A SEARCH FOR NEW CLASSES OF CONDUCTING POLYMERS USING 4d AND 5d METAL PORPHYRIN COMPLEXES

James P. Collman[†], John T. McDevitt[†], Kimoon Kim[†], James M. Garner[†], Matthew B. Zisk[†], Charles R. Leidner[†], Gordon T. Yee[†], Jacques W. Prodolliet[†], and William A. Little[‡]

Departments of [†]Chemistry and [‡]Physics
Stanford University
Stanford, California 94305

Introduction

Studies of molecular superconductivity have centered around a class of closely-related charge transfer salts. The synthesis, structural characterization, and especially the physics of this class of molecular superconductors dominate this conference.

There are many other synthetically accessible materials which are candidates for molecular conductors. Some of these substances may be found to exhibit superconductivity, perhaps through novel mechanisms. Herein we discuss the possibility of finding superconductivity among polymeric substances which contain 4d or 5d transition metal porphyrin complexes. This paper is divided into three parts: the "shish-kebab" polymers, metalloporphyrin dimers joined by multiple metal-metal bonds, and approaches to polymers having multiple metal-metal bonds along the polymer backbone.

The Shish-Kebab Polymers

This name was given to a class of coordination polymers derived from ruthenium and osmium octaethyl porphyrin derivatives linked together via bidentate unsaturated ligands such as pyrazine (Figure 1) [1]. The concept underlying these polymers is that their valence electrons (HOMOs) are metal-centered; charge transfer conductivity should not directly involve the porphyrin π-electrons. Such a situation could result in Coulombic interaction between the conducting core electrons and the polarizable, radial π-electrons.

In contrast, the classical types of conducting solids derived from porphyrins and other unsaturated macrocycles [2,3] principally involve π-electrons of the stacked macrocycle in the conduction process. However, in one notable exception [4], CoPcI (Pc = phthalocyanine), the metal spine supports the conduction.

The synthesis and physical characterization of the shish-kebab porphyrin polymers has been studied in considerable detail. Partially oxidized polymers exhibit varying degrees of conductivity which has been principally studied by means of charge transfer electronic absorption spectra in the mid-infrared range. The conductive, electrochemical, and optical properties were examined for Fe, Ru, and Os octaethylporphyrin

Organic Superconductivity, Edited by V.Z. Kresin and W.A. Little, Plenum Press, New York, 1990

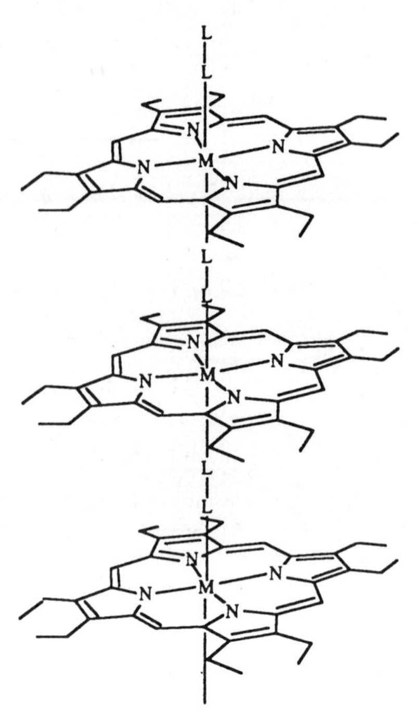

Figure 1. Structure of the octaethylporphyrin coordination polymers where M = {Fe, Ru, Os} and L-L = {pyrazine, 4,4'-bipyridine, dabco}.

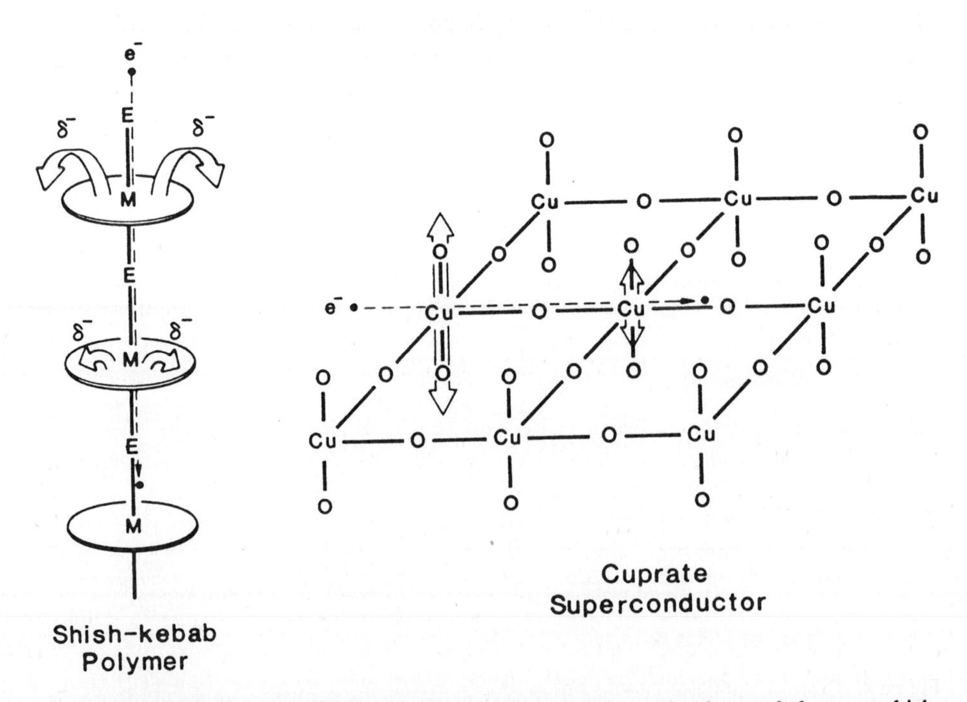

Figure 2. Possible Coulombic interactions between conducting and orthogonal electrons which could lead to the formation of Cooper pairs.

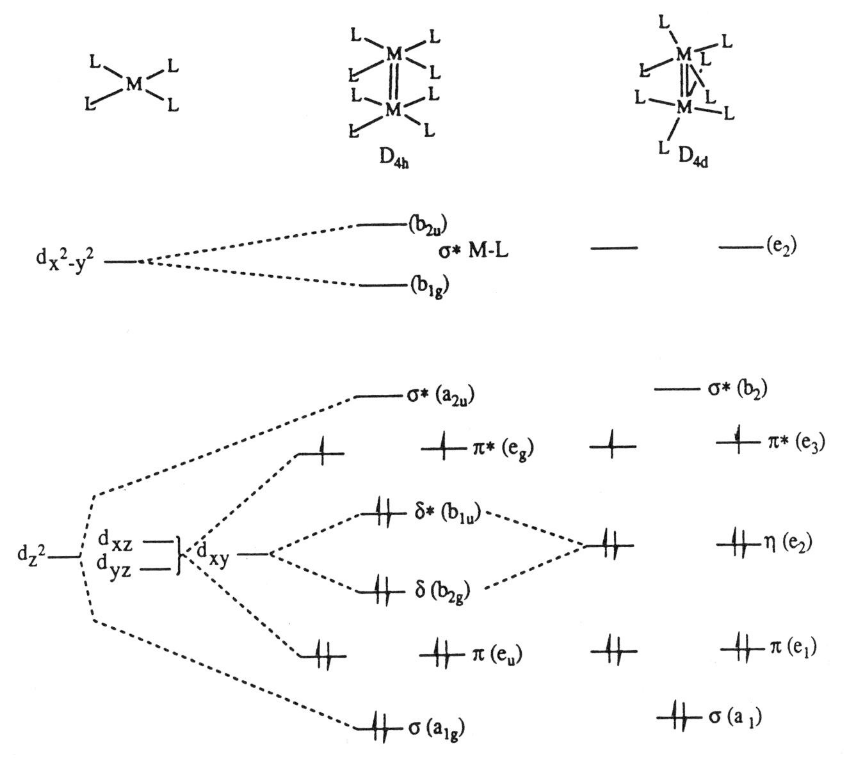

Figure 3. Molecular orbital diagrams showing the possible combinations d-orbitals in the formation of metal-metal bonds between 4d and 5d metalloporphyrins. Both conformations are shown: eclipsed (D_{4h}) and staggered (D_{4d}).

Cases Prepared Thus Far:

BO	4	3.5	3.0	2.5	2.0	1.5	1.0
	Mo-Mo	$[Re\text{-}Re]^+$	Re-Re	$[Ru\text{-}Ru]^+$	Ru-Ru	—	Rh-Rh
	W-W		$[Os\text{-}Os]^{2+}$	$[Os\text{-}Os]^+$	Os-Os	—	Ir-Ir
	$[Re\text{-}Re]^{2+}$		$[Ru\text{-}Ru]^{2+}$		Ru-Os	—	
S	0	1/2	0	1/2	1	(1/2)	0

BO = bond order; **S** = spin state

Figure 4. Examples of dimers derived from octaethylporphyrin derivatives of 4d and 5d metals.

coordination polymers bridged by dabco (a saturated bidentate ligand), bipy (4,4'-bipyridyl), and pyz (pyrazine). Partial oxidation of these polymers electrochemically and with diverse chemical oxidants leads to substantial increases in electrical conductivities which are also manifest by charge transfer absorption bands in the mid-infrared region (for Ru and Os). The doped conductivities and charge transfer absorption follow the trend Os > Ru > Fe and pyz > bipy > dabco. Electrochemical studies of the polymers using a novel carbon cloth technique demonstrate the presence of metal-centered anodic waves at potentials much less positive than that required for oxidation of the porphyrin ring. The broad, intense, low-energy spectroscopic transition is most pronounced at a mole fraction of oxidation intermediate between Os(II) and Os(III). A prominent antiresonance at 1580 cm^{-1} for $[Os(OEP)pyz)](PF_6)_{0.3}$ shifts to 1532 cm^{-1} upon substitution of deuterium for hydrogen in the bridging pyrazine ligand. This was explained by coupling between mixed-valence electronic transitions and an infrared silent pyz breathing mode. Such a result is strong evidence for the role of the bridging ligand, pyz, in the conduction process and shows that this is core-conducting.

Studies of the shish-kebab polymer family were suspended to examine the newly-discovered high T_c superconducting cuprate ceramic materials [5–7]. We suggested that the latter systems may form superconducting Cooper pairs by utilizing an electronic "excitonic" interaction such as the one we had hoped to find in the 4d and 5d metalloporphyrin polymers. The possible relationship is depicted in Figure 2.

We are now turning some attention back to molecular polymers derived from the heavy metal porphyrins, but using a different approach.

Multiple Metal-Metal Bonds in 4d and 5d Metal Porphyrin Dimers

In the absence of axial ligands, second- and third-row transition metal porphyrin complexes form dimers joined by multiple metal-metal bonds. The bond order and spin state of these multiple metal-metal bonds are accurately forecast from simple molecular orbital diagrams using only the d-electrons (Figure 3) [8]. We have developed techniques for the synthesis and characterization of an entire family of such dimers [9]; examples of octaethylporphyrin derivatives which have been prepared and characterized to date are displayed in Figure 4.

Note that some of the compounds shown in Figure 4 are cations; these metal-metal bonded dimers exhibit facile outer-sphere electron transfer reactions. Removal of valence d-electrons increases the bond order (cf. the MO diagram in Figure 3). All of the redox couples we have examined thus far undergo very rapid, outer-sphere electron transfer reactions (Figure 5). This characteristic makes these dimers excellent candidates for highly-conducting, charge transfer salts of a previously unknown type.

Polymers Having Multiple Metal-Metal Bonds Along the Backbone

The combination of the above metal-metal bonded dimers with a class of cofacial porphyrins, "Pacman-B" (DPB, Figure 6) and "Pacman-A" (DPA, Figure 7), leads to situations which can be used to create polymeric arrays. The fixed cofacial geometry of DPB enforces an intramolecular metal-metal bond The constraint within DPA does not allow intramolecular bonding; a polymer whose backbone is joined by multiple metal-metal bonds results. One example has been prepared but is not well-characterized. The partial oxidation of such a polymer should result in a conducting, mixed-valence compound such as the osmium derivative shown in Figure 7.

An alternative method for enforcing a discotic array of metal-metal bonding is shown in Figure 8. Synthetic routes to such mesodiscotic liquid crystalline porphyrins are presented in the poster by Matthew Zisk. Such systems will probably result in charge transfer waves because of their one-dimensional character.

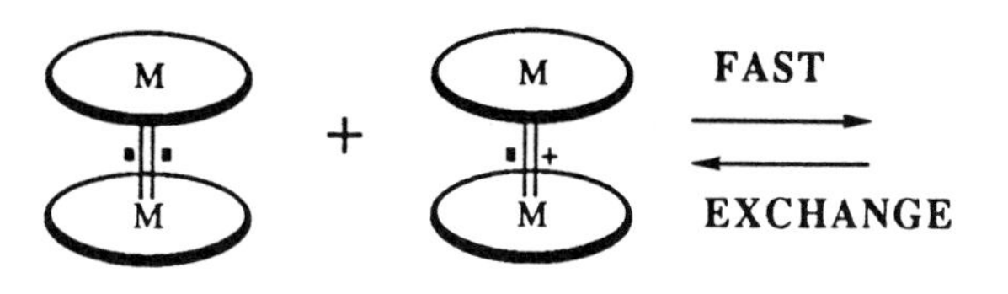

Figure 5. These metal-metal dimers exhibit rapid, outer-sphere electron transfer reactions.

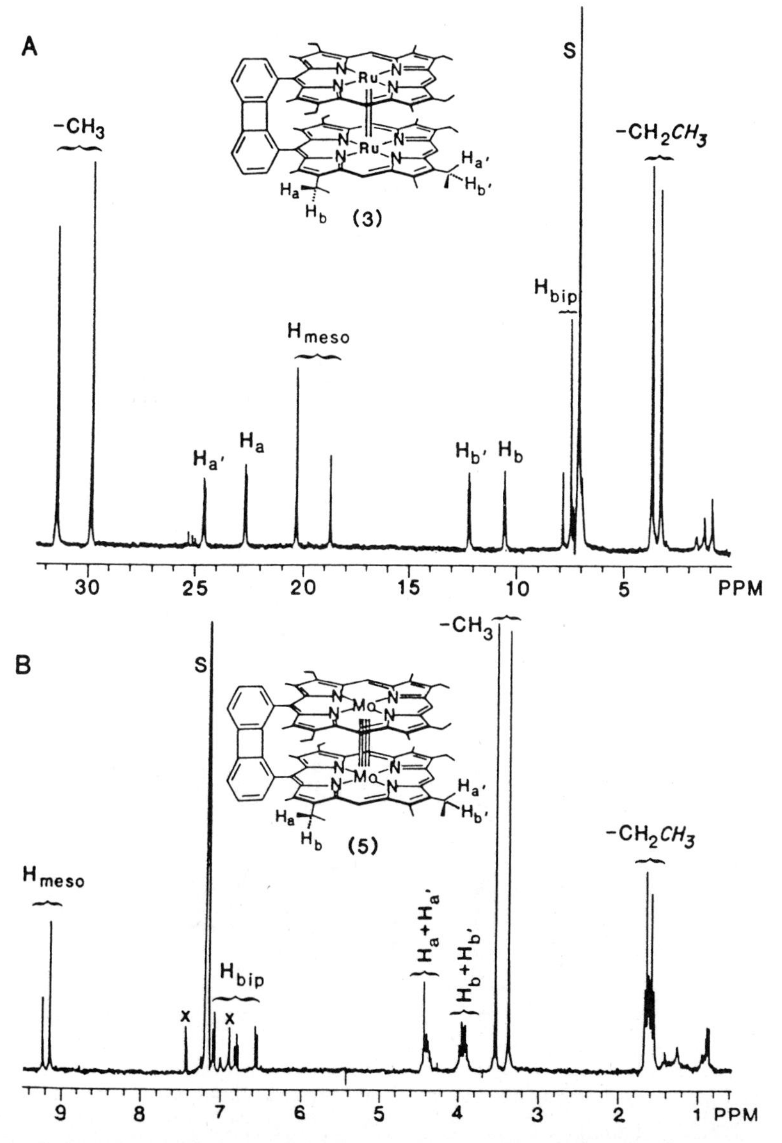

Figure 6. ^1H NMR spectra of the Ru(II) and Mo(II) Pacman-B dimer; the former is a ground state triplet and the latter is a singlet as predicted from the MO diagram in Figure 3.

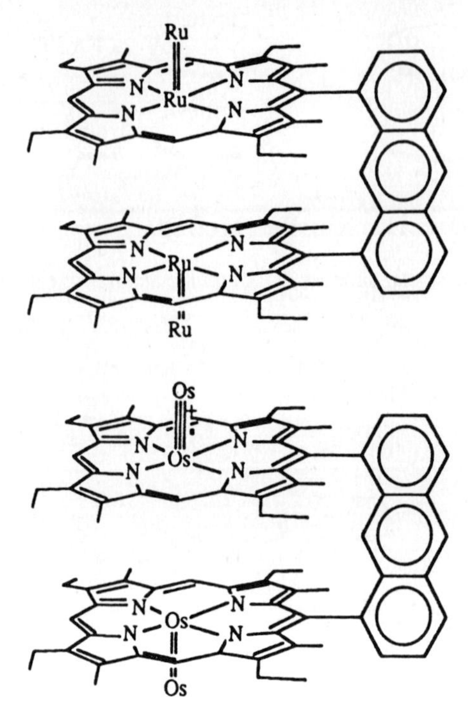

Figure 7. Polymers containing ruthenium and osmium multiple metal-metal bonds along the backbone. The Ru compound is known; the mixed-valence osmium derivative has not been prepared.

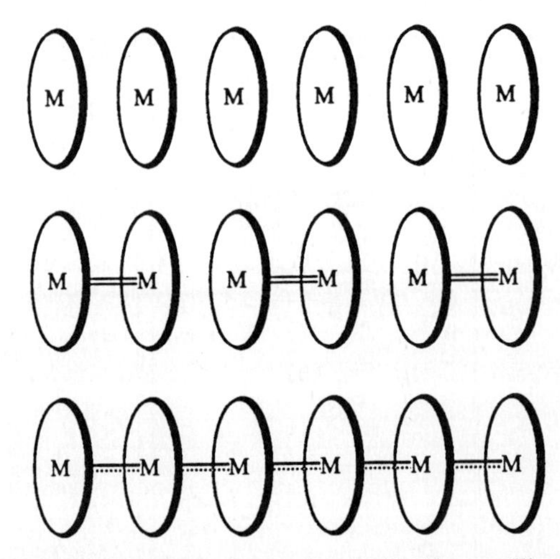

Figure 8. Postulated metal-metal bonding interactions in liquid crystalline 3d metal porphyrins.

Top: Nonbonding interactions (except magnetic dipole).
Middle: Pairing
Bottom: Cumulative bonding

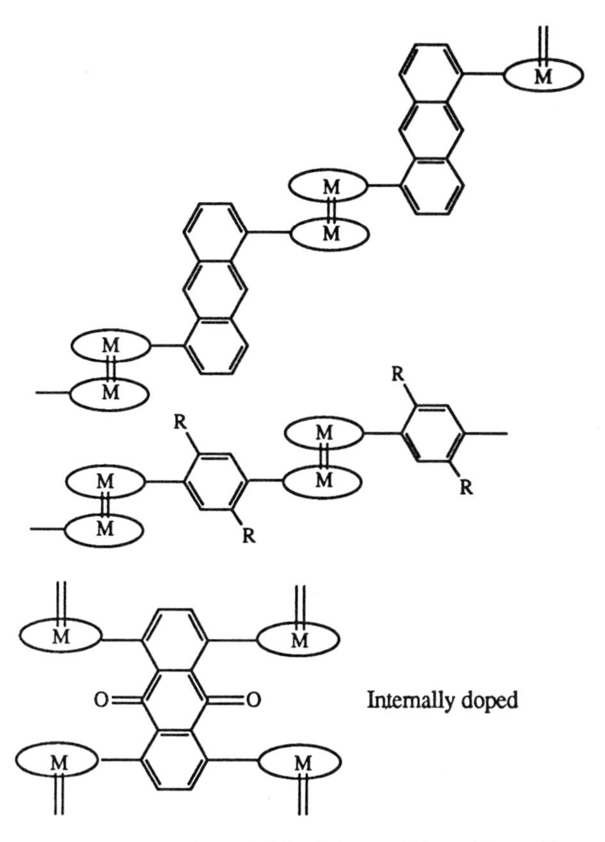

Figure 9. Schemes for preparing polymeric M≡M assemblies with non-linear morphologies.

Other strategies for preparing polymeric arrays of metal-metal bonded dimers are outlined in Figure 9. Although none of these polymers has yet been prepared, the proposed substances are based on straightforward synthetic sequences. Note that these polymers are designed to pack in the solid state in such a way as to encourage interchain electron transfer which might confer two- or three-dimensional character to the partially oxidized charge transfer polymers.

All of the charge transfer polymers suggested in this section have in common the possibility of Coulombic interactions between the metal-centered conduction electrons and the orthogonal π-electrons within the porphyrin rings as is illustrated in Figure 2.

Acknowledgment

The support of the U.S. Department of Energy (grant number DE-FG03-86ER45245) and the National Science Foundation (grant number CHE88-14949) is gratefully acknowledged.

REFERENCES

1. Collman, J.P.; McDevitt, J.T.; Leidner, C.R.; Yee, G.T.; Torrance, J.B.; Little, W.A. *J. Am. Chem. Soc.* **1987**, *109*, 4606.

2. Hoffman, B.M.; Ibers, J.A. *Acc. Chem. Res.* **1983**, *16*, 15.

3. (a) Dirk, C.W.; Inabe, T.; Schoch, K.F., Jr.; Marks, T.J. *J. Am. Chem. Soc.* **1983**, *105*, 1539.
(b) Joyner, R.D.; Kenney, M.E. *J. Am. Chem. Soc.* **1960**, *82*, 5790.
(c) Nohr, R.S.; Kuznesof, P.M.; Wynne, K.J.; Kenney, M.E.; Siebenman, P.G. *J. Am. Chem. Soc.* **1981**, *103*, 4371.
(d) Wynne, K.J.; Nohr, R.S. *Mol. Cryst. Liq. Cryst.* **1981**, *81*, 243.

4. Martinsen, J.; Stanton, J.L.; Greene, R.L.; Tanaka, J.; Hoffman, B.M.; Ibers, J.A. *J. Am. Chem. Soc.* **107**, 6915.

5. Collman, J.P.; Yee, G.T.; Holcomb, M.J.; McDevitt, J.T.; Little, W.A.; Brown, G.E. *J. Am. Chem. Soc.* **1988**, *110*, 1301.

6. Little, W.A.; Collman, J.P. *Proc. Natl. Acad. Sci. U.S.A. 1988*, *85*, 4596.

7. Yee, G.T.; Collman, J.P.; Little, W.A. *J. Superconductivity* in press.

8. Collman, J.P.; Barnes, C.E.; Woo, L.K. *Proc. Natl. Acad. Sci. U.S.A.* **1983**, *80*, 7684.

9. (a) Collman, J.P.; Barnes, C.E.; Collins, T.J.; Brothers, P.J.; Gallucci, J.; Ibers, J.A. *J. Am. Chem. Soc.* **1981**, *103*, 7030.
(b) Collman, J.P.; Woo, L.K. *Proc. Natl. Acad. Sci. U.S.A.*, **1984**, *81*, 2592.
(c) Collman, J.P.; Barnes, C.E.; Sweptson, P.N.; Ibers, J.A. *J. Am. Chem. Soc.* **1984**, *106*, 3500.
(d) Collman, J.P.; Prodolliet, J.W.; Leidner, C.R. *J. Am. Chem. Soc.* **1986**, *108*, 2916.
(e) Collman, J.P.; Kim, K.; Garner, J.M. *J. Chem. Soc., Chem. Commun.* **1986**, 1711.
(f) Tait, C.D.; Garner, J.M.; Collman, J.P.; Sattelberger, A.P.; Woodruff, W.H. *J. Am. Chem. Soc.* **1989**, *111*, 7806.
(g) Collman, J.P.; Garner, J.M.; Woo, L.K. *J. Am. Chem. Soc.* **1989**, *111*, 8141.
(h) Asahina, H.; Zisk, M.B.; Hedman, B.; McDevitt, J.T.; Collman, J.P.; Hodgson, K.O. *J. Chem. Soc., Chem. Commun.* **1989**, 1360.
(i) Tait, C.W.; Garner, J.M.; Collman, J.P.; Sattelberger, A.P.; Woodruff, W.H. *J. Am. Chem. Soc.* **1989**, *111*, 9072.
(j) Collman, J.P.; Garner, J.M. *J. Am. Chem. Soc.* **1990**, *112*, 166.

CONDUCTING AND SUPERCONDUCTING CRYSTALS BASED ON SOME UNSYMMETRICAL DONOR MOLECULES

G.C.Papavassiliou, D.Lagouvardos, V.Kakoussis, G.Mousdis, A.Terzis*, A.Hountas*, B.Hilti**, C.Mayer**, J.Zambounis**, J.Pfeiffer** and P. Delhaes***

Theor.Phys.Chem.Institute, National Hellenic Research Foundation, 48, Vassileos Constantinou Ave., Athens 116 35, Greece.* Inst.Mat.Sci., "Democritos", N.R.C. Ag.Paraskevi Attikis, Athens 153 10, Greece.** Central Research Laboratories, CIBA-GEIGY AG, CH-4004, Basel, Switzerland.***Centre de Recherche Paul Pascal, CNRS, Château Brivazac, Avenue A.Schweitzer, 33600 PESSAC, France

INTRODUCTION

Recently, a number of conducting and superconducting salts based on unsymmetrical π-donor molecules have been prepared and studied (see [1-5]). In this paper the preparations, crystal structures and physical properties of some salts based on the unsymmetrical π-donors ethylenedioxyethylenedithiotetrathiafulvalene (1), ethylenedioxymethylenedithiotetrathiafulvalene (2), ethylenedioxyvinylenedithiotetrathiafulvalene (3) and methylenedithiotetrathiafulvalene (4) are described.

(1): (EDOEDTTTF), (2): (EDOMDTTTF)

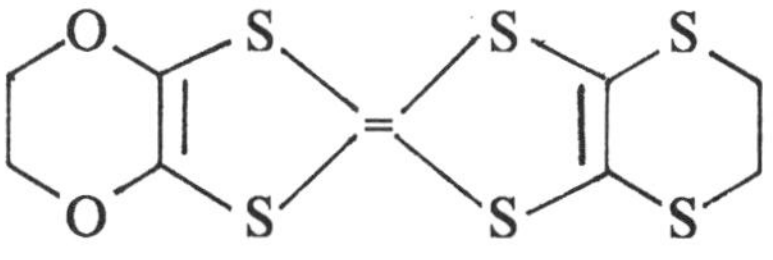

(3): (EDOVDTTTF), (4): (MDTTTF)

EXPERIMENTAL

Compounds (1), (2) and (3) were prepared by coupling of 4,5-ethylenedioxy-1,3-dithio-2-one with 4,5-ethylenedithio-1,3-dithiole -2-one, 4,5-methylenedithio-

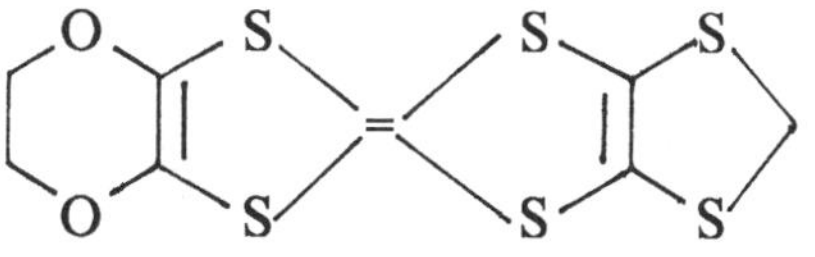

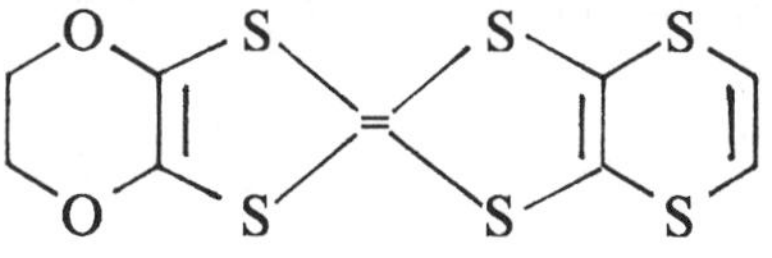

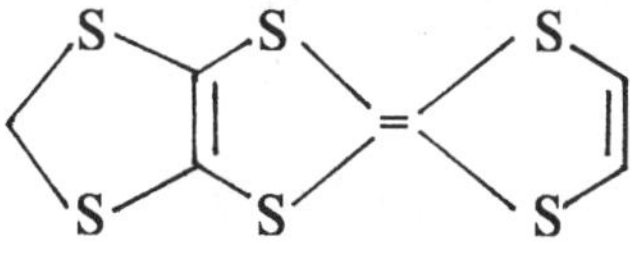

Organic Superconductivity, Edited by V.Z. Kresin and W.A. Little, Plenum Press, New York, 1990

1,3-dithiole-2-one, and 4,5-vinylenedithio-1,3-dithiole-2-one, respectively, via triethyl phosphite at ca 150 °C, followed by column-chromatography separation [3,6]. Compound (4) was prepared by a two-steps sequence: coupling of 4,5-methylenedithio-1,3-dithiole-2-one with 4,5-bis(methylcarboxy)-1,3-dithiole-2-thione, chromatography-separation of the cross-coupling product from the self-coupling products and then demethoxycarboxylation with LiBr in hexamethylphosphoramide, followed by chromatography-separation [1,7]. It was found that the values of half-wave oxidation potentials of (3) are close to those of BEDTTTF, while those of (1), (2) and (4) are lower. This means that substitution of an alkylenedithio-group or vinylenedithio-group by an ethylenedioxy-group increases the donor-capacity of the compound [1]. However, the difference between the second and first oxidation potential of (3), (4) and BEDTTTF is almost the same (see also [8]). The compounds (1)-(4) gave charge transfer complexes with TCNQ. This is a consequence of their low oxidation potentials. Also, a number of cation radical salts was prepared in single crystal form, mainly, by electrochemical crystallization in CH_2Cl_2. β′-$(EDOVDTTTF)_2I_3$ was prepared by chemical reaction of (3) with Bu_4NI_3 in CH_2Cl_2 and slow evaporation of the solvent . Data collection for the crystal structure determination was performed on a compurer-controlled difractometer with MoKā Nb-filtered radiation [9]. Electrical conductivity measurements were performed by a four-probe method. Measurements of static susceptibility on a polycrystalline sample of ϰ-$(MDTTTF)_2AuI_2$ were performed with a standard Faraday balance between 4 and 290 K in presence of a magnetic field strength of 1.1 T. Magnetization measurements at low temperatures were performed with a SQUID magnetometer working with a variable magnetic field strength. Resonance Raman spectra of the three different phases of $(EDOVDTTTF)_xI_3$ were recorded at room temperature with a Jobin Yvon Ramanor Laser Raman spectrophotometer, using for excitation the 488 nm line.

RESULTS AND DISCUSSION

From a large number of conducting salts prepared from the donors (1)-(4) only eleven of them, which crystallized well in single crystal form, were studied. Most of the crystals obtained from (1), (2) and (3) crystallize in tetragonal system, which is rarely observed in organic conductors. They have the formula $D_2(X)_1(X)_y$, where D=donor and X=anion. In this case a new kind of side-by-side arrangement of the donor molecules was observed.

$(EDOEDTTTF)_2Cu(NCS)_2$

It crystallizes as thin dark-brown plates [10]. It is metallic (σ_{RT}=29 S/cm, MIT at ca25K). Details on the temperature dependence of the conductivity are given in Fig.1a.

τ-$(EDOEDTTTF)_2(AuI_2)_1(AuI_2)_y$

It crystallizes as black big crystals, in the tetragonal system, space group P$\bar{4}$2c, a=7.485 c=34.032 Å, Z=4. It is semiconducting (σ_{RT}=4.6x10^{-5} S/cm).

β′-$(EDOEDTTTF)_2AuI_2$

It crystallizes as black plates in the orthorhombic system, space group A222,

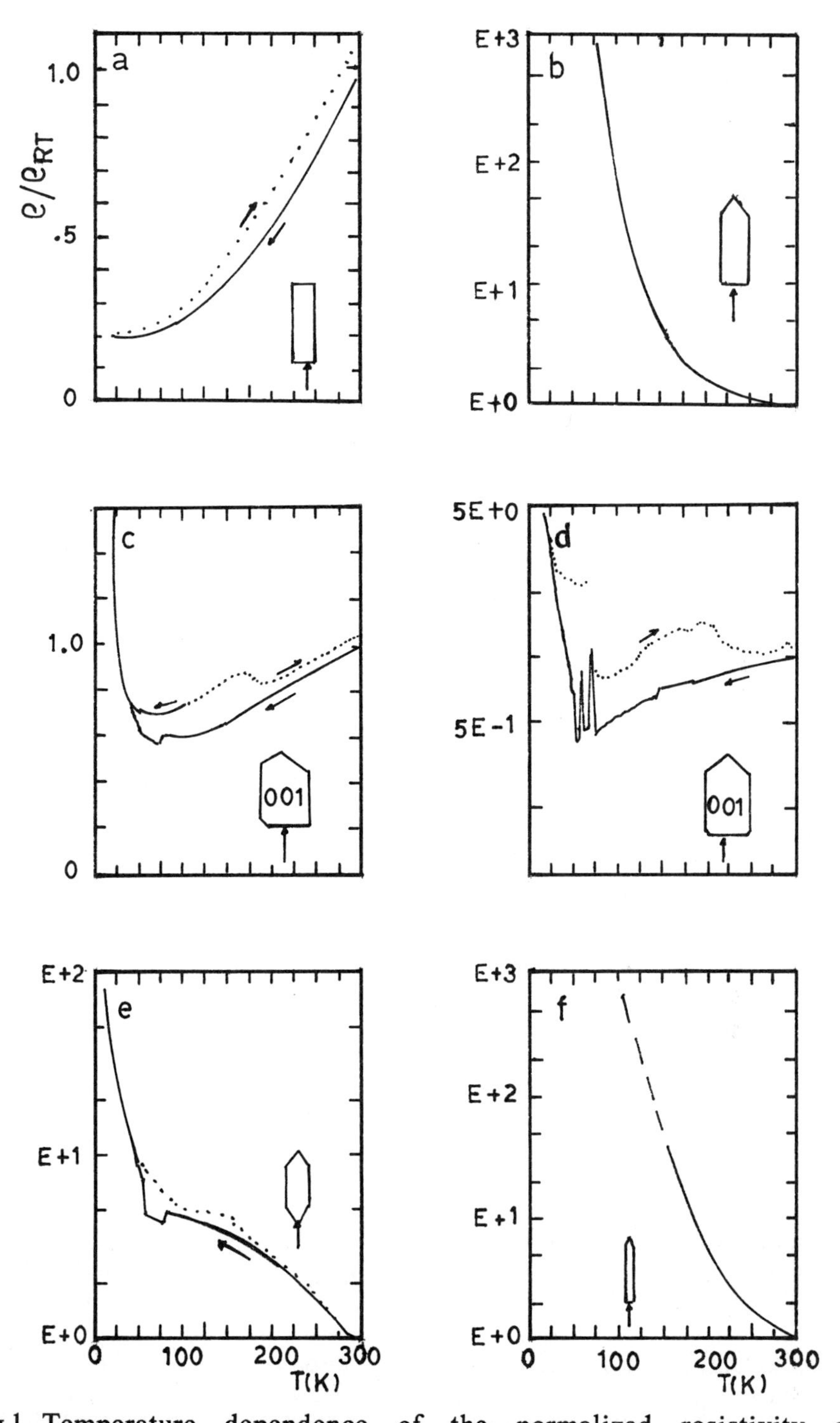

Fig.1. Temperature dependence of the normalized resistivity of $(EDOEDTTTF)_2Cu(NCS)_2$(a), τ-$(EDOMDTTTF)_2(I_3)_1(I_3)_y$, τ-$(EDOVDTTTF)_2(AuI_2)_1(AuI_2)_y$ (c), τ-$(EDOVDTTTF)_2(IBr_2)_1(IBr_2)_y$(d), $(EDOVDTTTF)_xAuBr_2$ (e), and β'-$(EDOVDTTTF)_2I_3$ (f) measured along the arrow-direction, in the temperature range from 15 to 300 K.

a=6.621, b=7.248, c=33.239 Å, Z=2. It is semiconducting (σ_{RT}=4x10^{-3} S/cm).

τ-$(EDOMDTTTF)_2(I_3)_1(I_3)_y$

It crystallizes as black plates in the tetragonal system, space group P$\bar{4}$2c. It is semiconducting (σ_{RT}=0.2 S/cm). Details on the temperature dependence of the conductivity are given in Fig.1b.

τ-$(EDOVDTTTF)_2(AuI_2)_1(AuI_2)_y$

It crystallizes as black plates in the tetragonal system, space group P$\bar{4}$2c, a=7.471, c=33.584 Å, Z=4. It is metallic (σ_{RT}=180 S/cm, MIT at ca 60K). Details on the temperatures dependence of the conductivity are given in Fig.1c.

τ-$(EDOVDTTTF)_2(IBr_2)_1(IBr_2)_y$

It crystallizes as black plates in the tetragonal system, space group P$\bar{4}$2c. It is metallic (σ_{RT}=60 S/cm, MIT at ca 70K). Details on the temperatures dependence of the conductivity are given in Fig.1d.

$(EDOVDTTTF)_xAuBr_2$

It crystallizes as black plates. It is semiconducting (σ_{RT}=11-22 S/cm). Details on the temperature dependence of the conductivity are given in Fig.1e.

γ'-$(EDOVDTTTF)_1I_3$

It crystallizes as black needles in the monoclinic system, space group $P2_1$. It is semiconducting. Resonance Raman spectra showed bands at 108, 217, 330 cm^{-1}.

β'-$(EDOVDTTTF)_2I_3$

It crystallizes as bronze needles. It is semiconducting (σ_{RT}=10.4 S/cm). Details on the temperature dependence of the conductivity are given in Fig.1f. Resonance Raman spectra showed bands at 107, 214, 329 cm^{-1}.

τ-$(EDOVDTTTF)_2(I_3)_1(I_3)_y$

It crystallizes as black plates with a golden lustre in the tetragonal system, space group P$\bar{4}$2c, a=7.563, c=33.883 Å, Z=4. A stereoview of the crystal structure is shown in Fig.2.There is no stacking of the donor-molecules. The large axis of the donor-molecules is parallel to the c-axis. The donor-molecules form extended layers parallel to the a- and b- axes at a=b=0.5. There is one I_3, which is well behaved, at the origin. The rest of I_3 is disordered. The whole y-axis at x=0.5 and z=0.25 is full of electronic density [10]. There is a disorder of the ethylenedioxo-group, as well, there are intermolecular (S---O)-contacts ($\geq$3.50Å) between the S and O atoms of vinylene-and ethylene-group, respectively, and (S---S)-contacts ($\geq$3.58 Å) between the sulfur atoms of TTF-cores, forming a 2-dimensional network perpendicular to the c-axis. The compound is metallic. The temperature dependence of the conductivity varies from crystal to crystal. Fig. 3 shows the temperature dependence of the conductivity for one crystal with σ_{RT}=45 S/cm. One

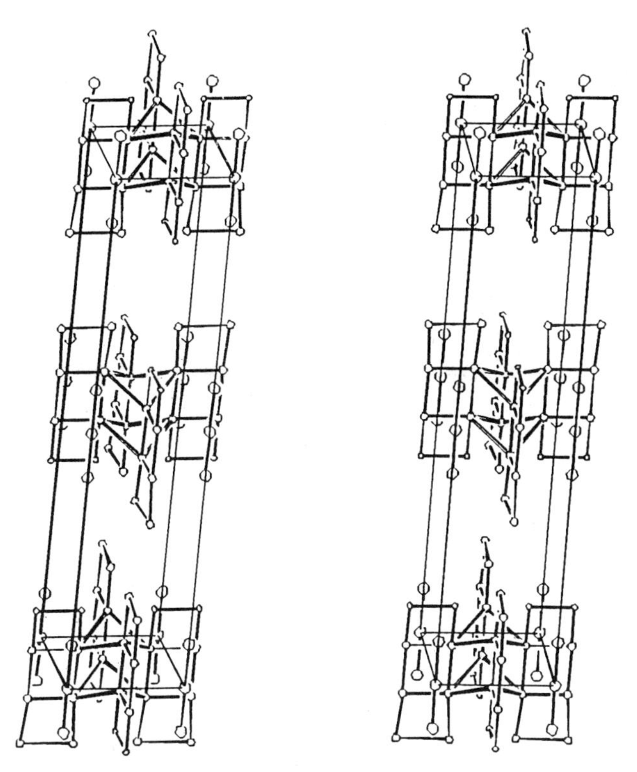

Fig.2. Stereoview of the molecular packing of τ-(EDOVDTTTF)$_2$(I$_3$)$_1$(I$_3$)$_y$. The disordered I$_3$ is not shown.

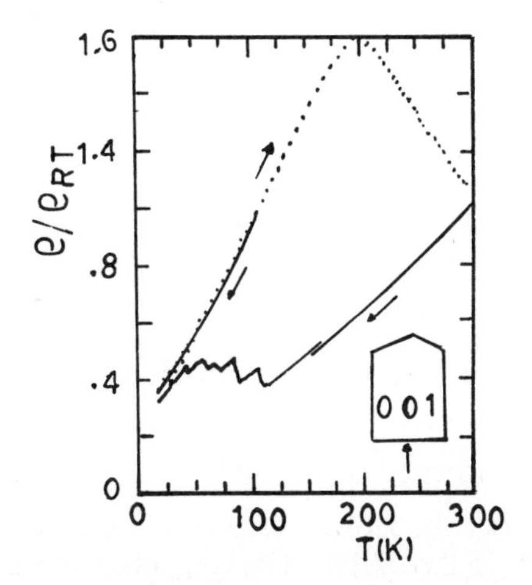

Fig.3. Same as in Fig.1, but for one crystal of τ-(EDOVDTTTF)$_2$(I$_3$)$_1$(I$_3$)$_y$.

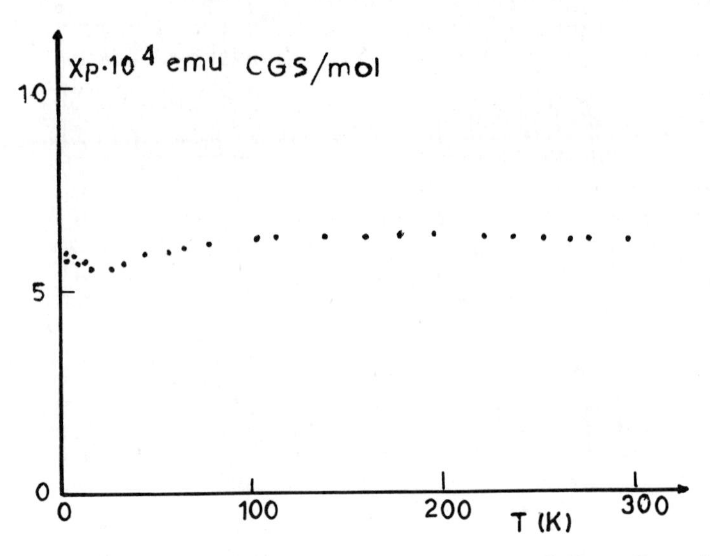

Fig.4. Temperature dependence of paramagnetic susceptibility of a polycrystalline sample of $(MDTTTF)_2AuI_2$ in the temperature range from 4K to 290 K.

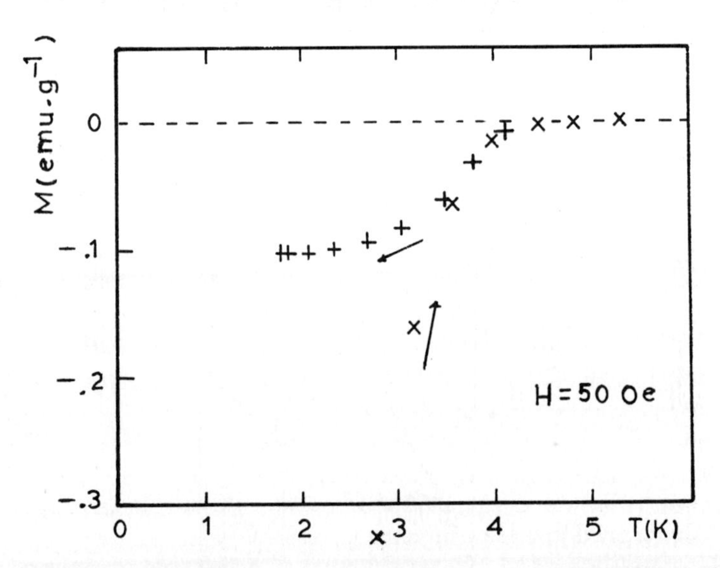

Fig.5. Temperature dependence of magnetization of $(MDTTTF)_2AuI_2$ at low temperatures (1.8-5K).

can see that on cooling down sudden drops in the conductivity began occuring around 150K. They disappeared after heating from 25K to 120 K and recooling to low temperatures. There is a 2-dimensional metallic conductivity, which is due to S---S interactions parallel to ab-plane. These preliminary results indicate that the crystal could remain metallic or become superconducting at very low temperatures. Resonance Raman spectrum showed bands at 108, 135(sh), 215-250(br), 335-345(br) cm^{-1}. The shoulder at 135 cm^{-1} indicates the disorder of I_3. Weissenberg photographs and Raman spectra showed that the disorder of I_3 decreases after heating the crystals at ca 70°C for three days. Perhaps, the observed metallic properties result from the presence of anion vacancies producing slightly nonstoichiometric crystals as in the case of ζ-$(ET)_2(I_3)(I_5)$ [11].

κ-$(MDTTTF)_2AuI_2$

It crystallizes as black-brown plates in the orthorombic system, space group Pbmm [9]. It is metallic (σ_{RT}=12-36 S/cm) and becomes superconducting at low temperature (Tc=5K) [4,12-14]. The temperature dependence of the paramagnetic susceptibility of a polycrystalline sample is shown in Fig.4. It indicates a Pauli paramagnetism. The temperature dependence of magnetization at lower temperatures (1.8-5K) is shown in Fig.5. For the observation of the diamagnetic shielding and Meissner effect the sample was first cooled in absence of any magnetic field down to 1.8K, then the field was applied and magnetization was measured upon heating and then it was cooled under magnetic field [15]. The Meissner phase starts to appear at ca 4.3 K (see also [12]), while from conductivity measurements it was found that Tc=5K [4,13]. Similar results have been obtained from other organic superconductors [16].

Conductivity measurements at lower temperatures under ambient and high pressure as well as magnetic properties of salts based on (1), (2), (3) and similar donors will be published in a future paper.

REFERENCES

1. G.C.Papavassiliou, in "Proc.NATO-ASI on Lower Dimensional Systems and Mol.Electronics", Spetses, Greece, June 1989, Ed.R.M.Metzger, Plenum (1990).
2. G.C.Papavassiliou, Pure and Appl.Chem., 62 , 483 (1990).
3. G.C.Papavassiliou, V.C.Kakoussis, D.J.Lagouvardos, and G.A.Mousdis, Mol.Cryst.Liq.Cryst. 181, 171 (1990).
4. G.C.Papavassiliou, G.Mousdis, V.Kakoussis, A.Terzis, A.Hountas, B.Hilti, C.W.Mayer, and J.S.Zambounis in "Proc.Int.Symp.Org.Superconductors" , Tokyo, August 1989, Ed.G.Saito, Springer-Verlag (1990).
5. I.Ikemoto, K.Kikuchi, K.Saito, K.Kanoda, T.Takahashi, K.Murata, and K.Kobayashi, Mol.Cryst.Liq.Cryst., 181, 185 (1990); H.Nakano, K.Miyawaki, T.Nogami, Y.Shirota, S.Harada and N.Kasai, Bull.Chem. Soc. Jpn. 62, 2604 (1989); Refs. cited in refs.[1-4] here.
6. G.C.Papavassiliou, D.J.Lagouvardos, V.C.Kakoussis and G.A.Mousdis, Z.Naturforsch., b, in press (1990).
7. G.C.Papavassiliou, J.S.Zambounis, G.A.Mousdis, V.Gionis, and S.Y.Yiannopoulos, Mol.Cryst.Liq.Cryst., 156 , 269 (1988).
8. G.C.Papavassiliou to be published: Experimental values of oxidation potentials of the unsymmetrical donors were found to be close to the average of potentials of the corresponding symmetrical donors.

9. A.Hountas, A.Terzis, G.C.Papavassiliou, B.Hilti, M.Burkle, C.W.Mayer and J.S.Zambounis, Acta Cryst. C, 46 , 228 (1990).
10. Elemental analysis of $(EDOETDTTF)_2Cu(NCS)_2$: C 29.77, H 2.02, N 3.17, Cu 3.15-3.19%. From the elemental analysis in one sample of τ-$(EDOVDTTTF)_2(I_3)_1(I_3)_y$ it was found that y=1; analysis: C 16.87, H 0.76, I 52.39%. Elemental analysis of similar compounds of the τ-phase will be published elsewhere.
11. M.A.Beno, Urs.Geiser, K.L.Kostka, H.H.Wang, K.S.Webb, M.A.Firestone, K.D.Carlson, L.Nunez, M.-H.Whangbo, and J.M.Williams, Inorg.Chem., 26, 1912 (1987).
12. G.C.Papavassiliou, G.A.Moudis, J.S.Zambounis, A.Terzis, A.Hountas, B.Hilti, C.W.Mayer and J.Pfeifer, Synth.Metals, 27 , B373(1988).
13. M.Freund et al, to be published.
14. A.M.Kini, M.A.Beno, D.Son, H.H.Wang, K.D.Carlson, L.C.Porter, V.Welp and J.E.Schirber, Sol.St. Commun. , 69 , 503(1989).
15. Details on the magnetic properties will be published elsewhere: P.Delhaes et al, to be published.
16. D.R.Harshman, R.N.Kleiman, R.C.Haddon, S.V.Chichester-Hicks, M.L.Kaplan, L.W.Rupp, Jr, T.Ptiz, D.L.Williams and D.B.Mitzi, Phys.Rev.Lett., 64 , 1293 (1990); G.Saito, Physica C ,162-164 , 577 (1989); C.P.Heidmann, H.Veith, K.Andres, H.Ruchs, K.Polborn and E.Polborn and E.Amberger, Sol.St.Commun., 57 , 161 (1982); H.Schwenk, S.S.P. Parkin, V.Y.Lee and R.L.Greene, Phys. Rev. B ,34, 3156 (1986).

CLOSING REMARKS

In closing this Conference I would like to comment on those aspects of the meeting which bore on the goals we originally set for holding it. These goals were to consider new classes of conductors and superconductors; to consider new syntheses; to present new developments; to present the present state-of-the-art of the field, the experimental techinques and capabilities of the various groups; and, to facilitate interactions among chemists, physicists and material scientists.

The meeting got off to an excellent start with the report from Jack Williams' group at Argonne of the discovery of a new variant in the ET family with a new high for the transition temperature for an organic superconductor. Clearly a great deal of thought and ingenuity went into the choice of the constituents and a study of trends in T_c in related compounds. The prospects for further modifications of the anions in the ET salts continues to look promising, with hints that with further expansion of the effective volume of the unit cell, the transition temperature might be edged up still higher, to the upper teens.

In regard to polymers, Elsenbaumer's report makes it clear that great strides have been made in achieving manageable, processable materials with excellent conductivities. As yet none of these is superconducting, but it can only be a matter of time for this further milestone to be achieved. What has changed in the area of conductive polymers in the past decade has been the growth in the number of monomers, and the number of donors and acceptors. This has increased geometrically the number of combinations, and in Fred Wudl's terms, the number of possible "hits", or successful combinations.

At the International Conference on Organic Superconductors held in Hawaii in 1969, Vitaly Ginzburg made the point that it was not important that the high temperature superconductor we were then seeking should be organic. It's composition would be irrelevent if its T_c was high enough. He encouraged us therefore not to limit ourselves purely to carbon based compounds but to broaden the study to other systems. On the otherhand, the power of organic chemistry lies in the deliberate, and precise manner in which structures are designed and synthesized, in contrast to the methods of the metallurgist and ceramicist, where much less control is exercised on the the structure and form of the final product. In the past several decades many of the leading organic chemists, several of whom are with us here today, appear to have ignored the elementary course desciption of organic chemistry, as the chemistry of compounds of carbon, nitrogen, hydrogen and oxygen, and have taken the methods and philosophy of the organic chemist and applied them to a host of systems built of other elements, including but not limited to S, Se, Te, Sn, Ge, Re, Os, W,.... This trend continues to enrich the field. Many of the suggestions of the panel participants, and of the contributed papers suggests that further growth in this area is likely to be substantial.

As a condensed matter physicist one cannot help but be impressed with the beautiful experimental detail revealed in the properties of many of the organic compounds - the Shubnikov - de Haas oscillations, the Quantum Hall Effect and the angle dependent magnetoresistance - all which give microscopic information needed for the development of a proper understanding of these materials. The variety of phases is rich - SDW, CDW, superconductivity, and field enhanced phases, to name a few. What is even more striking is that many or even most can be seen in the same material under suitable conditions of pressure and field! Whereas among conventional superconductors one has been limited to a subset of the 92 elements, their binary alloys and a few ternary alloys, among the organics the number of possible combinations is virtually limitless. This surely will prove to be one of the great contributions that the organics will make to condensed matter physics.

In looking back a few years, it is worth noting that the field of organic conductors and superconductors was not one which was welcomed by the condensed matter community with open minds and open arms. Indeed, as exemplified by the reception the first conference on this subject received in 1969 (Physics Today, p23, August1971; p11, April, 1972), it was rather one of scepticism and ridicule, at least from some quarters, not warmth. Part of this can be understood because the organics appeared to be so un-metal-like, part of it was a lack of imagination, and part was simply human nature. The position that we and others took at that time, was that if one was to believe what was then thought to explain the metallic state and its transport processes, then there was every reason to believe that some organics could exist in the same state. This has been amply borne out, particularly in the past decade, as microscopic information has accumulated on the band structures of these organics, their phases, and of their transport properties. The psychological barrier to the belief that *organics* could behave in such a manner, has disappeared. This is due partly to the older generation having been supplanted by the young, and partly due to evidence gathered over the past ten years or so which has convinced the remainder. It is clearly now no longer an issue.

Today's more open views of exotic conductors has been helped, of course, by the discovery of the high T_c superconducting cuprates. Indeed, one is struck by the similarities between them and the best superconducting organics. Both appear to be two-dimensional, both exhibit a mid-infrared excitation of some kind, both have highly anisotropic transport properties and both appear to have highly anisotropic gaps. The transition temperatures are still significantly different, but once the superconductivity of the cuprates is understood, then perhaps it may be possible to implant such a similar capability in the organics. It is an exciting possibility!

In the past, the presumption of the physicists had been that organics could not be prepared with a purity sufficient to meet the demands of the Fermiologists. It is a credit to the chemists that this presumption has clearly been shown to be false, as illustrated by the achievement of mobility figures in excess of a million, and the many Fermi Surface phenomena discussed at this meeting. These figures are so good, that one wonders whether perhaps in some organics, impurities, if present, are squeezed from the unit cell, sequestered at grain boundaries and eventually annealed from the crystal. The strikingly regular, STM pictures of Fainchtein, showing regions several hundreds of Angstoms on a side, apparently free of any inclusions, would support some such a view. Perhaps the organic chemists are not alone, with Nature herself helping clean their samples.

On the logistics side of the meeting, it appeared that the plan to hold mixed sessions of physics, chemistry and materials science worked rather well, with almost 100% attendance at every session over the four day period. This, too, must be something of a record!

The conference did succeed in identifying some areas where greater emphasis and, greater funding, would be warranted. One is the need for developing a better understanding of electrocrystallization. This powerful technique has been successful in the growth of charge transfer salts. It could make a significant impact on the preparation of high purity polymers if it was better understood. Perhaps some help from the funding agencies can be obtained here.

Finally, I was struck by Professor Collman's remark that the number of researchers involved in the development of new compounds such as those highlighted at this conference, compared to the number involved in natural product chemistry, is of the order of one in a thousand! This imbalance is so large, that it is hard to avoid the conclusion that a greater overall return on research investment would accrue with even a microscopic adjustment of this ratio.

In closing, I wish to thank the Department of Energy and the Office of Naval Research for their generous support of the conference. I would also like to thank the speakers for their excellent presentations that succeeded in holding our attention for these past, pleasant days; and to thank the chairmen of the sessions for controlling these speakers in so admirable a manner. Thank you !

W. A. Little
Stanford University
May, 1990

Attendees: International Conference on Organic Superconductivity / May 20 - 24th, 1990 / Fallen Leaf Lake, South Lake Tahoe, California, USA

1. Liaping Le 2. S. Yoon 3. S. D'Arcangelis 4. W. A. Little 5. D. Cowan 6. L. Montgomery 7. P. Erk
8. M.Holcomb 9. J. Brill 10. J. Schirber 11. R. L. Greene 12. J. Medrano 13. J. Novoa 14 .M. Weger
15 .B. Bush
16. W. Kang 17. O. Klein 18. G. Gruner 19. J. Ladik 20. T. Ishiguro 21. S. Brown 22. M. Zisk 23. F. Wudl
24. H.Morawitz 25. F. Higuchi 26. V. Kresin 27. A. Blumstein
28. G. Friedman 29. R. Kelley 30. M. Beno 31. M. Dreiling 32. E. Pope 33. J. Lagowski. 34. A. H. Alberts
35. T. Timusk 36. R. Elsenbaumer 37. S Wolf

38. C. Caley 39. Y. Uemura. 40. G. Saito 41. H. Wang 42. W. Smith 43. P. Chaikin 44. W. Fuller
45. S. Hannahs 46. R. Fainchtein 47. A. Hermann 48. J. Collman 49. H. Arnold 50. C. Agosta
51. A. Swanson 52. J. Williams 53. M. Whangbo 54. M. Tokumoto 55. G. Papavassiliou
56. K. Kornelson 57. J. Eldridge 58. H.Q. Lin

Absent: R. V. Kasowski

PARTICIPANTS

Charles Agosta
MIT, NW 14-1102D
Cambridge, MA 02139

Hillary Arnold
Chemistry Department
Stanford University
Stanford, CA 94305

Dr. Albert H. Alberts
Dept. Polymer Technology
P.O. Box 513
5600 MB Eindhoven
THE NETHERLANDS

Patrick Batail
Laboratoire de Physique des Solides
CNRS UAZ
Universite Parid (bat510)
91405 ORSAY FRANCE

Mark Beno
Argonne National Lab
9700 S. Cass Avenue
Argonne, IL 60439

Alexandre Blumstein
Dept. of Chemistry
Univ. of Lowell
1 University Avenue
Lowell, MA 01854

Joseph Brill
University of Kentucky
Dept. of Physics
Lexington, KY 40506-0055

Stuart Brown
U. of Florida
215 Williamson Hall
Gainesville, FL 32611

Benjamin Bush
Dept. of Chemistry
U. of Texas at Austin
Austin, TX 78712

Catherine Caley
Chemistry Dept.
Stanford University
Stanford, CA 94305

Paul Chaikin
Dept. of Physics
Princeton University
Princeton, NJ 08544

Jim Collman
Chemistry Dept.
Stanford University
Stanford, CA 94305

Dwaine Cowan
The Johns Hopkins University
Dept. of Chemistry
34th & Charles Streets
Baltimore, MD 21218

Samuel D'Arcangelis
The Johns Hopkins University
Dept. of Chemistry
34th & Charles Streets
Baltimore, MD 21218

Mark Dreiling
329 PL
Philips Petroleum Co.
Bartlesville, OK 74004

John Eldridge
Univ. of British Columbia
Physics Dept.
Vancouver B.C., V6T 2A6
Canada

Ronald Elsenbaumer
Allied Signal Inc.
P.O. Box 1021 R
Morristown, NJ 07962

Peter Erk
Dept. of Chemistry
Stanford University
Stanford, CA 94305

Raul Fainchtein
Johns Hopkins University
Applied Physics Lab
Johns Hopkins Road
Laurel, MD 20723

Richard L. Greene
Univ. of Maryland
Center for Superconductivity Research
Dept. of Physics
College Park, MD 20742

G. Gruner
Dept. of Physics, UCLA
405 Hilgard Avenue
Los Angeles, CA 90024

Scott Hannahs
Francis Bitter National Magnet Lab
MIT, NW14-1104
Cambridge, MA 02139

Allen Hermann
U. of Colorado
Dept. of Physics
Campus Box 390
Boulder, CO 80309-0390

Fumito Higuchi
3-17-8-203 Kamiyoga
Setagaya-ku, Tokyo 158
JAPAN

Matt Holcomb
Physics Department
Stanford University
Stanford, CA 94305

K. Holczer
Dept. of Physics
UCLA
405 Hilgard Avenue
Los Angeles, CA 90024

Takehiko Ishiguro
Kyoto University
Dept. of Physics/Fac. of Science
Kyoto, JAPAN

Woowon Kang
Princeton University
Dept. of Physics
P.O. Box 708
Princeton, New Jersey 08544

Robert Kasowski
Dupont Experimental Station
E356/245
Wilmington, DE 19890-0356

Richard Kelley
Dept. of Energy
Division of Material Sciences
Washington, D.C. 20545

Kevin Kornelsen
U. of British Columbia
Physics Dept.
6224 Agriculture Road
Vancouver B.C., Canada V6T 2A6

Vladimir Kresin
Lawrence Berkeley Laboratory
UC Berkeley
Berkeley, CA 94720

Prof. J. Ladik
Chair of Theoretical Chemistry
University Erlangen-Nurnberg
Egerlandstr. 3
D-8520 Erlangen, FRG

J.J. Lagowski
University of Texas
Dept. of Chemistry
Austin, TX 78712

Liaping Le
Box 44 Pupin
Columbia University
New York, NY 10027

Hai-Qing Lin
Los Alamos National Lab
CNLS MS-B258
Los Alamos, NM 87545

W.A. Little
Physics Department
Stanford University
STanford, CA 94305

Jorge Medrano
WRDC/MLBP
Wright-Patterson AFB
OH 45433-6533

Lawrence Montgomery
Indiana University
Dept. of Chemistry
Bloomington, IN 47405

Hans Morawitz
IBM Almaden
650 Harry Road
SanJose, CA 95120

Juan Novoa
N. Carolina University
Dept. of Chemistry
Raleigh, N. Carolina 27695-8204

Dr. G.C. Papavassiliou
Theoretical and Physical Chemistry Institute
National Hellenic Research Foundation
48, Vassileos Constantinou Avenue
Athens 116 35, GREECE

Edward Pope
Matech
31304 Via Colinas, Ste 102
Westlake Village, CA 91362

Gunzi Saito
Dept. of Chemistry, Faculty of Sci.
Kyoto University, Kitahirakawa
Sakyou-ku, Kyoto 606, JAPAN

Walter Smith
U. of Texas
Physics Dept.
Austin, TX 78712

Alka Swanson
9 Osborne Road
Arlington, MA 02174

Thomas Timusk
Dept. of Physics
McMaster University
Hamilton, Ontario
Canada L85 4MI

Madoka Tokumoto
Electrotechnical Laboratory
1-1-4 Umezono
Txukuba, Ibaraki 305
JAPAN

V. Uemura
Columbia University
Physics Dept.
Box 13
New York, NY 10027

Dr. Jurgen Voitlander
Insitut fur Physikalische Chemie der Universitat Munchen
Sophienstr. 11-8000
Munchen 2 FRG

Hau Wang
Argonne National Lab
C-105 Bldg. 200
Argonne, IL 60439

Meir Weger
Racah Institute of Physics
Hebrew University
Jerusalem, ISRAEL

M. Whangbo
N. Carolina State Univ.
Dept. of Chemistry
Raleigh, N. Carolina 27695-8204

Jack Williams
Argonne National Laboratory
Chemistry & Materials Science Divisions
9700 S. Cass Avenue
Argonne, IL 60439

Stu Wolf
Naval Research Lab
Code 6340
Washington, D.C. 20375-5000

Fred Wudl
Prof. Of Chemistry & Physics
Institute for Polymers & Organic Solids
UCSB
Santa Barbara, CA 93106

Seokwon Yoon
Dept. of Physics
U. of Texas
Austin, TX 78712

M. Zisk
Stanford University
Dept. of Chemistry
Stanford, CA 94305

AUTHOR INDEX

SUBJECT INDEX